COLLOQUIA MATHEMATICA
SOCIETATIS JÁNOS BOLYAI, 32.

NONPARAMETRIC
STATISTICAL INFERENCE

Edited by:

B.V. GNEDENKO
M.L. PURI
and
I. VINCZE

Vol. I.

NORTH-HOLLAND PUBLISHING COMPANY
AMSTERDAM — OXFORD — NEW YORK

© BOLYAI JÁNOS MATEMATIKAI TÁRSULAT

Budapest, Hungary, 1982

ISBN North-Holland: 0 444 86196 3
ISSN Bolyai: 0139 3383
ISBN Bolyai: 963 8021 53 5 Vol. I-II
ISBN Bolyai: 963 8021 54 3 Vol. I

Joint edition published by

JÁNOS BOLYAI MATHEMATICAL SOCIETY

and

NORTH-HOLLAND PUBLISHING COMPANY

Amsterdam — Oxford — New York

In the U.S.A. and Canada

NORTH-HOLLAND PUBLISHING COMPANY

52 Vanderbilt Avenue

New York, N.Y. 10017

Printed in Hungary
Szegedi Nyomda
Szeged

PREFACE

The field of nonparametric statistics continues to play an increasingly succesful role in statistical theory as well as in its applications. Non-parametric methods are not only mathematically elegant, they also provide a variety of applications in several fields such as engineering, economics, agriculture, meteorology and biometrics among others. To stimulate further research and provide an opportunity for personal contacts among scholars whose fields of specialization cover broad spectra, a colloquium on nonparametric statistical inference was held in Budapest from June 23-27, 1980. It was organized by the János Bolyai Mathematical Society and was sponsored by the Bernoulli Society for Mathematical Statistics and Proba-bility and by the Hungarian Central Statistical Office. Professors D.R. Cox and I. Vincze were kind enough to make the opening remarks on behalf of the Bernoulli Society and the Bolyai Society respectively.

The program committee of the colloquium consisted of Professors D.R. Cox, B.V. Gnedenko, J. Jurečková, P. Révész, K. Sarkadi, I. Vincze (Chairman), W.R. Van Zwet and myself. The national organizing com-mittee consisted of Professors M. Arató, P. Bártfai, E. Csáki, B. Gyires, A. Marton and G. Tusnády with A. Krámli, T. Móri and L. Rejtő as secretaries. It is a pleasure to express our sincere thanks to the members of both these committees for the overall help they gave in making the colloquium a great success. Evidence of the results of their efforts was the rich and active participation of about 150 scholars from 19 different countries. The members of the national organizing committee are also to be thanked for taking care of problems connected with local arrangements. Our Hungarian hosts deserve our thanks not only for the superb organiza-tion of the conference but also for providing an excellent atmosphere of warmth, friendliness and generous hospitality.

The proceedings of the colloquium are divided into two volumes containing a broad range of topics such as testing and estimation, ranking and selection, empirical distributions and stochastic processes among others. The papers in the volumes were presented at the colloquium and were reviewed by the referees. It is a pleasure to express our appreciation to them as well as the authors for their help and cooperation.

Special thanks are due to the Hungarian Central Statistical Office and to its President Mrs. Vera Nyitrai for their help as well as encouragement in organizing the colloquium.

We hope that the proceedings of the colloquium will be of interest to mathematicians as well as applied statisticians.

Madan L. Puri

CONTENTS
Volume I

Volume II

SCIENTIFIC PROGRAM

June 23, 1980
Morning session (Congress Hall)
Chairman: P. Révész

$9^{30} - 10^{00}$ Opening session

$10^{15} - 11^{00}$ W.R. van Zwet: On the Edgeworth expansion for the simple linear rank statistic

$11^{15} - 12^{00}$ D.R. Cox: The randomization theory of experimental design

Afternoon session

Session A
Chairman: Z. Govindarajulu

$14^{30} - 14^{50}$ P.K. Sen: The UI principle and LMP rank tests

$14^{55} - 15^{15}$ I.R. Savage: Lehmann models

$15^{20} - 15^{40}$ M. Csörgő — P. Révész: An invariance principle for N.N. empirical density functions

Interval

$16^{00} - 16^{20}$ R.J. Does — R. Helmers: Edgeworth expansions for functions of uniform spacings

$16^{25} - 16^{45}$ K. Sarkadi: A direct proof for a ballot type theorem

$16^{50} - 17^{10}$ O.I. Toskin: Control of multidimensional hypothesis about quantity of fractiles of continuous distributions

Session B
Chairman: V.S. Koroljuk

$14^{30} - 15^{15}$ J. Blum – V. Susarla – G. Walter: Estimation of the prior distribution using differential operators

Interval

$16^{00} - 16^{20}$ B. Epstein: Some nonparametric estimation techniques in reliability theory

$16^{25} - 16^{45}$ R. Zmyślony: On admissible estimation in linear models

$16^{50} - 17^{10}$ G. Leunbach: Parameter-free inference in models with sufficient estimation

June 24, 1980
Morning session

Session A
Chairman: V. Mammitzsch

$9^{00} - 9^{45}$ E. Hmaladze: The martingale approach in the theory of nonparametric goodness of fit tests

$9^{50} - 10^{10}$ S.S. Gupta – G.C. McDonald: Nonparametric procedures in multiple decisions (ranking and selection procedures)

Interval

$10^{30} - 10^{50}$ H.S. Konijn: Evaluation of a new treatment

$10^{55} - 11^{15}$ R.S. Sudakov: A confidence bound for function with preventing binomial parameters

$11^{20} - 11^{40}$ H. Callaert – P. Janssen: The convergence rate of fixed width sequential confidence intervals

$11^{45} - 12^{05}$ C. Stepniak: On testability of statistical hypotheses

Session B
Chairman: B. Penkov

$9^{00} - 9^{45}$ J. Jurečková: Tests of location and criterion of tails

$9^{50} - 10^{10}$ B. Gyires: Linear rank statistics generated by uniformly distributed sequences

Interval

$10^{30} - 10^{50}$ Z. Prášková: A local limit theorem and an asymptotic expansion for a two-sample rank test

$10^{55} - 11^{15}$ M. Hušková: On bounded length sequential confidence interval for parameter in regression model based on ranks

$11^{20} - 11^{40}$ D. Vorličková: Exact slopes for rank statistics for a two-sample problem under discrete distribution

$11^{45} - 12^{05}$ T. Gerstenkorn: On some statistical problem joining the inflated Polya distribution to Bayes rule

Afternoon session

Session A
Chairman: D. Rasch

$14^{00} - 14^{20}$ B. de Finetti: Aspects of heuristics

$14^{25} - 14^{45}$ V.K. Rohatgi: Operator self-similarity

$14^{50} - 15^{10}$ H. Grimm: Transformation of variables versus nonparametrics

Interval

$15^{30} - 15^{50}$ G. Enderlein: Hierarchical clustering methods

$15^{55} - 16^{15}$ M. Ghosh − R. Dasgupta: Berry − Esseen theorems for U-statistics in the non i.i.d. case

$16^{20} - 16^{40}$ Z. Grabos: Some applications of spectral decomposition of a matrix in analysis of variance

$16^{45} - 17^{05}$ C. Domański: Notes on the Theil test for the hypothesis of linearity for the model with two explanatory variables

Session B
Chairman: H. Witting

$14^{00} - 14^{20}$ K. Jogdeo: Nonparametric methods for gross error models involving location parameters

$14^{25} - 14^{45}$ V.S. Koroljuk – Y.V. Borovskih: Analytical method in theory of rank statistics

$14^{50} - 15^{10}$ Z. Govindarajulu: Asymptotic normality of linear combinations of functions of order statistics in one and several samples

Interval

$15^{30} - 15^{50}$ U. Müller-Funk: On contiguity and weak convergence with an application to sequential analysis

$15^{55} - 16^{15}$ W. Schlee: Nonparametric tests of the monotony and convexity of regression

$16^{20} - 16^{40}$ S.A. Hashimov: An effective estimate of nonlinear functionals

$16^{45} - 17^{05}$ G. Sadasivan: Within pair order effects in paired comparisons

June 26, 1980

Morning session

Session A

Chairman: J. Blum

$9^{00} - 9^{20}$ J.K. Ghorai — A. Susarla — V. Susarla — J. Van Ryzin: Nonparametric estimation of mean residual life time with censored data

$9^{25} - 9^{45}$ E.G. Phadia: Nonparametric Bayesian inference based on censored data — an overview

$9^{50} - 10^{10}$ V.N. Nair: Goodness of fit tests for multiply right censored data

Interval

$10^{30} - 10^{50}$ N. Reid: Nonparametric estimation for censored data

$10^{55} - 11^{15}$ R. Thrum: Convergence of quadratic forms and applications in mixed models

$11^{20} - 11^{40}$ B. Levit: On asymptotic optimality of nonparametric estimators

$11^{45} - 12^{05}$ B. Prasad — R.S. Singh: Nonparametric kernel estimates of a density function alongwith its derivatives

Session B

Chairman: V.K. Rohatgi

$9^{00} - 9^{20}$ G. Halász: Nonparametric regression and density estimations for smooth functions

$9^{25} - 9^{45}$ B. Kim -- J. Van Ryzin: On the asymptotic distribution of a histogram density estimator

$9^{50} - 10^{10}$ M.A. Mirzahmedov: Nonparametric estimation of probability density

Interval

$10^{30} - 10^{50}$ R.J. Serfling: Properties and applications of metrics on nonparametric density estimators

$10^{55} - 11^{15}$ A. Kozek -- W. Wertz: Estimators with values in Banach spaces under convex loss and applications to curve estimation problems

$11^{20} - 11^{40}$ H.L. Koul: Some weighted empirical inferential procedures for a simple regression model

$11^{45} - 12^{05}$ P. Deheuvels: Some applications of the dependence functions to statistical inference: nonparametric estimates of extreme values distributions, and a Kiefer type universal bound for the uniform test of independence

Session C
Chairman: B. Epstein

9^{00} – 9^{20} K.A. Borovkov: On the rate of convergence in the gener-
                 alized invariance principle

9^{25} – 9^{45} E. Schumacher: Kendall's Tau, used as a coefficient of
                 disarray between permutations with unoccupied
                 places

9^{50} – 10^{10} P. Stolarski: A nonparametric method for classification
                 based on orthonormal series

Interval

10^{30} – 10^{50} W.R. Allen: On Kolmogorov bounds for survival functions
                 when there are losses as well as deaths

10^{55} – 11^{15} V. Mammitzsch: Is the distribution function of
                 $f(X + Y) - X$ continuous, if X and Y are inde-
                 pendent and have continuous distribution func-
                 tions?

11^{20} – 11^{40} M.L. Puri – N.S. Rajaram: Stochastic integrals and rank
                 statistics

Afternoon session

Session A
Chairman: P.K. Sen

$14^{00} - 14^{20}$ E. Csáki: On the standardized empirical distribution function

$14^{25} - 14^{45}$ I.S. Borisov: An approximation of empirical fields

$14^{50} - 15^{10}$ A. Sahanenko: On unimprovable estimates of the rate of convergence in invariance principle

Interval

$15^{30} - 15^{50}$ J.-P. Raoult: Some remarks on generalized Skorohod topology, in connection with weak convergence of multi-dimensional empirical processes (non-stationary φ-mixing case)

$15^{55} - 16^{15}$ S. Csörgő: The empirical moment generating function

$16^{20} - 16^{40}$ H.-G. Müller: Uniformly consistent kernel estimates of a multivariate regression function and of its mixed partial derivatives

Session B
Chairman: W. Wertz

$14^{00} - 14^{45}$ K.B. Athreya – P. Ney: Some aspects of ergodic theory and laws of large numbers for Harris-recurrent Markov chains

$14^{50} - 15^{10}$ W. Jahn: Regression analysis under the condition that the determinant of the covariance matrix is small

Interval

$15^{30} - 15^{50}$ D. Dabrowska: Parametric and nonparametric models with special schemes of stochastic dependence

$15^{55} - 16^{40}$ D. Rasch – G. Herrendörfer: Optimum experiment design and analysis – estimating the regression function

Session C
Chairman: B. Gyires

$14^{00} - 14^{20}$ B. Ceranka – S. Mejza: A proposition for new classification of block design

$14^{25} - 14^{45}$ S. Mejza: On the estimation of treatment contrasts in the linear mixed model for an incomplete block design

$14^{50} - 15^{10}$ P.G. Miasinkov – I.A. Stepanenko: On the nonparametric approach to planning regression experiment

Interval

$15^{30} - 15^{50}$ L. Tar: A generalized model of the Latin square design

$15^{55} - 16^{15}$ S. Molnár: The application of geostatistics in mining

June 27, 1980
Morning session

Session A
Chairman: I.R. Savage

$9^{00} - 9^{20}$ G.R. Shorack – J.A. Wellner: Limit theorems for the uniform empirical process indexed by intervals

$9^{25} - 9^{45}$ H. Koul – V. Susarla – J. Van Ryzin: Regression analysis with randomly censored data

$9^{50} - 10^{10}$ B.B. Winter – A. Földes: Nonparametric failure rate estimation

$10^{15} - 10^{20}$ M. Hollander – R. Korwar: Nonparametric Bayesian estimation of horizontal distance between two populations (presented by I.R. Savage)

Interval

$10^{30} - 10^{50}$ A. Földes – L. Rejtő: On a property of the Kaplan – Meier estimator on the whole line

$10^{55} - 11^{15}$ M.D. Burke: Tests for exponentiality based on randomly censored data

$11^{20} - 11^{40}$ G. Campbell – A. Földes: Large-sample properties of nonparametric bivariate estimators with censored data

$11^{45} - 12^{05}$ L. Horváth: Two-sample problems under random censorship

Session B
Chairman: J. Jurečková

$9^{00} - 9^{20}$ M. Arató: Dynamic placement of records and order statistics

$9^{25} - 9^{45}$ R.-D. Reiss: One-sided test for quantiles in certain nonparametric models

$9^{50} - 10^{10}$ R. Ahmad: On tests of independence for the multiway Markovian contingency tables

Interval

$10^{30} - 10^{50}$ R. Bentkus: On optimal estimates of spectral density

$10^{55} - 11^{15}$ M.L. Tiku: Robust parametric tests versus nonparametric tests

$11^{20} - 11^{40}$ U. Müller-Funk – H. Witting: On the rate of convergence in the CLT for signed linear rank statistics

$11^{45} - 12^{05}$ M. Krzysko: Nonparametric estimation associated with discriminant analysis

Afternoon session

Session A
Chairman: R. Van Zwet

$14^{00} - 14^{20}$ S.G. Mohanty: On some computational aspects of rectangular probabilities

$14^{25} - 14^{45}$ L. Gerencsér – T. Lengyel: A derivative free stochastic Newton method

$14^{50} - 15^{10}$ S. Geman: An application of the method of sieves: functional estimator for the drift of a diffusion

$15^{15} - 15^{35}$ I. Vincze: On the Cramér – Fréchet – Rao inequality

LIST OF PARTICIPANTS

AHMAD, R., Department of Mathematics, University of Strathclyde, Livingstone Tower, 26 Richmond Street, GB-Glasgow G1 1XH, UK.

ALLEN, W.R., 68 Magnolia Lane, Princeton, NJ 08540, USA.

ARATÓ, M., Institute for Coord. of Comp. Techn., P.O. Box 227, 1536, Budapest, Hungary.

BÁNYAI, GY., H-1125 Budapest, Hadik A. u. 7. Hungary.

BARABÁS, B., Dept. of Math., Civil Engineering Faculty, Univ. of Technology, H-1111 Budapest, Egry J. u. 2, Hungary.

BENTKUS, R., Institute of Math. and Cyb., SU-232600 Vilnius, Ul. Poželos 54, USSR.

BERNT, H., Karl-Marx-Universität, Bereich Medizin, Lehrstuhl für Med. Stat. und Dokumentation, 701 Leipzig, Liebigstr. 27, GDR.

BLUM, J.R., Division of Statistics, Univ. of California, Davis, CA 95616, USA.

BORISOV, I.S., Institute of Mathematics, AN USSR, SU-630090 Novosibirsk-90, USSR.

BOROVKOV, A.A., Institute of Mathematics, AN SSR, SU-630090 Novosibirsk-90, USSR.

BURKE, M.D., Dept. of Math. and Stat., University of Calgary, Calgary, Alberta T2N 1N4, Canada.

CAMPBELL, G., Dept. of Stat., Purdue University, West Lafayette, IN 47907, USA.

ČERANKA, B., Akademia Rolnicza, ul. Wojska Polskiego 28, 60-637 Poznań, Poland.

COX, D.R., Dept. of Math., Imperial College, Queen's Gate, London SW7 2BZ, UK.

CSÁKI, E., Math. Inst. of Hung. Acad. Sci., Reáltanoda u. 13-15, H-1053 Budapest, Hungary.

CSÖRGŐ, S., Bolyai Institute, Aradi vértanúk tere 1, 6720 Szeged, Hungary.

DABROWSKA, D., Inst. of Comp. Sci., Polish Acad. Sci., PL-00-901 Warszawa, P.O. Box 22, Poland.

DECKER, A., Institute of Oncology, H-1525 Budapest, P.O. Box 21, Hungary.

DEHEUVELS, P., Inst. de Stat. des Univ. de Paris, Tour 45.55-3e Étage, 4 Place Jussieu, F-75230 Paris, France.

DOBOSI, E., H-1195 Budapest, Petőfi u. 26, Hungary.

DOES, R.J.M.M., University of Limburg, Data Processing Dept., P.O. Box 616, NL-6200 MD. Maastricht, The Netherlands.

DOMAŃSKI, C., University of Łódź, Inst. of Econ. and Stat., PL-90-214 Łódź, ul. Rewolucji 1905 r. 41, Poland.

ENDERLEIN, G., ZI f. Arbeitsmedizin der DDR, DDR-1134 Berlin, Nöldner strasse 40-42, GDR.

EPSTEIN, B., Stat. Area, Israel Inst. of Technology, Haifa, Israel.

de FINETTI, B., piazza Filattiera 49, I-00139 Roma, Italy.

FISCHER, J., AdW der DDR, ZI f. Molekularbiologie, Rechenzentrum, DDR-1115 Berlin, Lindenberger Weg 70, GDR.

GEMAN, S., Div. of Appl. Math., Brown University, Providence, RI 02912, USA.

GERENCSÉR, L., Res. Inst. for Comp. and Automat. Sci., Victor Hugo u. 18-22, 1132 Budapest, Hungary.

GERSTENKORN, T., Inst. of Math., Łódź Univ., PL-90-238 Łódź, ul. Banacha 22, Poland.

GHOSH, M., Dept. of Stat., Iowa State Univ., Ames, Iowa 50011, USA.

GOVINDARAJULU, Z., Dept. of Stat., Univ. of Kentucky, Lexington, Kentucky 40506, USA.

GÖTZE, F., Math. Inst. d. Univ. zu Köln, Weyertal 86, D-5000 Köln 41, GFR.

GRABOŠ, Z., Inst. of Appl. Math., Agricult. Acad., Akademicka 13, PL-20-934 Lublin, Poland.

GRIMM, H., ZI f. Mikrobiologie u. exp. Therapie, AdW der DDR, DDR-69 Jena, Beutenbergstr. 11, GDR.

GUPTA, S.S., Dept. of Stat., Purdue Univ., West Lafayette, IN 47907, USA.

GÜNTHER, R., Friedrich-Schiller-Universität, Sekt. Math., DDR-6900 Jena, Schillerstr., GDR.

GYIRES, B., Math. Inst., Kossuth L. Univ. of Debrecen, P.O. Box 10, 4010 Debrecen, Hungary.

GYŐRFFY, J., H-1016 Budapest, Zsolt u. 7, Hungary.

GYŐRFI, Z., Univ. of Technology, H-1111 Budapest, Stoczek u. 2, Hungary.

HAJTMAN, B., H-1114 Budapest, Szabolcska M. u. 9, Hungary.

HASHIMOV, S.A., Mathematical Institute, SU-700000 GSP Tashkent, USSR.

HEINHOLD, J., Römerstr. 49, D-8035 Gauting, GFR.

HMALADZE, E.V., Math. Inst. AN USSR, Math. Stat. Dept., SU-117333 Moscow, USSR.

HORVÁTH, L., Bolyai Institute, Aradi vértanúk tere 1, 6720 Szeged, Hungary.

HUŠKOVÁ, M., Dept. of Stat., Charles Univ., Sokolovská 83, CS-186 00 Prague 8, Czechoslovakia.

JAHN, W., Karl-Marx-Universität, Sekt. Math., DDR-701 Leipzig, Karl-Marx-Platz 10, GDR.

JANECEK, Z., Prague, Czechoslovakia.

JANSSEN, P., Limburgs Univ. Centrum, Universitaire Campus, B-3610, Belgium.

JEKOVA, V.I., Comm. for Unified System of Social Information at the Council of Ministers, BG-Sofia, Volov. str. 2, Bulgaria.

JOGDEO, K., Dept. of Math., Univ. of Illinois, Urbana, ILL 61801, USA.

JÓZSEF, S., H-1038 Budapest, Márton u. 7, Hungary.

JUREČKOVÁ, J., Dept. of Stat., Charles Univ., Sokolovská 83, CS-186 00 Prague 8, Czechoslovakia.

KABOS, S., H-1025 Budapest, Őzgida u. 3, Hungary.

KARDOS, J., H-1064 Budapest, Izabella u. 70, Hungary.

KELEMEN, Z., H-1055 Budapest, Honvéd u. 38, Hungary.

KIM, B.K., Dept. of Math. and Stat., Memorial Univ. of Newfoundland, St. John's, Canada-Nfld. A1B 3X7, Canada.

KLAJNMIC, H., Electricite de France, Etudes et Recherches, Service IMA, 1 Ave. du Général de Gaulle, F-92141 Clamart, France.

KNIJPSTRA, S., Roghorst 206, NL-6708 KT Wageningen, The Netherlands.

KOLTAY, L., Math. Dept., Chemical Univ. Veszprém, H-8200 Veszprém, P.O. Box 28, Hungary.

KONIJN, H.S., Tel-Aviv University, Fac. of Social Sciences, Department of Statistics, Ramat-Aviv, Tel-Aviv, Israel.

KOROLJUK, V.S., Math. Inst. AN USSR, SU-Kiev, GSP-1, ul. Repina 3, USSR.

KOUL, H.L., Dept. of Stat. and Prob., M.S.U., East Lansing, MI 48824, USA.

KRÁMLI, A., Res. Inst. for Comp. and Automat. Sci., Victor Hugo u. 18-22, 1132 Budapest, Hungary.

KRZYŚKO, M., Inst. of Math., Mickiewicz Univ., Matejki 48/49, PL-60-769 Poznań, Poland.

KUCZYŃSKI, M., Inst. of Appl. Math., Agricult. Acad., Akademicka 13, PL-20-934 Lublin, Poland.

LEUNBACH, G., Danish Inst. f. Educ. Res., Hermodsgade 28, DK-2200 Copenhagen N, Denmark.

LEVIT, B., Institute "Orgenergostroy", SU-113 105 Moscow, Varshavskoe Shosse 17, USSR.

MAMMITZSCH, V., FB Math., Univ. Marburg, Lahnberge, D-3550 Margurg/Lahn, GFR.

MARTON, Á., Hungarian Central Statistical Office, H-1024 Budapest, Keleti K. u. 5-7, Hungary.

MEJZA, S., Akademia Rolnicza, ul. Wojska Polskiego 28, PL-60-637 Poznań, Poland.

MIASINKOV, D.G., Mashinostroitelnij Inst. Kaf. VM, SU-640669 Kurgan, pl. Lenina, USSR.

MIKOS, H., Dept. of Appl. Math., Agricult. Acad., Akademicka 13, PL-20-934 Lublin, Poland.

MIRZAHMEDOV, M.A., Gosuniversitet, Mat. Fak., SU-Tashkent, USSR.

MOHANTY, S.G., Dept. of Math. Sci., McMaster Univ., Hamilton, Ont. L8S 4K1, Canada.

MOLLIERE, J.L., Electricite de France, Etudes et Recherches, Service IMA, 1 Ave. du Général de Gaulle, F-92141 Clamart, France.

MOLNÁR, L.G., Univ. of Technology, H-1111 Budapest, Stoczek u. 2, Hungary.

MOLNÁR, S., KBFI, H-1025 Budapest, Varsányi I. u. 40-44, Hungary.

MÓRI, T., H-1096 Budapest, Hámán K. u. 44, Hungary.

MÜLLER, H.-G., Universität Heidelberg, Im Neuenheimer Feld 293, D-6900 Heidelberg 1, GFR.

MÜLLER, W., ZI f. Herz- u. Kreislauf-Regulationsforschung der AdW der DDR, DDR-1115 Berlin, Wiltbergstr. 50, GDR.

MÜLLER-FUNK, U., Institut für Math. Stoch. der Albert-Ludwig-Universität, Hebelstr. 27, D-7800, Freiburg, GFR.

NAIR, V.N., WB 1J-304, Bell Telephone Laboratories, Holmdel, NJ 07733, USA.

NEY, P., Math. Dept., University of Wisconsin, Madison, WI 53706, USA.

NOVOVIĆOVÁ, J., Inst. of Inf. Theory and Automation, Pod. Vodárenskou Věží 4, CS-Prague 8, Czechoslovakia.

NYITRAI, V., Hungarian Central Statistical Office, H-1024 Budapest, Keleti K. u. 5-7, Hungary.

PÁL, M., H-1111 Budapest, Lágymányosi u. 14/a, Hungary.

PENKOV, B., P.O. Box 373, BG-1090 Sofia, Bulgaria.

PESTI, L., Hungarian Central Statistical Office, H-1024 Budapest, Keleti K. u. 5-7, Hungary.

PHADIA, E.G., Dept. of Math., W. Paterson College of New Jersey, Wayne, N.J. 07470, USA.

PRÁŠKOVÁ, Z., Dept. of Stat., Charles Univ., Sokolovská 83, CS-186 00 Prague 8, Czechoslovakia.

PRŐHLE, T., H-1113 Budapest, Villányi u. 38, Hungary.

PURI, M.L., Dept. of Math., Indiana Univ., Bloomington, IN 47405, USA.

RAISZ, P., H-3530 Miskolc, Vándor S. u. 6, Hungary.

RASCH, D., DDR 25, Rostock, Karl-Marx-str. 15, GDR.

RAOULT, J.-P., Faculté des Sciences et des Techniques de l'Université de Rouen, B.P. 67, 76130 Mont-Saint-Aignan, France.

REID, N., Dept. of Math., Imperial College, Queen's Gate, London SW7 2BZ, UK.

REISS, R.-D., University of Siegen, FB6, D-5900 Siegen 21, Hölderlinstr. 3, GFR.

REJTŐ, L., Math. Inst. of Hung. Acad. Sci., Reáltanoda u. 13-15, H-1053 Budapest, Hungary.

RÉVÉSZ, P., Math. Inst. of Hung. Acad. Sci., Reáltanoda u. 13-15, H-1053 Budapest, Hungary.

ROHATGI, V.K., Bowling Green State Univ., Bowling Green, Ohio 43403, USA.

ROSTÁS, J., H-1113 Budapest, Dávid F. u. 6, Hungary.

RUDAS, T., Eötvös L. Univ., H-1052 Budapest, Pesti B. u. 1, Hungary.

RUJBANI, S.M., Libanon.

Van RYZIN, J., Div. of Biostat., Columbia Univ., School of Public Health, 600 W. 168th St., New York, NY 10032, USA.

SADASIVAN, G., I.A.S.R.I., Library Avenue, 110012-New-Delhi, India.

SAHANENKO, A., Institute of Mathematics, AN SSR, SU-630090 Novosibirsk-90, USSR.

SARKADI, K., Math. Inst. of Hung. Acad. Sci., Reáltanoda u. 13-15, H-1053 Budapest, Hungary.

SAVAGE, Yale University, Dept. of Stat., Box 2179 Yale Station, New Haven, Connecticut 06520, USA.

SCHLEE, W., TU, Inst. f. Stat. u. Unterf., D-8000 München 2, Arcisstr. 21, GFR.

SCHUMACHER, E., Universität Hohenheim, Inst. f. Angew. Math. und Stat., D-7000 Stuttgart 70, Schloss, Westhof-Süd, GFR.

SEN, P.K., Dept. of Biostat., 201 H Univ. of North Carolina, Chapel Hill, NC 27514, USA.

SERFLING, R.J., Dept. of Math. Sci., Johns Hopkins Univ., Baltimore, Maryland 21218, USA.

SIMAHIN, V.A., Mashinostroitelnij Inst., Kaf. VM, SU-640669, pl. Lenina, USSR.

SINGH, R.S., University of Guelph, Dept. of Math. and Statist., Guelph, Ontario, N1G 2W1, Canada.

SOLT, GY., H-1027 Budapest, Mártirok ú. 56, Hungary.

STEPNIAK, C., Inst. of Appl. Math., Agricult. Acad., Akademicka 13, PL-20-934 Lublin, Poland.

STOLARSKI, P., Inst. of Math., A. Mickiewicz Univ., Matejki 48/49, PL-60-769 Poznań, Poland.

SUDAKOV, R.S., SU-129010 Moscow, ul. Bolshaja, Spassija 32, kv. 181, USSR.

SUSARLA, V., Dept. of Stat. and Prob., M.S.U., East Lansing, MI 48824, USA.

SZÉKELY, G.J., Dept. of Prob., Eötvös L. Univ., H-1088 Budapest, Múzeum krt. 6-8, Hungary.

SZÜTS, I., H-1426 Budapest, Buzogány u. 10-12, Hungary.

TAR, L., H-4027 Debrecen, Ujkert u. 13, Hungary.

TEJMAN, A.J., USSR.

TIKU, M.L., Dept. of Math. Sci., McMaster Univ., Hamilton, Ontario, L8S 4K1, Canada.

TIOSKIN, O.I., SU-131090 Moskovskaja obl., Bolsevo-1, d. 6, kv. 57, USSR.

TUSNÁDY, G., Math. Inst. of Hung. Acad. Sci., Reáltanoda u. 13-15, H-1053 Budapest, Hungary.

VARGHA, M., H-1107 Budapest, Fokos u. 12, Hungary.

VIGASSY, J., Central Physical Res. Inst. Hung. Acad. Sci., H-1525 Budapest, 114, P.O. Box 49, Hungary.

VINCZE, I., Math. Inst. of Hung. Acad. Sci., Reáltanoda u. 13-15, H-1053 Budapest, Hungary.

VISY, K., H-1112 Budapest, Kérő u. 20, Hungary.

VORLÍČKOVÁ, D., Dept. of Stat., Charles Univ., Sokolovská 83, CS-186 00 Prague 8, Czechoslovakia.

WANDL, H., ZI f. Math. u. Mech. der AdW der DDR, DDR-108 Berlin, Mohrestr. 39, GDR.

WEIMA, J., Agricultural University, De Dreijen 8, NL-Wageningen, The Netherlands.

WELLNER, J.A., University of Rochester, Dept. of Stat., Rochester, NY 14627, USA.

WERTZ, W., Institut f. Statistik, TU Wien, Karlsplatz 13, A-1040 Wien, Austria.

WESELOWSKA-JACZAREK, M., Dept. of Appl. Math., Agricult. Acad., Akademicka 13, PL-20-934 Lublin, Poland.

WIEGHARDT, M., H-1014 Budapest, Bécsikapu tér 1, Hungary.

WINTER, B.B., Math. Dept., Univ. of Ottawa, Ottawa, Ontario, K1N 9B4, Canada.

WITTING, H.,   Institut für Math. Stoch. der Albert-Ludwig-Universität,
   Hebelstr. 27, D-7800, Freiburg, GFR.
WÓJCIK, A.R.,   Inst. of Comp. Sci., Polish Acad. Sci., PL-00-901
   Warszawa, P.O. Box 22, Poland.
ZMYŚLONY, R.,   ul. Bacciarellego 63/10, PL-51-649 Wrocław, Poland.
Van ZWET, W.R.,   University of Leiden, Dept. of Math., Wassenaarseweg
   80, P.O. Box 9512, NL-2300 Leiden, The Netherlands.

COLLOQUIA MATHEMATICA SOCIETATIS JÁNOS BOLYAI

32. NONPARAMETRIC STATISTICAL INFERENCE,

BUDAPEST (HUNGARY), 1980.

ON TESTS OF INDEPENDENCE FOR THE MULTIWAY MARKOVIAN CONTINGENCY TABLES

R. AHMAD

1. INTRODUCTION AND PRELIMINARIES

The statistical inference structure on Markov chains and processes has been studied by B a r t l e t t [3], H o e l [15], C o c h r a n [8], [9], A n d e r s o n and G o o d m a n [2], B i l l i n g s l e y [5], R a o [23], [24], J o h n s o n and R o u s s a s [16], R o u s s a s [25], A k r i t a s and R o u s s a s [1], B a s a w a and R a o [4] among others. Recently, some investigations in contingency tables were carried out by D a r r o c h [10], B i r c h [6], P l a c k e t t [21], [22], G o o d m a n [12], [13], [14], M a n t e l [18], B i s h o p et al. [7], W i l l i a m s and G r i z z l e [27], W i l l i a m s [28], F i e n b e r g [11] and others.

In a paper P a t e l and A h m a d [20] dealt with testing the hypothesis of independence of two attributes in $r \times c$ Markovian contingency tables. In essence, most of the above results fall in the family of asymptotic χ^2-test class and its variants. The objective of this investigation is to give a generalization of tests of independence from $r \times c$ to the multiway multidimensional Markovian contingency tables. In practice, the situations which give rise to multidimensional Markovian contingency tables exist in medical diagnoses, sociopsychological analysis, mental and physiological tests. Furthermore, the area of computer tomography (the study of making acoustic vibrations visible by using computers) in connection with measurements of human brain volumes to study intellect and emotion etc., and various similar problems in social and behavioural sciences give rise to multidimensional Markovian contingency tables.

2. MULTIDIMENSIONAL MARKOVIAN CONTINGENCY TABLES: A GENERAL STRUCTURE

Let $A^1, A^2, \ldots, A^m$ be a set of m attributes with $a_1, a_2, \ldots, a_m$ mutually exclusive states, respectively, that is we have the following situation:

$$A^1 = (A_1^1, A_2^1, \ldots, A_{a_1}^1)$$

$$A^2 = (A_1^2, A_2^2, \ldots, A_{a_2}^2)$$

(2.1)
$$\cdot$$
$$\cdot$$
$$\cdot$$

$$A^m = (A_1^m, A_2^m, \ldots, A_{a_m}^m)$$

Assume a random sample of n individuals is drawn from a given population at time $t = 0$ and each of them is observed with regard to attributes $\{A^j, j = 1, 2, \ldots, m\}$ at time $t = 0, 1, 2, \ldots, T$. The data consists of $(T + 1)$ $a_1 \times a_2 \times \ldots \times a_m$ time dependent contingency tables. Now let

(2.2) $\{X_i^1(t), X_i^2(t), \ldots, X_i^m(t); \ i = 1, 2, \ldots, n\}$

be a m-dimensional random vector corresponding to the i-th individual at time t defined as

$$X_i^1(t) = \begin{cases} j^1 & \text{if the } i\text{-th individual at time } t \text{ belongs to the} \\ & j\text{-th category of attribute } A^1, \\ 0 & \text{otherwise;} \end{cases}$$

$$X_i^2(t) = \begin{cases} k^2 & \text{if the } i\text{-th individual at time } t \text{ belongs to the} \\ & k\text{-th category of attribute } A^2, \\ 0 & \text{otherwise;} \end{cases}$$

(2.3)

.
.
.

$$X_i^m(t) = \begin{cases} r^m & \text{if the } i\text{-th individual at time } t \text{ belongs to the} \\ & r\text{-th category of attribute } A^m, \\ 0 & \text{otherwise,} \end{cases}$$

where $j^1 = 1, 2, \ldots, a_1$; $k^2 = 1, 2, \ldots, a_2$; $\ldots$; $r^m = 1, 2, \ldots, a_m$; and $t = 0, 1, 2, \ldots, T$. Note that the superscript on j, k, etc. refers to the 1-st, 2-nd, etc. attribute, and we shall drop the superscript if no confusion arises. The successive observations at various times of the stochastic process generated by the above random vector are assumed to follow a stationary Markov chain of first order for each i. All the individuals are supposed to behave independently of each other. Now, we define the related parameters as below:

(2.4) $\quad p_{jk \ldots r} = P\{X_i^1(t) = j, X_i^2(t) = k, \ldots, X_i^m(t) = r\}$,

(∗∗) $\quad \begin{array}{l} (1 \leqslant j \leqslant a_1 ; 1 \leqslant k \leqslant a_2 ; \ldots ; 1 \leqslant r \leqslant a_m ; \\ t = 0, 1, 2, \ldots, T; \ i = 1, 2, \ldots, n) \end{array}$

$$p_{j \ldots r(q \ldots s)} =$$

(2.5) $\quad = P\{X_i^1(t) = q, \ldots, X_i^m(t) = s \,|\, X_i^1(t-1) =$

$\quad = j, \ldots, X_i^m(t-1) = r\}$,

(over ∗∗ with $q \leqslant a_1, \ldots, s \leqslant a_m$)

(2.6) $\qquad \gamma_j^1 = P\{X_i^1(t) = j\}, \ldots, \gamma_r^m = P\{X_i^m(t) = r\},$

$$\gamma_{j(q)}^1 = P\{X_i^1(t) = q \mid X_i^1(t-1) = j\},$$

(2.7) $\qquad \vdots$

$$\gamma_{r(s)}^m = P\{X_i^m(t) = s \mid X_i^m(t-1) = r\},$$

(all over **). From the above definitions it is clear that the following equalities hold:

$$(2.8) \qquad \sum_{j=1}^{a_1} \gamma_j^1 \gamma_{j(q)}^1 = \gamma_q^1, \ldots, \sum_{r=1}^{a_m} \gamma_r^m \gamma_{r(s)}^m = \gamma_s^m,$$

$$(2.9) \qquad \sum_{j=1}^{a_1} \sum_{k=1}^{a_2} \cdots \sum_{r=1}^{a_m} P_{jk\ldots r} P_{jk\ldots r(q\ldots s)} = P_{q\ldots s}.$$

For the above general structure the null hypothesis can be written as $H_0 \equiv H_0(t=0) \cap H_0(t=1,2,\ldots,T)$, where

$$(2.10)(i) \quad H_0(t=0): \quad p_{jk\ldots r} = \gamma_j^1 \gamma_k^2 \cdots \gamma_r^m \quad \text{for all} \quad j, k, \ldots, r;$$

$$(2.11)(ii) \qquad \begin{aligned} H_0(t=1,2,\ldots,T): \quad & p_{jk\ldots r(qp\ldots s)} = \gamma_{j(q)}^1 \gamma_{k(p)}^2 \cdots \gamma_{r(s)}^m \\ & \text{for all} \quad j, k, \ldots, r \quad \text{and} \quad q, p, \ldots, s. \end{aligned}$$

The first part of the null hypothesis means that the initial random sample has come from a population with m independent attributes, and in part (ii) one considers that the sequence of changes in the classifications are mutually stochastically independent. In fact, one can construct several variants (both realistic and practicable) of the above hypothesis by introducing pairwise independence of the classifications. In these situations there are many interesting combinations of null hypotheses. Clearly, the analysis in such cases is more intricate but conceptually very simple. However, in the sequel, we restrict our attention to the null nypothesis H_0 as defined above. The alternative hypothesis of interest is that at least one of the above two conditions is not valid.

To get the appropriate maximum likelihood estimates for the unknown parameters we define the following statistics:

(2.12) $n_{jk...r}^{(0)}$ = the number of individuals in state $(j, k, \ldots, r)$ at time 0,

(2.13) $n_{jk...r(qp...s)}^{(t)}$ = the number of individuals in state $(j, k, \ldots, r)$ at time $t-1$, and in state $(q, p, \ldots, s)$ at time t,

where $1 \leqslant j, q \leqslant a_1; 1 \leqslant k, p \leqslant a_2; \ldots; 1 \leqslant r, s \leqslant a_m;$ and $t = 1, 2, \ldots, T$. Obviously, $n_{jk...r}^{(0)}$ and $n_{jk...r(qp...s)}^{(t)}$ are sufficient statistics for $p_{jk...r}$ and $p_{jk...r(qp...s)}$, respectively. Hence, the likelihood function in the unrestricted parameter space Ω is given by

$$
L(\Omega) = \prod_{j,q=1}^{a_1} \prod_{k,p=1}^{a_2} \cdots \prod_{r,s=1}^{a_m} [p_{jk...r(qp...s)}]^{n_{jk...r(qp...s)}} \times
$$

(2.14)

$$
\times \prod_{j=1}^{a_1} \prod_{k=1}^{a_2} \cdots \prod_{r=1}^{a_m} p_{jk...r}^{n_{jk...r}^{(0)}}
$$

where

(2.15) $n_{jk...r(qp...s)} = \sum_{t=1}^{T} n_{jk...r(qp...s)}^{(t)}$.

On the other hand, the likelihood function in the restricted parameter space ω, under the null hypothesis is given by

$$
L(\omega) = \prod_{j,q=1}^{a_1} [\gamma_{j(q)}^1]^{u_{j(q)}^1} \cdots \prod_{r,s=1}^{a_m} [\gamma_{r(s)}^m]^{u_{r(s)}^m} \times
$$

(2.16)

$$
\times \prod_{j=1}^{a_1} [\gamma_j^1]^{v_j^1} \cdots \prod_{r=1}^{a_m} [\gamma_r^m]^{v_r^m},
$$

where

$$v_j^1 = \sum_{k=1}^{a_2} \cdots \sum_{r=1}^{a_m} n_{jk\ldots r}(0), \ldots, v_r^m = \sum_{j=1}^{a_1} \cdots \sum_{w=1}^{a_{m-1}} n_{jk\ldots wr}(0)$$

$$u_{j(q)}^1 = \sum_{t=1}^{T} \sum_{k=1}^{a_2} \cdots \sum_{r=1}^{a_m} \sum_{p=1}^{a_2} \cdots \sum_{s=1}^{a_m} n_{jk\ldots r(qp\ldots s)}(t)$$

(2.17)

$$\cdot$$
$$\cdot$$
$$\cdot$$

$$u_{r(s)}^m = \sum_{t=1}^{T} \sum_{j=1}^{a_1} \cdots \sum_{z=1}^{a_{m-1}} \sum_{q=1}^{a_1} \cdots \sum_{w=1}^{a_{m-1}} n_{jk\ldots zr(qp\ldots ws)}(t).$$

The maximum likelihood estimates for the parameters in the equations (2.14) and (2.16) are given by

$$\hat{p}_{jk\ldots r(qp\ldots s)} = \frac{n_{jk\ldots r(qp\ldots s)}}{\sum\limits_{q=1}^{a_1} \cdots \sum\limits_{s=1}^{a_m} n_{jk\ldots r(qp\ldots s)}},$$

$$\hat{p}_{jk\ldots r} = \frac{n_{jk\ldots r}(0)}{n},$$

(2.18)

$$\hat{\gamma}_{j(q)}^1 = \frac{u_{j(q)}^1}{\sum\limits_{q=1}^{a_1} u_{j(q)}^1}, \ldots, \hat{\gamma}_{r(s)}^m = \frac{u_{r(s)}^m}{\sum\limits_{s=1}^{a_m} u_{r(s)}^m},$$

$$\hat{\gamma}_j^1 = \frac{v_j^1}{n}, \ldots, \hat{\gamma}_r^m = \frac{v_r^m}{n}.$$

Under the unrestricted parameter space Ω, we estimate, say $d(\Omega)$, parameters: $\{p_{jk\ldots r}, p_{jk\ldots r(qp\ldots s)}\}$ with the side restriction

(2.19) $$\sum_{j=1}^{a_1} \cdots \sum_{r=1}^{a_m} p_{jk\ldots r} = 1 = \sum_{q=1}^{a_1} \cdots \sum_{s=1}^{a_m} p_{jk\ldots r(qp\ldots s)}.$$

Similarly, for the restricted parameter space ω, that is under the null hypothesis, we estimate, say $d(\omega)$, parameters $\{\gamma_j^1, \ldots, \gamma_r^m; \gamma_{j(q)}^1, \ldots, \gamma_{r(s)}^m\}$ with the restriction that

$$(2.20) \quad \sum_{j=1}^{a_1} \gamma_j^1 = \ldots = \sum_{r=1}^{a_m} \gamma_r^m = \sum_{q=1}^{a_1} \gamma_{j(q)}^1 = \ldots = \sum_{s=1}^{a_m} \gamma_{r(s)}^m = 1.$$

By substituting the maximum likelihood estimates given by (2.18), one gets the likelihood ratio test statistic $\lambda = \dfrac{L(\hat{\omega})}{L(\hat{\Omega})}$. This gives us the well-known statistic $-2 \log \lambda$ which has an asymptotic χ^2-distribution with $d = d(\Omega) - d(\omega)$ degrees of freedom.

Remark 2.1. In the null hypothesis defined by (2.10) and (2.11), one has

$$d(\Omega) = \left(\prod_{i=1}^{m} a_i \right)^2 - 1 \quad \text{and} \quad d(\omega) = \sum_{i=1}^{m} (a_i^2 - 1).$$

For various other hypotheses, these degrees of freedom will be different.

To construct the χ^2-statistic of the form used in contingency tables to test the hypothesis of independence, first we let

$$(2.21) \quad n_{jk\ldots r} = \sum_{q=1}^{a_1} \ldots \sum_{s=1}^{a_m} n_{jk\ldots r(q\ldots s)}.$$

Now, define the following two measures of discrepancy:

$$(2.22) \quad D_1^* = \sum_{j=1}^{a_1} \ldots \sum_{r=1}^{a_m} \frac{[n_{jk\ldots r}(0) - n\hat{\gamma}_j^1 \hat{\gamma}_k^2 \ldots \hat{\gamma}_r^m]^2}{n\hat{\gamma}_j^1 \hat{\gamma}_k^2 \ldots \hat{\gamma}_r^m}$$

$$(2.23) \quad D_2^* = \sum_{j=1}^{a_1} \sum_{q=1}^{a_1} \ldots \sum_{r=1}^{a_m} \sum_{s=1}^{a_m} \frac{1}{n_{jk\ldots r}\hat{\gamma}_{j(q)}^1 \hat{\gamma}_{k(p)}^2 \ldots \hat{\gamma}_{r(s)}^m} \times$$

$$\times [n_{jk\ldots r(qp\ldots s)} - n_{jk\ldots r}\hat{\gamma}_{j(q)}^1 \hat{\gamma}_{k(p)}^2 \ldots \hat{\gamma}_{r(s)}^m]^2.$$

The above two statistics D_1^* and D_2^*, respectively, correspond to the part (i) and part (ii) of the null hypothesis. By using similar assumptions, conditions and arguments as in the Theorem 4.1 of Patel and Ahmad [20], one can obtain the following general result.

Lemma 2.1. *Under the null hypothesis the three statements given below are true:*

(i)   *For large* n, D_1^* *has an asymptotic* χ^2-*distribution with* $d_1 =$

$$= \prod_{i=1}^{m} (a_i - 1) \quad \text{degrees of freedom;}$$

(ii)   *For large* n *or* T, D_2^* *has an asymptotic* χ^2-*distribution with* $d_2 = d - d_1$ *degrees of freedom;*

(iii)   *The statistics* D_1^* *and* D_2^* *are asymptotically independent.* *Consequently,* $T^* = D_1^* + D_2^*$ *has an asymptotic* χ^2-*distribution with* $d = d_1 + d_2$ *degrees of freedom.*

Thus to test for independence of a multiway Markovian contingency table either we can use $-2 \log \lambda$, the likelihood ratio statistic, or the χ^2-statistic T^* as developed above. Both of these statistics have the same asymptotic χ^2-distribution with d degrees of freedom.

The above results, with some essential modifications, can be extended to r-th order Markovian contingency tables, sub-Markovian contingency tables and strong Markovian contingency tables.

Furthermore, various other null hypotheses of interest can be tested through this approach. In essence, the results are similar to those developed for the classical large sample multiway contingency tables, with multinomial structure. Of course, here the sources of variation are complex and large in number, which somewhat complicates the enumeration of the number of unknown parameters to be estimated and the corresponding degrees of freedom.

Acknowledgements. The author is grateful to a referee whose helpful comments on an earlier version of this paper led to improvements in both the content and the exposition.

REFERENCES

[1] M.G. Akritas – G.G. Roussas, Asymptotic inference in continuous time semi-Markov processes, *Scand. J. Statist.*, 7 (1979), 73-79.

[2]   T.W. Anderson – L.A. Goodman,   Statistical inference
       about Markov chains, *Ann. Math. Statist., 28* (1957), 89-110.

[3]   M.S. Bartlett,   The frequency goodness of fit test for prob-
       ability chains, *Proc. Cambridge, Phil. Soc., 47* (1951), 86-95.

[4]   I.V. Basawa – B.L.S.P. Rao,   *Statistical inference for
       stochastic processes,* Academic Press, 1980.

[5]   P. Billingsley,   Statistical methods in Markov chains, *Ann.
       Math. Statist., 32* (1961), 12-40.

[6]   M.W. Birch,   Maximum likelihood in three way contingency
       tables, *J.R. Statist. Soc. Ser. B, 25* (1963), 220-233.

[7]   Y.M.M. Bishop – S.E. Fienberg – P. Holland, *Dis-
       crete multivariate analysis,*   M.I.T. Press, Cambridge, Mass., 1975.

[8]   W.G. Cochran,   The χ^2-test of goodness of fit, *Ann. Math.
       Statist., 23* (1952), 315-346.

[9]   W.G. Cochran,   Some methods of strengthening the common
       χ^2-tests, *Biometrics, 10* (1954), 417-451.

[10]   J.N. Darroch,   Interactions in multifactor contingency tables,
       J.R. Statist. Soc. Ser. B, 24 (1962), 251-263.

[11]   S.E. Fienberg, The use of chi-squared statistics for categorical
       data problems, *J.R. Statist. Soc. Ser. B, 41* (1979), 54-64.

[12]   L.A. Goodman,   The analysis of multidimensional contingency
       tables: Stepwise procedures and direct estimation methods for
       building models for multiple classifications,   *Technometrics, 12*
       (1970), 33-61.

[13]   L.A. Goodman,   Partitioning of chi-square, analysis of mar-
       ginal contingency tables, and estimation of expected frequencies
       in multidimensional contingency tables, *J. Amer. Statist. Assoc.,
       66* (1971), 339-344.

[14] L.A. Goodman, The analysis of multiple contingency tables when some variables are posterior to others: a modified path analysis approach, *Biometrika,* 60 (1973), 179-192.

[15] P.G. Hoel, A test for Markov chains, *Biometrika,* 41 (1954), 430-433.

[16] R.A. Johnson — G.G. Roussas, Asymptotically optimal tests in Markov processes, *Ann. Math. Statist.,* 41 (1970), 918-938.

[17] M.G. Kendall — A. Stewart, *Advanced theory of statistics,* Vol. 2 and 3, Hafner, London, 1961, 1966.

[18] N. Mantel, Incomplete contingency tables, *Biometrics,* 26 (1970), 291-304.

[19] V.K. Murthy — A.V. Gafarian, Limiting distribution of some variations of the chi-square statistics, *Ann. Math. Statist.,* 41 (1970), 188-194.

[20] H.I. Patel — R. Ahmad, On tests of independence for $r \times c$ Markovian contingency tables, *Ann. Inst. Statist. Math.,* 25 (1973), 355-361.

[21] R.L. Plackett, Multidimensional contingency tables: a survey of models and methods, *Bull. Inter. Statist. Inst.,* 43 (1969), 133-142.

[22] R.L. Plackett, *The analysis of categorical data,* Griffin, London, 1974.

[23] M.M. Rao, Inference in stochastic processes, II and III., *Z. Wahrscheinlichkeitstheorie verw. Geb.,* 5 (1966), 317-335, 8 (1967), 49-72.

[24] M.M. Rao, Inference in stochastic processes IV: predictions and projections, *Sankhya Ser. A,* 36 (1974), 63-120.

[25] G.G. Roussas, Asymptotic distribution of the log-likelihood function for stochastic processes, *Z. Wahrscheinlichkeitstheorie verw. Geb.*, 47 (1979), 31-46.

[26] A.S. Whittemore, Collapsibility of multidimensional tables, *J.R. Statist. Soc. Ser. B*, 40 (1978), 328-340.

[27] O.D. Williams − J.E. Grizzle, Analysis of contingency tables, *Biometrics*, 28 (1972), 177-202.

[28] D.A. Williams, Improved likelihood ratio tests for complete contingency tables, *Biometrika*, 63 (1976), 33-37.

R. Ahmad

Dept. of Math., University of Strathclyde, Livingstone Tower, 26 Richmond Street, GB-Glasgow G1 1XH.

SOME ASPECTS OF ERGODIC THEORY AND LAWS OF LARGE NUMBERS FOR HARRIS-RECURRENT MARKOV CHAINS

K.B. ATHREYA — P. NEY

1. INTRODUCTION

The theory of Markov chains on a general state space was originated by Doeblin [4]. Significant contributions to its development have been made by Harris [5] and Orey [11]. The latter's 1971 monograph contains an excellent account of the state of the theory at that time, as do the books of Neveu [8] and Revuz [12]. In these works the reader can find complete bibliographies and references to other contributors to the literature.

In recent years techniques have been developed which have greatly simplified the proofs of the principal theorems, and led to several improvements and generalizations. The basic idea in this approach is to show that in a suitable sense *all* Harris-recurrent chains have a recurrent atom, and then to take advantage of the embedded renewal sequence of return times to this state. This device was discovered (concurrently and independently) by E. Nummelin [9] and by ourselves [1] and has been used to prove such limit laws as Orey's ergodic theorem, weak (Cesaro mean) ratio theorems, individual (strong) ratio theorems; ergodic theorems for semi-Markov processes; as well as extensions to general non-negative kernels including the Chacon – Ornstein theorem (e.g. Nummelin [10], Athreya, McDonald, Ney [2], Athreya, Ney [3]).

In this lecture we present a brief survey of these developments, focusing in some detail on the strong ratio theorems. These results are "somewhat" new in that the hypotheses and proofs are more transparent. Also the proof of the weak ergodic theorem given here is a further simplification of our earlier argument (via a martingale Wald's identity argument).

2. THE WEAK ERGODIC THEOREM AND LAW OF LARGE NUMBERS

Let $\{X_n\}_0^\infty$ be a Harris chain on a measurable state space $(S, \mathscr{S})$ with transition function $P(\cdot, \cdot)$ and reference measure $\varphi(\cdot)$. In [1], we showed that there is a regeneration scheme for such chains in the following sense: there exists an integer $n_0 \geqslant 1$ for which the skeleton chain $Y_n = \{X_{nn_0}\}_{n=0}^\infty$ can be extended to a chain $\widetilde{Y}_n$ on an enlarged state space $\widetilde{S} = S \cup \{\Delta\}$ with transition function $\widetilde{P}$, where Δ is a *recurrence point* for $\widetilde{Y}_n$; namely $\widetilde{P}_x(\widetilde{Y}_n = \Delta$ for some $n \geqslant 1) \equiv 1$, and for any bounded measurable f on $(S, \mathscr{S})$ there exists an extension $\widetilde{f}$ to $\widetilde{S}$ such that $(P^n f)(x) \equiv E_x f(Y_n) = E_x \widetilde{f}(\widetilde{Y}_n) \equiv \widetilde{P}^n \widetilde{f}(x)$ for all x in S. Thus, for the Harris chain, the study of $(P^n f)(x)$, and in particular that of the sums $\sum_{j=0}^n (P^j f)(x)$ or of their ratios, is easily deduced from the case when there exists a recurrence point. For the details of this analysis the reader is referred to Athreya, Ney [1], [3]. We will assume in Sections 2 and 3 (without loss of generality) that such a recurrence point exists. The results are stated in full generality in Section 4.

We recall the following result from [1].

Theorem 1. *The measure*

$$(2.1) \qquad v(A) = E_\Delta \Big(\sum_{j=0}^{T_\Delta - 1} \chi_A(X_j) \Big),$$

where $T_\Delta = \inf\{n: n \geqslant 1, X_n = \Delta\}$, is an invariant measure for P, namely $\nu P = \nu$. If $\lambda(\cdot)$ is another σ-finite measure satisfying $\lambda P \leqslant \lambda$, then $\lambda(\cdot) = \lambda(\Delta)\nu(\cdot)$.

The following ergodic theorem is a special case of Theorem 3.1 in [3], but the probabilistic proof here is somewhat easier.

Theorem 2. *Let $u_n = P_\Delta(X_n = \Delta)$, $U_n = \sum_{j=0}^{n} u_j$. Then for all $f \in L_1(\nu)$*

$$(2.2) \qquad \lim_{n \to \infty} \frac{1}{U_n} \sum_{j=0}^{n} E_x f(X_j) = \int f d\nu$$

for almost all x with respect to $\nu(\cdot)$.

Proof. Let $N_0 = 0$, $N_{j+1} = \inf\{n: n \geqslant N_j + 1, X_n = \Delta\}$ for $j \geqslant 0$. By the Markov property the excursions $\{X_i; N_j \leqslant i < N_{j+1}\}$, $j \geqslant 1$, are independent and identically distributed. Let $K_n = \sup\{j: N_j \leqslant n\}$ and for given $f \geqslant 0$, let $\eta_j(f) = \sum_{i=N_j}^{N_{j+1}-1} f(X_i)$. Then

$$(2.3) \qquad \sum_{j=0}^{K_n} \eta_j(f) \leqslant \sum_{i=0}^{n} f(X_i) \leqslant \sum_{j=0}^{K_n+1} \eta_j(f).$$

By Wald's identity (applied to K_n, and then to $K_n + 1$)

$$E_\Delta \left(\sum_{j=0}^{K_n} \eta_j(f) \right) = (E_\Delta \eta_0(f))(E_\Delta K_n) = \left(\int f d\nu \right) E_\Delta K_n.$$

But $E_\Delta K_n = E_\Delta$ (# returns to Δ by time n) $= U_n$, and hence by (2.3)

$$(2.4) \qquad U_n \left(\int f d\nu \right) \leqslant E_\Delta \left(\sum_{j=0}^{n} f(X_i) \right) \leqslant (U_n + 1) \left(\int f d\nu \right),$$

implying

$$\lim_{n \to \infty} \frac{1}{U_n} \sum_{j=0}^{n} E_\Delta f(X_j) = \int f d\nu.$$

(This fact could also have been derived by a renewal argument.)

But from (2.4) we see that

$$\mathsf{E}_x(\eta_0(f)) + (U_n - 1)\left(\int f d\nu\right) \leqslant \mathsf{E}_x\left(\sum_{j=0}^{n} f(X_j)\right) \leqslant$$

(2.5)

$$\leqslant \mathsf{E}_x(\eta_0(f)) + U_n\left(\int f d\nu\right).$$

Thus to prove (2.2) it suffices to observe that

(2.6) $\mathsf{E}_\Delta \eta_0(f) < \infty$ implies $\mathsf{E}_x \eta_0(f) < \infty$ a.e. $[\nu]$.

But

$$\mathsf{E}_\Delta \eta_0(f) = \mathsf{E}_\Delta\left(\sum_{j=0}^{T_\Delta - 1} f(X_j)\right) \geqslant \mathsf{E}_\Delta\left(\sum_{j=r}^{T_\Delta - 1} f(X_j); \ T_\Delta > r\right) =$$

$$= \int \mathsf{E}_x \eta_0(f) \nu_r(dx)$$

where $\nu_r(\cdot) = P(X_n \in \cdot; \ T_\Delta > r)$. Thus (2.6) holds a.e. $[\nu_r]$, and hence a.e. $[\nu]$ since $\nu(\cdot) = \sum_{r=0}^{\infty} \nu_r(\cdot)$. ∎

Obviously Theorem 2.2 implies

Corollary 2.1 (Weak ratio theorem). *For $f, g \in L_1(\nu)$*

(2.7) $$\frac{\sum_{j=0}^{n} \mathsf{E}_x f(X_j)}{\sum_{j=0}^{n} \mathsf{E}_x g(X_j)} \to \frac{\int f d\nu}{\int g d\nu} \quad a.e. \ [\nu].$$

(This is deduced from the Chacon — Ornstein theorem in Orey's book.)

Also from the proof of the theorem we get the following law of large numbers

Corollary 2.2. *If f and $g \in L_1(\nu)$, then for any initial distribution of X_0*

(2.8) $$\lim_{n \to \infty} \frac{\sum_{j=0}^{n} f(X_j)}{\sum_{j=0}^{n} g(X_j)} = \frac{\int f d\nu}{\int g f \nu} \quad a.e.$$

(where the a.e. is with respect to the measure on the Markov chain).

Proof. Follows from (2.3). Note

$$\frac{\sum\limits_{j=0}^{n} f(X_j)}{\sum\limits_{j=0}^{n} g(X_j)} \leqslant \frac{K_n}{K_n+1} \frac{\left(\frac{1}{K_n} \sum\limits_{j=0}^{K_n} \eta_j(f)\right)}{\left(\frac{1}{K_n+1} \sum\limits_{j=0}^{K_n+1} \eta_j(g)\right)} \to \frac{\int f dv}{\int g dv},$$

since

$$\frac{1}{K_n} \sum\limits_{j=0}^{K_n} \eta_j(f) \to \mathsf{E}\eta_j(f) = \mathsf{E}_\Delta \sum\limits_{i=0}^{N-1} f(X_i) = \int f dv;$$

and similarly for g, yielding lim sup LHS $\leqslant$ RHS. Interchanging f and g yields lim inf LHS $\geqslant$ RHS. ∎

Remark 2.1. As observed before, although these results appear to depend on the existence of an atom Δ, one can use the extension arguments of [1] and [3] mentioned above to show that they remain valid for *any* Harris recurrent chain. See Section 4 for detailed statements.

Remark 2.2. If the chain is furthermore *positive* recurrent and aperiodic, then (2.2) can be strengthened to

(2.9) $\mathsf{E}_x f(X_n) \to \int f dv,$

and in fact to $\|\mathsf{P}_x\{X_n \in \cdot\} - v(\cdot)\| \to 0$ (where $\|\cdot\| =$ total variation norm).

Remark 2.3. In the aperiodic, null-recurrent case a further argument, using Ornstein's coupling, can be used to prove Orey's theorem: for any initial measures η, λ,

$$\|\mathsf{P}_\mu(X_n \in \cdot) - \mathsf{P}_\lambda(X_n \in \cdot)\| \to 0.$$

3. THE STRONG RATIO THEOREM

Having seen that for all Harris chains

$$(3.1) \qquad \frac{\sum\limits_{j=0}^{n} E_x f(X_j)}{\sum\limits_{j=0}^{n} u_j} \to \int f d\nu \quad \text{a.e.,}$$

we proceed to look for conditions for the stronger result

$$(3.2) \qquad \frac{1}{u_n} E_x f(X_n) \to \int f d\nu \quad \text{a.e.}$$

We will see that further hypotheses on the chain will be needed, as well as some extra conditions on f and on the initial measure. This section is devoted to clarifying these matters.

As before, it is no loss of generality to assume that there esists a recurrence point Δ for the chain and we do so throughout this section. The form of the theorems in the general case will be indicated at the end of the paper.

We say that the strong ratio limit property (SRLP) holds for (B, λ), where $B \in \mathscr{S}$ and λ is an "initial" measure on $(S, \mathscr{S})$, if

$$(3.3) \qquad \lim_{n \to \infty} \frac{P_\lambda(X_n \in B)}{u_n} = \nu(B),$$

where

$$\nu(B) = E_\Delta \sum_{j=0}^{T_\Delta - 1} \chi_B(X_j)$$

is the invariant measure introduced in the previous section. For SRLP to hold, two kinds of conditions are needed.

The first is a condition on the chain, and is customarily stated as

$$(3.4) \qquad \frac{u_{n+1}}{u_n} \to 1 \quad \text{as} \quad n \to \infty,$$

(recall, $u_n = P_0(X_n = \Delta)$). This is not an entirely satisfactory hypothesis, since what the result then in effect says is that if one property involving the asymptotic behavior of P^n holds (namely $u_{n+1} \sim u_n$) then so does another (namely SRLP). It is desirable to have reasonable conditions on $P(\cdot, \cdot)$ itself which assure $u_{n+1} \sim u_n$. In the discrete state-space case a strong sufficient condition is given by the Kingman – Orey theorem (see [11]), namely that for some integer n_0 and $\epsilon > 0$

(3.5) $P^{n_0}(x, x) \geqslant \epsilon$ for all $x \in S$.

An improved sufficient condition was given by K e s t e n [7] as

(3.6) $\displaystyle\sum_y P^{n_0 + m_0}(x, y) \wedge P^{n_0}(x, y) \geqslant \epsilon$

for some integers m_0, n_0, some $\epsilon > 0$, and all $x \in S$ $(a \wedge b = \min(a, b))$. Whereas the proof of the Kingman – Orey theorem required only a simple mixing argument, Kesten's proof of sufficiency was more delicate. (Professor K e s t e n sent us this proof in a private communication, which we gratefully acknowledge.)

The Kesten condition has the advantage that it suggests a natural extension to general state spaces, and one of our students at Wisconsin, J . K i n g [6] has proved such a result. He has shown that the existence of integers m_0, n_0, an $\epsilon > 0$ and a stochastic transition function $Q(\cdot, \cdot)$ on $(S, \mathscr{S})$ such that

(3.7) $P^{n_0}(x, E) \wedge P^{n_0 + m_0}(x, E) \geqslant \epsilon Q(x, E)$

for all $x \in S$, $E \in \mathscr{S}$ (in conjunction with the further requirements on (B, λ) to be discussed) is sufficient for SRLP. His proof (which will be contained in his thesis) does not attempt to generalize Kesten's argument, but involves a more complicated version of the original mixing scheme in Kingman – Orey.

We will not go further into this aspect of the problem, but assume from now on that $u_{n+1} \sim u_n$ or a sufficient condition like (3.7) holds.

Turning to the second set of hypotheses, namely those on (B, λ)

we will show that

(i)   there is a decomposition of S into a countable number of sets $\{B_i: i \geqslant 1\}$ of finite ν-measure,

(ii) a decomposition of ν into finite measures $\{\lambda_j: j \geqslant 1\}$: $\nu = \sum \alpha_j \lambda_j$ $(0 \leqslant \alpha_j \leqslant 1)$, such that if B is contained in a finite union of B_i's and λ is a finite convex combination of λ_j's, then SRLP holds for (B, λ).

We now proceed to make this precise.

Proposition 3.1. *Let* $A \in \mathscr{S}$ *be such that for some constant* c *and integer* r, *and for all* $k \geqslant 1$,

$$(3.8) \qquad P_\Delta(X_k \in A, T_\Delta > k) \leqslant c P_\Delta(k < T_\Delta \leqslant k + r).$$

Then "$u_{n+1} \sim u_n$" *implies that*

$$(3.9) \qquad \frac{1}{u_n} P_\Delta(X_n \in A) \to \nu(A) \quad as \quad n \to \infty.$$

Proof. The following argumnet is patterned after O r e y 's [11]. By a standard decomposition based on the first passage time to Δ,

$$P_\Delta(X_n \in A) =$$
$$(3.10)$$
$$= P_\Delta(X_n \in A, T_\Delta > n) + \sum_{k=0}^{n} P_\Delta(X_{n-k} \in A) P_\Delta(T_\Delta = k),$$

and hence

$$P_\Delta(X_n \in A) = \sum_{k=0}^{n} P_\Delta(X_k \in A, T > k) u_{n-k}.$$

Write

$$\frac{1}{u_n} P_\Delta(X_n \in A) = S_{n,M} + R_{n,M},$$

where

$$S_{n,M} = \frac{1}{u_n} \sum_{k=0}^{M} P_\Delta(X_k \in A, T > k) u_{n-k};$$

$$R_{n,M} = \frac{1}{u_n} \sum_{k=M+1}^{n} P_{\Delta}(X_k \in A, \; T > k) u_{n-k}.$$

Now $u_{n+1} \sim u_n$ implies that

$$\lim_{M \to \infty} \lim_{n \to \infty} S_{n,M} = \sum_{k=0}^{\infty} P_{\Delta}(X_k \in A, \; T > k) = v(A),$$

and it remains to show that

(3.11)
$$\lim_{M \to \infty} \lim_{n \to \infty} R_{n,M} = 0.$$

But (3.8) implies that

$$R_{n,M} \leqslant \frac{c}{u_n} \sum_{k=M+1}^{n} u_{n-k} \sum_{j=1}^{r} P_{\Delta}(T_{\Delta} = k+j) \leqslant$$

$$\leqslant c \sum_{j=1}^{r} \frac{1}{u_n} \sum_{k=M+1+j}^{n+j} u_{n+j-k} P_{\Delta}(T_{\Delta} = k) \leqslant$$

$$\leqslant c \sum_{j=1}^{r} \frac{1}{u_n} \Big\{ u_{n+j} - \sum_{k=0}^{M+j} u_{n+j-k} P_{\Delta}(T_{\Delta} = k) \Big\}.$$

Hence

$$\limsup_{M \to \infty} \limsup_{n \to \infty} R_{n,M} \leqslant c \sum_{j=1}^{r} (1 - P_{\Delta}(T_{\Delta} \leqslant M+j)) = 0.$$

Corollary 3.1. *If for some* $r < \infty$

$$\inf_{y \in A} P_y(T_{\Delta} \leqslant r) \equiv \epsilon > 0,$$

then A satisfies (3.8) and hence (3.9) holds.

Proof.

$$P_{\Delta}(k < T_{\Delta} \leqslant k+r) = \int_{S} P_{\Delta}(X_k \in dy, \; T_k > k) P_y(T_{\Delta} \leqslant r) \geqslant$$

$$\geqslant \int_{A} P_{\Delta}(X_k \in dy, \; T_k > k) P_y(T_{\Delta} \leqslant r) \geqslant$$

$$\geqslant \epsilon P_{\Delta}(X_k \in A, \; T_{\Delta} > k). \qquad \blacksquare$$

Remark. We record also the following corollary of the above proof, which will be needed in the proof of the next proposition. Namely, for A satisfying (3.8)

$$(3.12) \qquad \lim_n \frac{1}{u_n} P_\Delta (X_n \in A, \ T_\Delta > n) = 0.$$

To see this, observe that by (3.10) the left side of (3.12)

$$\leqslant \lim \Big\{ \frac{1}{u_n} P_\Delta (X_n \in A) -$$

$$- \sum_{k=0}^{M} P_\Delta (T_\Delta = k) \Big[\frac{1}{u_n} P_\Delta (X_{n-k} \in A) \Big] \Big\} =$$

$$= \nu(A) [1 - P_\Delta (T_\Delta \leqslant M)] \quad \text{for all} \quad M < \infty,$$

$$\to 0 \quad \text{as} \quad M \to \infty.$$

Turning to the requirements on the initial measure, we start with

Proposition 3.2. *Let* $A \in \mathscr{S}$ *satisfy* (3.8), *and* λ *be a probability measure such that for some* $c, r,$ *and* j_0

$$(3.13a) \quad P_\lambda (T_\Delta = k) \leqslant c P_\Delta (k < T_\Delta \leqslant k + r)$$

and

$$(3.13b) \quad P_\lambda (X_n \in A, \ T_\Delta > n) \leqslant c P_\Delta (X_{n+j_0} \in A, \ T_\Delta > n + j_0)$$

for all $n \geqslant 1.$ *Then*

$$(3.14) \qquad \lim_{n \to \infty} \frac{1}{u_n} P_\lambda (X_n \in A) = \nu(A).$$

Proof. By the usual decomposition

$$\frac{1}{u_n} P_\lambda (X_n \in A) = V_n + S_{n,M} + R_{n,M},$$

where

$$V_n = \frac{1}{u_n} P_\lambda (X_n \in A, \ T_\Delta > n),$$

$$S_{n,M} = \frac{1}{u_n} \sum_{k=0}^{M} P_\lambda(T_\Delta = k) P_\Delta(X_{n-k} \in A)$$

$$R_{n,M} = \frac{1}{u_n} \sum_{k=M+1}^{n} P_\lambda(T_\Delta = k) P_\Delta(X_{n-k} \in A).$$

By Proposition 3.1, $\displaystyle\lim_{M\to\infty} \lim_{n\to\infty} S_{n,M} = \nu(A)$, and hence it is enough to show that $R_{n,M} \to 0$ and $V_n \to 0$. Applying (3.13), a change of variables, (3.10), (3.12) and Proposition 3.1

$$\lim_n R_{n,M} \leqslant \lim_n c \sum_{j=1}^{r} \sum_{k=M+1}^{n} P_\Delta(T_\Delta = k+j) P_\Delta(X_{n-k} \in A) \leqslant$$

$$\leqslant \lim_n c \sum_{j=1}^{r} \frac{1}{u_n} \sum_{k=M+1+j}^{n+j} P_\Delta(T_\Delta = k) \times$$

$$\times P_\Delta(X_{n+j-k} \in A) \leqslant$$

$$\leqslant \lim_n c \sum_{j=1}^{r} \frac{1}{u_n} \Big[P_\Delta(X_{n+j} \in A) -$$

$$- \sum_{k=0}^{M+j} P_\Delta(X_{n+j-k} \in A) P_\Delta(T_\Delta = k) -$$

$$- P_\Delta(X_{n+j} \in A, T_\Delta > n+j) \Big] \leqslant$$

$$\leqslant c\nu(A)[1 - P_\Delta(T_\Delta \leqslant M + j)] \to 0 \quad \text{as} \quad M \to \infty.$$

Finally by (3.13b) and then (3.8)

$$V_n \leqslant \frac{c}{u_n} P_\Delta(X_{n+j_0} \in A, T_\Delta > n + j_0) \leqslant$$

$$\leqslant \frac{c^2}{u_n} P(n + j_0 < T_\Delta \leqslant n + j_0 + r).$$

But

$$P_\Delta(T_\Delta = n) = u_n - \sum_{k=0}^{n-1} P_\Delta(T_\Delta = k) u_{n-k} \leqslant$$

$$\leqslant u_n - \sum_{k=0}^{M} P_\Delta(T_\Delta = k) u_{n-k}, \quad M < \infty$$

and hence

$$\limsup_{n \to \infty} \frac{1}{u_n} P_\Delta (T_\Delta = n) \leqslant 1 - P_\Delta (T_\Delta \leqslant M) \to 0 \quad \text{as} \quad M \to \infty .$$

Thus $\limsup \dfrac{1}{u_n} P_\Delta (T_\Delta = n + j) = 0$ for all j and $\limsup V_n = 0. \blacksquare$

Corollary 3.2. *If* A *satisfies* (3.8) *and*

$$(3.15) \quad \lambda_j (\cdot) = P_\Delta (X_j \in \cdot \mid T_\Delta > j)$$

then (3.13) *holds and hence*

$$(3.16) \quad \lim_n \frac{1}{u_n} P_{\lambda_j} (X_n \in A) = \nu(A).$$

Proof. Just verify (3.13). $\blacksquare$

These propositions lead at once to the decompositions of the state space S and the invariant measure $\nu(\cdot)$ discussed at the beginning of this section.

Theorem 3.1. *If* $u_{n+1} \sim u_n$ *and a recurrence point* Δ *for* $P(\cdot, \cdot)$ *exists, then there exists a sequence of sets* $\{B_i; i = 1, 2, \ldots\}$ *and of measures* $\{\lambda_j : j = 1, 2, \ldots\}$, *such that*

(i) $\displaystyle\bigcup_{i=1}^{\infty} B_i = S$ *a.e.* $[\nu]$, *and* $\nu(B_i) < \infty$.

(ii) $\nu(E) = \displaystyle\sum_{j=1}^{\infty} \alpha_j \lambda_j (E)$, *where* $\alpha_j = P_\Delta (T_\Delta > j)$.

(iii) $\dfrac{1}{u_n} P_{\lambda_j} (X_n \in B_r) \to \nu(B_r)$ *as* $n \to \infty$, *for each* j, r.

Remark. The series in (ii) may diverge for some sets E and converge for others, but the individual measures λ_j are all probability measures.

Proof. For the B_i's we take the sets

$$B_{r,s} = \left\{ y : P_y (T_\Delta \leqslant r) \geqslant \frac{1}{s} \right\} \quad (r, s = 1, 2, \ldots),$$

and the λ_j's are given by (3.15). By Corollary 3.1 the B's satisfy (3.8). Also $\nu(B_{r,s}) < \infty$ since the number of visits to $B_{r,s}$ between returns to Δ is bounded by a geometric random variable. Thus (i) holds.

By definition $\lambda_i(S) = 1$ and thus (ii) holds, while (iii) follows from the fact that $B_{r,s}$ satisfies (3.8) and λ_j satisfies (3.13). ∎

Finally, from Theorem 3.1 we can extract the following convergence in measure result (Orey [11]).

Theorem 3.2. *If* $u_{n+1} \sim u_n$ *and a recurrence point exists, and if* A *satisfies (3.8), then for every sequence of integers* $n_k \uparrow \infty$, *there exists a subsequence* $\{n'_k\}$ *such that*

(3.17)
$$\lim_{n'_k \to \infty} \frac{1}{u_{n'_k}} P_x(X_{n'_k} \in A) = \nu(A) \quad a.e. \quad [\nu].$$

Proof. For any x, let $f_n(x) = \frac{1}{u_n} P_x(X_n \in A)$. Then as before we can write $f_n(x) = S_{n,M}(x) + R_{n,M}(x)$ where

$$S_{n,M} = \frac{1}{u_n} \sum_{k=1}^{M} P_x(T_\Delta = k) P_\Delta(X_{n-k} \in A)$$

and

$$R_{n,M} = \frac{1}{u_n} P_x(X_n \in A, \ T_\Delta > M).$$

Clearly $\lim_{M \to \infty} \lim_{n \to \infty} S_{n,M} = \nu(A)$ and hence for any sequence $N_k \uparrow \infty$ we have

(3.18)
$$\liminf_{n_k \to \infty} f_{n_k}(x) \geq \nu(A).$$

We also know by Proposition 3.2 that

$$\lim_{n_k \to \infty} \int f_{n_k}(x) \lambda_j(dx) = \nu(A)$$

and hence by Fatou's lemma

$$\nu(A) \geqslant \liminf \int f_{n_k} \, d\lambda_j \geqslant \int (\liminf f_{n_k}) \, d\lambda_j.$$

This, with (3.18) implies (since $\lambda_j(S) = 1$)

(3.19) $\liminf f_{n_k}(x) = \nu(A)$ a.e. $[\lambda_j]$.

Hence there is a subsequence $\{m_k\} \subset \{n_k\}$ such that

$$\lim_{m_k \to \infty} f_{m_k}(x) = \nu(A) \quad \text{a.e.} \quad [\nu_j].$$

This sequence is of course also dependent on j (via λ_j), but by diagonalization we can find a single subsequence $\{n'_k\} \subset \{n_k\}$ such that

(3.20) $\lim_{n'_k \to \infty} f_{n'_k}(x) = \nu(A)$ a.e. $[\nu]$.

Arguing as in O r e y [11], this leads to

Corollary 3.3. *Under the hypothesis of Theorem* 3.2

(3.21) $\displaystyle \lim_{n \to \infty} \varphi \left\{ x : \left| \frac{P_x(X_n \in A)}{u_n} - \nu(A) \right| > \epsilon \right\} = 0$

for every $\epsilon > 0$, *and every finite measure* φ *that is absolutely continuous with respect to* ν.

4. STATEMENT OF GENERAL RESULTS FOR HARRIS CHAINS

All the above proofs have been carried out under the hypothesis of the existence of a recurrence point Δ. We have stressed that this involves no loss of generality, and that by the methods of [1] and [3] an extension to arbitrary Harris chains can be accomplished. Since these extension arguments are by now fairly routine, we will not supply the proofs, but state the results.

Recurrence hypothesis. *There exists a set* A, *an integer* n_0, *a number* $0 < \lambda < 1$, *and a probability measure* φ *on* A, *such that*

(i) $P_x(X_n \in A \text{ for some } n \geqslant 1) = 1$ *for all* $x \in S$

(ii) $P_x(X_{n_0} \in E) \geqslant \lambda \varphi(E)$ *for* $x \in A$, $E \subset A$.

This can be shown to be equivalent to Harris recurrence. We then have

Theorem 4.1. *Under the recurrence hypothesis there exists an invariant measure v for* **P.** *Furthermore for every* $f \in L_1(v)$

$$(4.1) \qquad \lim_{n \to \infty} \frac{1}{U_n} \sum_{j=0}^{n} E_x f(X_j) = \int f dv \quad a.e. \quad [v],$$

where $U_n = \sum_{i=0}^{n} u_i$ *and*

$$(4.2) \qquad u_n = \int_A \varphi(ds) P^n(s, A).$$

The weak ratio theorem (Corollary 2.1) and law of large numbers (Corollary 2.2) remain valid as they are, as do the remarks about the positive recurrent case and Orey's theorem.

To obtain the extension of the strong ratio theorem, we strengthen the recurrence hypothesis.

Theorem 4.2. *If the recurrence hypothesis holds with* $n_0 = 1$, *and* $u_{n+1} \sim u_n$ *(see (4.2)), then there exists a sequence of sets* $\{B_i\}$, *measures* $\{\lambda_j\}$, *and constants* $\{\alpha_j\}$, *satisfying the conclusions of Theorem* 3.1.

Remark. All the above $\{B_i\}, \{\lambda_j\}, \{\alpha_j\}$, as well as v, can be represented explicitly in terms of a so-called regeneration time N, which replaces T_Δ in the above arguments (see [1]). Theorem 3.2 and its corollary similarly carry over.

REFERENCES

[1] K.B. Athreya – P. Ney, A new approach to the limit of theory of recurrent Markov chains, *TAMS,* 245 (1978), 493-501.

[2] K.B. Athreya – D. McDonald – P. Ney, Limit theorems for semi-Markov processes and renewal theory for Markov chains, *Annals of Prob.,* 6 (1978), 788-797.

[3]   K.B. Athreya – P. Ney,   A renewal approach to the Perron – Frobenius theory of non-negative kernels on general state spaces, University of Wisconsin Technical Report.

[4]   W. Doeblin,   Elements d,une theorie generale des chaines simples constantes de Markoff, *Ann. Sci. de l'Ec. Norm. Sup. de Paris,* 57 (1940), 61-111.

[5]   T.E. Harris,   The existence of stationary measures for certain Markov processes, *3rd Berkeley Symposium,* Vol. II (1956), 113-124.

[6]   J. King,   Strong ratio theorems,   University of Wisconsin Ph. D. Thesis, (under preparation).

[7]   H. Kesten,   Existence and uniqueness of countable one-dimensional Markov random fields, *Annals of Prob.,* 4 (1976), 557-569.

[8]   J. Neveu,   *Mathematical foundations of the calculus of probability,* Holden Day, San Francisco, 1965.

[9]   E. Nummelin,   A splitting technique for Harris recurrent Markov chains, *Z. Wahrscheinlichkeitstheorie verw. Geb.,* 43 (1978), 309-318.

[10]   E. Nummelin,   Strong ratio limit theorems for φ-recurrent Markov chains, *Annals of Prob.,* 7 (1979), 639-650.

[11]   S. Orey,   *Limit theorems for Markov chain transition probabilities,* Van Nostrand, New York, 1971.

[12]   D. Revuz,   *Markov chains,*   North Holland, Amsterdam, 1975.

K.B. Athreya

Department of Mathematics, University of Wisconsin, Milwaukee, WI, USA.

P. Ney

Department of Mathematics, University of Wisconsin, Madison, WI, USA.

ESTIMATION OF THE PRIOR DISTRIBUTION USING DIFFERENTIAL OPERATORS

J. BLUM — V. SUSARLA* — G. WALTER

1. INTRODUCTION

The empirical Bayes decision problem of R o b b i n s [12] naturally leads to the problem of estimating an unknown prior distribution given an i.i.d. sample $(\Theta_1, X_1), (\Theta_2, X_2), \ldots, (\Theta_n, X_n)$ of random vectors in which the $X_1, X_2, \ldots, X_n$ are observable but the $\Theta_1, \Theta_2, \ldots, \Theta_n$ are not. The problem is to determine the distribution of the latter from $X_1, \ldots, X_n$. The distribution functions of X, and Θ, are related by an equation of the form

$$(1.1) \qquad F(x) = \int F_\theta(x)\, dG(\theta)$$

where $F_\theta(x)$ is the conditional distribution function of X given $\Theta = \theta$. The function F_θ in (1.1) is assumed known, while F and thence G must be estimated from the observations.

*Research supported in part by NIH Grant No. 7-R01-GM28405, and the revision was done while visiting Wayne State University.

Various methods have been proposed for estimating G. In addition to Robbins, Blum and Susarla [1], [3], Choi [4], Deely and Kruse [6], Maritz [11], Torterella and O'Bryan [16], and O'Bryan and Walter [10], among others, have obtained estimators for G which converge to it in some weak sense. More recently, Blum and Susarla [3] have shown that in certain cases, the prior distribution G may be obtained from the marginal F by means of an appropriate differential operator. Their results were an extension of previous work by Fox [7] and lead to an estimator converging to G in law. They considered two cases in which the conditional distribution function F_θ was respectively of the form

(1.2) $\qquad F_\theta(x) = \left(\frac{x}{\theta}\right)^m \qquad (0 < x < \theta, \; m, \theta > 0),$

and of the form of an exponential distribution with θ a location parameter. In both cases they found a differential operator $P(D)$ such that

(1.3) $\qquad P(D)F = G.$

In this work, we generalize these two cases to certain classes of conditional distributions which, considered as the kernel of an integral operator, are invertible by means of a differential operator. Such kernels are widely used in the theory of differential equations and are called Green's functions. In fact each linear differential operator has, under quite general hypotheses, associated with it a Green's function. (See Coddington and Levinson [5], p. 192). However, not all Green's functions can be interpreted as conditional distribution functions nor do all distribution functions form Green's functions for some differential operator. One of our purposes in this work will be to determine which ones do.

Once it is known that the integral operator of (1.1) may be inverted with a differential operator it is an easy matter to estimate G given a sample of F. One needs merely to use nonparametric estimation procedures estimating densities and their derivatives. For this purpose a Fourier transform estimator, a Hermite series estimator, or one of various kernel estimators are most appropriate. Such estimators $\hat{G}$ of G can then be used to obtain confidence bands for G on any finite interval using the recent results of R o s e n b l a t t [13].

As an example where our results can be applied, consider a situation in which a component is manufactured at k sites and that the life of the component given that it is manufactured at site i is exponential with location parameter $\theta_i \geqslant 0$. (θ_i can be interpreted as the minimum period of time this component will work.) If the proportions of the components manufactured at various sites is unknown, then the theory mentioned in the above paragraph can be applied to estimate G, i.e. to estimate θ_i and the jump at θ_i under G. Since, as will be noted later on in the paper, a differential equation as in (1.3) is satisfied in this situation. This example can be generalized to a gamma family with unknown origin instead of exponential distribution with unknown location.

In Section 2 we present some of the background from differential equations and the associated Laplace transform theory. In Section 3 a general theory for infinite intervals is developed. The conditional distributions which correspond to constant coefficient differential operators are characterized. In Section 4 the case of bounded intervals is considered for constant coefficient differential operators. In Section 5 the case when θ is a scale parameter and in Section 6 the case of a normal location parameter are investigated.

The distribution function we consider will all be absolutely continuous. Thus we may replace F, G and F_θ by their derivatives and may rewrite 1.1) and (1.3) in terms of the density functions

(1.4) $\qquad f(x) = \int k(x, \theta) g(\theta)\, d\theta$

(1.5) $\qquad P(D)f = g.$

If the differential operator $P(D)$ has constant coefficients then clearly the operators in (1.3) and (1.5) are the same. If not, then the modifications needed to convert one to the other are straightforward.

2. BACKGROUND

In this section we present some elements of the theory of differential operators and Laplace transforms which we need in subsequent sections. More details may be found in the books by Coddington and Levinson [5], Stakgold [14] and Widder [19].

Let $P(D)$ be a differential operator given by

$$(2.1) \qquad P(D) = \sum_{i=1}^{n} a_i(x) \frac{d^i}{dx^i}$$

on an open bounded interval I, $a_i(x)$ functions in class C^n on the closure $\bar{I}$ of I and $a_0(x) \neq 0$ on $\bar{I}$.

Let $k(x, \theta)$ be a function that satisfies

$$(2.2) \qquad P(D) k(x, \theta) = \delta(x - \theta)$$

where δ is the Dirac "delta function" with unit mass at zero. Such a function is called a Green's function. The associated integral operator K satisfies

$$(2.3) \qquad (Kg)(x) = \int k(x, \theta) g(\theta)\, d\theta$$

and is the inverse of $P(D)$, i.e.,

$$(2.4) \qquad P(D) Kg = g \qquad (g \in C(\bar{I})).$$

Such Green's functions exist for all such operators (see p. 188 of Coddington and Levinson [5]). They are not necessarily unique but may be made so by the addition of appropriate boundary conditions. The same is true for infinite intervals except that the class of densities g must be restricted to ensure that the integral (2.3) exists.

In general the Green's functions are *not* densities since they are not necessarily nonnegative nor can they be normalized to L^1 norm 1. However, some are, and the problem we need to consider therefore, is to identify pairs $(P(D), k(x, \theta))$ of differential operators and density k with parameter θ such that

(2.5) $P(D)k(x, \theta) = \delta(x - \theta)$ $(x, \theta \in I)$,

where I is a finite or an infinite open interval.

Example 2.1. Let n be a nonnegative integer; let $k(x, \theta)$ be given by

$$k(x, \theta) = \frac{(x - \theta)^n}{n!} e^{-(x-\theta)} U(x - \theta)   (x, \theta \in (0, \infty)),$$

where U is the indicator function of $(0, \infty)$, and let $P(D)$ be given by

$$P(D) = \sum_{m=0}^{n+1} \binom{n+1}{m} D^m.$$

This pair satisfies (2.5) on $I = (0, \infty)$.

Example 2.2. Let m, n be positive integers, and let k be given by

$$(2.6)\qquad k(x, \theta) = \frac{\left(\frac{x}{\theta}\right)^{n-1}\left(1 - \frac{x}{\theta}\right)^{m-1}}{\beta(n, m)} U\left(1 - \frac{x}{\theta}\right) U\left(\frac{x}{\theta}\right)   (\theta, x \in (0, 1))$$

and let $P(D)$ be given by

$$(2.7)\qquad P(D) = \frac{x^{-1}(xD - n)(xD - n - 1) \ldots (xD - m - n + 1)x}{n(n+1) \ldots (n + m - 1)(-1)^m}$$

where $\beta(n, m) = \dfrac{(n-1)!(m-1)!}{(n+m-1)!}$. It will be shown in Section 6 that this pair satisfies (2.5).

Example 2.1 is of course the Gamma distribution with θ a location parameter, while (2.2) is the Beta distribution with θ a scale parameter.

One of the principal tools in studying linear differential equations with constant coefficients is the Laplace transform, i.e. the operator Lf given by

$$(2.8)\qquad F(s) = (Lf)(s) = \int_{-\infty}^{\infty} e^{-sx} f(x)\, dx   (f \in L^1_{loc}).$$

The transform exists for certain values of $s \in \mathscr{C}$ which depend on the behavior of f at $\pm \infty$.

Definition 2.1. A function of $f \in L^1_{loc}$ is *admissible* if there exist real constants C and C', $C < C'$ such that $f(x)$ is dominated by e^{Cx} as $x \to +\infty$ and by $e^{C'x}$ as $x \to -\infty$.

The function of $f(x) = 1$ is not admissible while $f(x) = e^{-|x|}$ is. (Take $C = -1$, $C' = 1$.) Clearly the Laplace transform exists for admissible functions if $C < \text{Re } s < C'$.

Definition 2.2. Let $b = \inf C$ and $b' = \sup C'$ taken over all C and C' for which f is admissible. Then f is said to be of *type* (b, b').

Note. The definition of Laplace transform as well as the definitions of admissible and type may be extended to measures or Stieltje's integrals in the obvious way.

Definitions 2.3. Let $f, f^{(1)}, \ldots, f^{(p)}$ be functions or measures which are admissible of type $(b_0, b_0'), (b_1, b_1'), \ldots, (b_p, b_p')$ respectively. Then they are said to be of *common* type (b, b') if $b = \max\{b_0, \ldots, b_p\}$ and $b' = \min\{b_0', \ldots, b_p'\}$.

Remark 2.1. A function with support on a half line (a, ∞) is of type (b, ∞) while one with support on $(-\infty, a)$ is of type $(-\infty, b)$. The Gamma distribution

$$f(x) = \frac{x^{\alpha-1}}{\Gamma(\alpha)} e^{-x} U(x)$$

is of type $(-1, \infty)$ while the normal distribution

$$f(x) = \frac{1}{\sqrt{2\pi}} e^{-\frac{x^2}{2}}$$

is of type $(-\infty, \infty)$. A measure with compact support is of type $(-\infty, \infty)$.

The power of the Laplace transform lies in its ability to transform differentiation into multiplication. Thus if $f^{(k)}$ $(k = 0, 1, \ldots, p)$ are admissible and $P(D)$ is an operator of the form (2.1) with *constant coefficients*, then $L(P(D)f) = P(s)F(s)$. The transformed function $F(s)$ will be holomorphic in the strip $b < \text{Re } s < b'$, and will come from a unique admissible function of the same type.

The definitions presented here are those of the two-sided Laplace transform. The more common one-sided transform involving the integral from 0 to ∞ may be considered as a special case.

3. KERNELS WITH EXTENSIONS TO R

In this section we study those kernels (i.e. conditional density functions) which either are defined on all of R or have a continuously differentiable extension from I to R. (See Example 2.1.) We shall characterize those that are invertible by means of constant coefficient differential operators, i.e. those that satisfy (2.5) for some $P(D)$. We first show that they must be of location parameter type and then that the polynomial must have certain properties. A complete characterization is given for second order differential operators and partial characterizations for higher orders.

Lemma 3.1. *Let* $\{P(D); k(x, \theta)\}$ *be a pair satisfying* (2.5) *on* R^1 *where* $P(D)$ *is a constant coefficient differential operator and* $D_x^i k(x, \theta)$ $(i = 0, 1, \ldots, n - 1)$ *are of common type* (b, b') $(-\infty \leqslant b < b' \leqslant \infty)$. *Then*

(i) $k(x, \theta) = k((x - \theta), 0)$,

(ii) $P(0) = 1$,

(iii) $P(s)$ *has no zeros in* $b < \operatorname{Re} s < b'$,

(iv) *if the degree of* $P(s) \leqslant 2$, *all zeros are real.*

Remark 3.1. The requirement that the interval be R^1 is not a very stringent one. Any conditional density defined on I may be considered to be defined on R^1 provided only that $k(x, \theta)$ and its first $(n - 1)$ derivatives have continuous extension across the boundary of I for each $\theta \in I$. For example the exponential density

$$k(x, \theta) = e^{-(x - \theta)} U(x - \theta) \qquad (x, \theta \in (0, \infty)),$$

is identically zero in a neighborhood of zero and hence may be extended to R'.

Proof.

(i) The Laplace transform of (2.5) is given by

(3.1) $P(s)K(s, \theta) = e^{-s\theta}$ $b < \mathrm{Re}\, s < b'$,

where $K(x, \theta)$ is the Laplace transform of $k(x, \theta)$. Hence

$$K(s, \theta) = \frac{e^{-s\theta}}{P(s)} = e^{-s\theta} K(s, 0)$$

except possibility at the zeros of $P(s)$. But both sides are holomorphic in $b < \mathrm{Re}\, s < b'$, and hence identical. The conclusion follows from the uniqueness theorem and the fact that the Laplace transform of $k(x - \theta, 0)$ is $e^{-s\theta} K(s, 0)$.

(ii) Since k is a density it must be in $L^1(R^1)$, and $K(s, 0)$ can be extended continuously to $s = 0$. Indeed

$$K(0, 0) = \int_{-\infty}^{\infty} k(x, 0)\, dx = 1.$$

Since P is a polynomial and $P(s)K(s, 0) = 1$, it follows that $P(s)$ can be extended to $s = 0$ and

$$P(0)K(0, 0) = P(0) = 1.$$

(iii) Since $K(s, 0)$ is holomorphic in $b < \mathrm{Re}\, s < b'$, $P(s)$ can have no zeros in this region.

(iv) If $P(s)$ had a pair of non real zeros then $k(x, 0)$ would have the form

$$A\,(\cos \gamma x)e^{-\beta x} + B\,(\sin \gamma x)e^{-\beta x}$$

which would not be nonnegative no matter what choice of A and B on any infinite interval.

Corollary 3.2. *Let $k(x, \theta)$ be a normal density as a function of x; then there is no differential operator with constant coefficients $P(D)$ such that $\{P(D), k\}$ satisfying (2.5).*

Proof. Since k is normal it is of type $(-\infty, \infty)$. If there were such a $P(D)$, by (iii) $P(s)$ would have no zeros in the entire plane. Hence it would be constant. Then it is clearly impossible that (2.5) be satisfied.

Remarks 3.2. While there is no finite order differential operator which works for the conditional normal distribution, we shall see in Section 6 that an infinite order differential operator can be used.

Since by the Lemma $k(x, \theta)$ must be of the form $k(x - \theta, 0)$ we may suppose it to be a function of one variable only. Then (2.5) may be replaced by similar conditions

(3.2) $P(D)k(x) = \delta(x) \qquad (x \in I).$

We first characterize those kernels for which $P(D)$ is a second degree polynomial.

Propositions 3.3. *Let* $P(D) = a_0 D^2 + a_1 D + 1$ *where* a_0, a_1 *are real constant; then* $k(x)$ *satisfies (3.2) for some choice of* a_0 *and* a_1 *if and only if it is given by one of the forms:*

(i) $k(x) = |\alpha| e^{-\alpha x} U(x \operatorname{sgn} \alpha), \quad \alpha \neq 0,$

(ii) $k(x) = \alpha^2 x e^{-\alpha x} U(x \operatorname{sgn} \alpha), \quad \alpha \neq 0,$

(iii) $k(x) = \dfrac{\alpha\beta}{\alpha - \beta} \{\operatorname{sgn} \beta e^{-\beta x} U(x \operatorname{sgn} \beta) - \operatorname{sgn} \alpha e^{-\alpha x} U(x \operatorname{sgn} \alpha)\},$

$\alpha\beta \neq 0, \quad \alpha \neq \beta.$

Note 3.2. It can be noted that the function k described in (iii) above has for $\alpha, \beta > 0$ constant hazard in the right tail, and zero near the origin. Such a k can be thought of as representing the density of the lifetime of a component which works for a minimum period of time, and in addition, if it has worked for a long period of time, then it works as if its lifetime from then on has an exponential tail.

Proof. If $k(x)$ has one of the forms shown then it follows easily that (3.2) is satisfied for polynomials whose zeros are at $-\alpha$ and/or $-\beta$. If on the other hand $P(D)$ is such a polynomial its zeros must be real. If $a_0 = 0$ then the solution to (3.2) is obtained by first integrating (3.2).

$$(a_1 D + 1) k(x) = \delta(x)$$

to obtain

$$k(x) = \frac{1}{a_1} e^{-\frac{x}{a_1}} U(x) + C e^{\frac{x}{a_1}}.$$

If $a_1 > 0$, C must be 0 in order for k to be in L'. If $a_1 < 0$ then C must be $\frac{1}{a_1}$. Hence $k(x)$ has the form given by (i). If $a_0 \neq 0$, then k may be found similarly to have the form (ii) or (iii) according to whether the zeros are the same or differ. Note that $\alpha\beta = \frac{1}{a_0}$ because of the form of $P(D)$.

Remark 3.3. The function k has support on a positive half line if $\alpha > 0$ and $\beta > 0$; on a negative half line if $\alpha < 0$ and $\beta < 0$; and is never zero if $\alpha\beta < 0$. If $\alpha = -\beta$, it is the double exponential distribution.

Proposition 3.4. *Let*

$$P(D) = \prod_{i=1}^{m} (D - \alpha_1) \prod_{j=1}^{n} (D + \beta_j)$$

$$(0 < \alpha_1 < \alpha_2 < \ldots < \alpha_m, \ 0 < \beta_1 < \beta_2 < \ldots < \beta_n)$$

be a polynomial operator; then $k(x)$ given by

$$k(x) = \sum_{i=1}^{m} \frac{e^{\alpha_i x}}{P'(\alpha_i)} U(-x) + \sum_{j=1}^{n} \frac{e^{-\beta_j x}}{P'(-\beta_j)} U(x)$$

is the unique solution to (3.2) for $P(D)$.

Note 3.4. It can be seen that the above proposition is a generalization of a particular case of Proposition 3.3 and just as in the note succeeding it, one can interpret some special case of this family of densities as having the same interpretation as in Note 3.2.

The proof of the above Proposition 3.4 is similar to the previous case. Each of the constants obtained in solving the differential equation has only one possible choice consistent with the fact that k is a density.

Remark 3.3. This is only one of the many possible cases since multiple zeros and nonreal zeros were excluded. The former leads to certain linear combinations of Gamma densities (see Example 2.1), while the latter leads to oscillatory behavior.

Example 3.1. Let

$$P(s) = \frac{1}{2}(s^3 + 3s^2 + 4s + 2),$$

$$k(x) = 2\left(\cos\frac{x}{2}\right)^2 e^{-x}U(x).$$

Then $\{P(D), k(x)\}$ are a pair satisfying (3.2). The zeros of P are at $s = -1$, $s = -1 \pm i$.

4. KERNELS AND BOUNDED INTERVALS

We now consider the case of kernels $k(x, \theta)$ in which both x and θ are restricted to a bounded interval which we take to be $(0, b)$. Lemma 3.1 no longer is valid and indeed $k(x, \theta)$ will not be expressible as $k(x - \theta, 0)$ even in simple cases. Indeed, the differential operators with constant coefficients are associated with $k(x, \theta)$ which do not correspond to the standard distributions. One such case is given by a $k(x, \theta)$ which is related to a truncated exponential distribution.

Proposition 4.1. Let $P(D) = (D + \alpha)$ on $(0, b)$ and let $\alpha > 1$ or $\alpha > 0$ and $b < 1$; then $k(x, \theta)$ is a density satisfying (2.5) for $I = (0, b)$ if and only if it is given by

(4.1) $k(x, \theta) = e^{-\alpha(x - \theta)}U(x - \theta) + e^{-\alpha x}C(\theta)$

where

$$C(\theta) = \begin{cases} \dfrac{\theta - b}{b} & \text{if } \alpha = 0 \\[3mm] \dfrac{1 - \alpha - e^{-\alpha(b - \theta)}}{e^{-\alpha b} - 1} & \text{if } \alpha > 0. \end{cases}$$

Proof. If k as the given form then clearly (2.5) and $\int k = 1$ are satisfied. To show that $k(x, \theta) \geqslant 0$ we first consider $k(0, \theta) = C(\theta)$ which

– 67 –

is a monotone increasing function for $\theta \in (0, b)$. Its value at $0+$ is $-1 + \dfrac{\alpha}{1 - e^{-\alpha b}}$ which is nonnegative since

$$\frac{1 - e^{-\alpha b}}{\alpha} = \int_0^b e^{-\alpha x} \, dx \leqslant 1$$

under either hypothesis. Hence $C(\theta)$ is nonnegative and $k(x, \theta)$ being the sum of two nonnegative terms must be nonnegative. The converse follows by showing that (4.1) is the solution to the differential equation.

5. SCALE PARAMETERS

In the previous sections we have considered equations with constant coefficient differential operators $P(D)$. On infinite intervals these were associated with $k(x, \theta)$ of the form $k(x - \theta)$, that is the case in which θ is a location parameter. In this section we consider scale parameters and operators with polynomial coefficients. The scale parameters involve $k(x, \theta)$ of the form $h\left(\dfrac{x}{\theta}\right)$.

Fortunately there is a simple relation between densities for certain location parameters and those for some scale parameters. This family of scale distributions includes the family of beta distributions which are conjugates to the family of binomial distributions.

Lemma 5.1. *Let $P(D)$ be an operator with constant coefficients, $k(x)$ a density function on $(-\infty, \infty)$ such that (3.2) is satisfied. Then $h(y) = \dfrac{k(\ln y)}{y}$ is a density function on $(0, \infty)$ and*

$$P\left(y \frac{d}{dy}\right) yh(y) = \delta(y - 1).$$

The proof is immediate from the change of variable $x = \ln y$. Then

$$P\left(y \frac{d}{dy}\right) yh(y) = P\left(\frac{d}{dx}\right) k(x) = \delta(x) = \delta(\ln y) = \delta(y - 1).$$

This may be used to obtain the equation for a location parameter since

$$P\left(y\frac{d}{dy}\right)yh\left(\frac{y}{\theta}\right) = P\left(\frac{y}{\theta}\frac{d}{dy}\right)\frac{y}{\theta}\,\theta h\left(\frac{y}{\theta}\right) =$$

(5.1)

$$= \theta\delta\left(\frac{y}{\theta}-1\right) = \theta^2\delta(y-\theta) = y^2\delta(y-\theta).$$

Hence we have

Corollary 5.1. *Let* $h(y) = \dfrac{k(\ln y)}{y}$. *Then*

$$(5.2) \qquad y^{-2}P\left(y\frac{d}{dy}\right)yh\left(\frac{y}{\theta}\right) = \delta(y-\theta).$$

Example 5.1. Let

$$k(x) = \sum_{i=1}^{n}\frac{e^{\alpha_i x}}{P'(\alpha_i)}\,U(-x) \qquad (\alpha_i > 0).$$

Then $P(D) = \displaystyle\prod_{i=1}^{n}(D-\alpha_i)$ and

$$h(y) = \sum_{i=1}^{n}\frac{y^{\alpha_i-1}}{P'(\alpha_i)}\,U(-\ln y) \qquad (y > 0),$$

or

$$h(y) = \sum_{i=1}^{n}\frac{y^{\alpha_i-1}}{P'(\alpha_i)}\,U(-y+1)U(y).$$

The Beta distribution is given by

$$h(y) = \frac{y^{p-1}(1-y)^{q-1}}{\beta(p,q)}.$$

If q is an integer, then

$$h(y) = \sum_{k=0}^{q-1}\frac{\binom{q-1}{k}y^{p+k+1}}{\beta(p,q)}(-1)^k\,U(y)U(1-y)$$

and

$$k(x) = \sum_{k=0}^{q-1}\frac{\binom{q-1}{k}e^{(p+k)x}(-1)^k}{\beta(p,q)}\,U(-x).$$

Hence

$$P(s) = \frac{(s-p)(s-p-1)\dots(s-p-q+1)}{p(p+1)(p+q-1)(-1)^q}.$$

6. THE NORMAL CASE

In Section 3 we saw that the normal density with location parameter is not the Green's function of a differential operator with a finite number of constant coefficient terms. However, if we allow an infinite number of terms, we then can interpret the normal density as a generalization of a Green's function. The operator will involve the Hermite functions $\{h_n\}$

$$\text{(6.1)} \qquad h_n(x) = \frac{H_n(x)e^{-\frac{x^2}{2}}}{\pi^{\frac{1}{4}}\sqrt{n!2^n}} \qquad (x \in R)$$

where $H_n(x)$ is the Hermite polynomial given by

$$\text{(6.2)} \qquad H_n(x) = (-1)^n e^{x^2} D^n e^{-x^2} \qquad (x \in R).$$

(See Szegő [20] for details of these functions.) The idea of approximating as above can also be found in Gaffey [8]. However, the details to follow are not available there. The operator we use will be given by the expression

$$\text{(6.3)} \qquad P(D) = \sum_{n=0}^{\infty} \frac{h_n(0)(-1)^n D^n}{\pi^{\frac{1}{4}}\sqrt{2^n n!}}$$

and $P(D)f$ will mean the weak limit of the sequence of functions

$$\text{(6.4)} \qquad P_N(D)f = \sum_{h=0}^{N} \frac{h_n(0)(-1)^n D^n f}{\pi^{\frac{1}{4}}\sqrt{2^n n!}}.$$

Lemma 6.1. *Let* $P(D)$ *be given by* (6.3), *then*

$$P(D)e^{-x^2} = \delta(x).$$

Proof. By (6.2) we see that

$$P_N(D)e^{-x^2} = \sum_{h=0}^{N} \frac{h_n(0)e^{-x^2}H_n(x)}{\pi^{\frac{1}{4}}\sqrt{2^n n!}} = e^{-\frac{x^2}{2}} \sum_{n=0}^{N} h_n(0)h_n(x).$$

But the expansion in Hermite series of δ is just

$$\delta(x) \sim \sum_{n=0}^{\infty} h_n(0)h_n(x).$$

This series converges weakly to δ, i.e. for each element φ of the space of rapidly decreasing C^∞ functions,

$$\langle \delta, \varphi \rangle = \sum_{n=0}^{\infty} h_n(0)\langle h_n, \varphi \rangle.$$

But $\delta(x)e^{-\frac{x^2}{2}} = \delta(x)$. Hence $P_N(0)e^{-x^2} \to \delta(x)$ weakly.

Now we are ready to attack the problem of inverting

(6.5) $\qquad f(x) = \int_{-\infty}^{\infty} \frac{1}{\sigma\sqrt{2\pi}} e^{-\frac{(x-\theta)^2}{2\sigma^2}} dG(\theta).$

We first normalize the problem to $2\sigma^2 = 1$ and then consider the case when G is absolutely continuous and when it is not. In the former case, we have

$$g(x) = \int_{-\infty}^{\infty} \delta(x - \theta)g(\theta)\,d\theta =$$

(6.6) $\qquad = \int_{-\infty}^{\infty} \sum_{N=0}^{\infty} h_N(0)h_N(x - \theta)e^{-\frac{(x-\theta)^2}{2}} g(\theta)\,d\theta =$

$\qquad = \sum_{N=0}^{\infty} h_N(0) \int_{-\infty}^{\infty} h_N(x - \theta)e^{-\frac{(x-\theta)^2}{2}} g(\theta)\,d\theta$

where the convergence is again the sort of weak convergence mentioned above.

We suppose that g is in the class $C^p[-\infty, \infty]$, i.e. each of its first p derivatives is continuous and bounded. The integral in (6.6) may be estimated by using the fact (derivable from (6.2)) that

$$(6.7) \qquad D_x(e^{-\frac{x^2}{2}} h_n(x)) = -\sqrt{2(n+1)}\, h_{n+1}(x) e^{-\frac{x^2}{2}}.$$

Indeed we have

$$\left| \int_{-\infty}^{\infty} h_N(\theta) e^{-\frac{\theta^2}{2}} g(x-\theta)\, d\theta \right| =$$

$$= \left| \int D_\theta \frac{e^{-\frac{\theta^2}{2}} h_{N-1}(\theta)}{\sqrt{2N}} g(x-\theta)\, d\theta \right| =$$

$$(6.8) \qquad = \left| \int \frac{h_{N-1}(\theta)}{\sqrt{2N}} e^{-\frac{\theta^2}{2}} D_\theta(g(x-\theta))\, d\theta \right| =$$

$$= \ldots = \left| \int \frac{h_{N-p}(\theta) e^{-\frac{\theta^2}{2}}}{\sqrt{2N \cdot 2(N-1) \cdot \ldots \cdot 2(N-1)}} D_\theta^p g(x-\theta)\, d\theta \right| \leqslant$$

$$\leqslant \| g^{(p)} \|_\infty \frac{\| h_{N-p} \|_2}{(2N)^{\frac{p}{2}}} \left(\int_{-\infty}^{\infty} (e^{-\frac{\theta^2}{2}})^2\, d\theta \right)^{\frac{1}{2}} = \frac{\pi^{\frac{1}{4}} \| g^{(p)} \|_\infty}{(2N)^{\frac{p}{2}}}.$$

We now operate on (6.5) with $P_N(D)$ to obtain

$$P_N(D)f(x) = \frac{1}{\sqrt{\pi}} \int_{-\infty}^{\infty} P_N(D) e^{-\frac{(x-\theta)^2}{2}} g(\theta)\, d\theta =$$

$$(6.9)$$

$$= \frac{1}{\sqrt{\pi}} \sum_{l=0}^{N} h_l(0) \int_{-\infty}^{\infty} h_l(x-\theta) e^{-\frac{(x-\theta)^2}{2}} g(\theta)\, d\theta$$

by Lemma 6.1. Hence $\sqrt{\pi}\, P_N(D)f(x)$ converges to $g(x)$ as $N \to \infty$ in the weak sense. Hence, if this series converges uniformly, it must converge to $g(x)$.

Lemma 6.2. *Let* $g \in C^p[-\infty, \infty]$, $p > 1$, *and let*

$$f(x) = \frac{1}{\sqrt{\pi}} \int_{-\infty}^{\infty} e^{-\frac{(x-\theta)^2}{2}} g(\theta)\, d\theta.$$

Then, for $N \geqslant p$,

$$|\sqrt{\pi}\,P_N(D)f(x) - g(x)| \leqslant c_p N^{-\frac{p}{2}+\frac{3}{4}}$$

uniformly for $x \in R$ *where*

$$c_p = \frac{\pi^{\frac{1}{4}}\|g^{(p)}\|_\infty}{\frac{p}{2}-\frac{3}{4}}.$$

Proof. By (6.9) and (6.6) we have

$$|\sqrt{\pi}\,P_N(D)f(x) - g(x)| =$$

$$= \left| \sum_{l=N+1}^{\infty} h_l(0) \int_{-\infty}^{\infty} h_l(x-\theta)e^{-\frac{(x-\theta)^2}{2}} g(\theta)\, d\theta \right| \leqslant$$

(6.10)
$$\leqslant \sum_{l=N+1}^{\infty} l^{-\frac{1}{4}-\frac{p}{2}} \frac{\pi^{\frac{1}{4}}}{2^{\frac{p}{2}}} \|g^{(p)}\|_\infty \leqslant$$

$$\leqslant N^{-\frac{p}{2}+\frac{3}{4}} \frac{\pi^{\frac{1}{4}}}{\frac{p}{2}-\frac{3}{4}} \|g^{(p)}\|_\infty.$$

To estimate g, we first choose N, then estimate f and its first N derivatives by some convenient method. Either the Fourier transform method or the Hermite series method will give good rates for f and its derivatives under the same hypothesis for g.

REFERENCES

[1] J.R. Blum – V. Susarla, Estimation of a mixing distribution function, *Ann. Prob.*, 5 (1977), 200-209.

[2] J.R. Blum – V. Susarla, A Fourier inversion method for the estimation of a density and its derivative, *J. Austs. Math. Soc.*, 22 (1977), 166-171.

[3] J.R. Blum – V. Susarla, Confidence curves for a prior distribution function, Submitted.

[4]   K. Choi, Estimators for the parameters of a finite mixture of distributions, *Ann. Inst. Stat. Math.*, 21 (1969), 107-116.

[5]   E.A. Coddington – N. Levinson, *Theory of ordinary differential equations*, McGraw-Hill, New York, 1955.

[6]   J.J. Deely – R.L. Kruse, Construction of sequences estimating the mixing distribution, *Ann. Math. Statist.*, 39 (1968), 286-288.

[7]   R.J. Fox, Estimating the empiric distribution function of a parameter sequence, *Ann. Math. Statist*, 41 (1970), 1845-1852.

[8]   W. Gaffey, A consistent estimator of a component of a convolution, *Ann. Math. Statist.*, 30 (1959), 198-205.

[9]   W. Gaffey, Solutions to empirical Bayes squared sequence error loss estimation problems, *Ann. Statist.*, 6 (1978), 846-853.

[10]   T.E. O'Bryan – G. Walter, Meansquare estimation of the prior distribution, *Sankhya*, 41 (1979).

[11]   J. Maritz, *Empirical Bayes methods*, Methuen's Monographs on Applied Probability and Statistics, Methuen and Co., Ltd., London, 1970.

[12]   H. Robbins, The empirical Bayes approach to statistical decision problems, *Ann. Math. Statist.*, 35 (1964), 1-20.

[13]   M. Rosenblatt, Maximal deviation of k-dimensional density estimates, *Ann. Prob.*, 4 (1976), 1079-1085.

[14]   I. Stakgold, *Green's functions and boundary value problem*, Wiley, New York, 1975.

[15]   V. Susarla – G. Walter, Estimation of a multivariate density using delta sequences, *Ann. Stat.*, 9 (1981), 347-355.

[16]   M. Torterella – T.E. O'Bryan, Estimation of the prior distribution by best approximation in uniformly convex function spaces, *Bull. Inst. Math. Acad. Simca.*, 7 (1979), 69-85.

[17]   G. Walter,   Properties of Hermite series estimation of probability density, *Ann. Statist,* 5 (1977), 1258-1264.

[18]   G. Walter – J. Blum,   Probability density estimation using delta sequences, *Ann. Statist.,* 7 (1979), 328-340.

[19]   D.L. Widder,   *The Laplace transform,*   Princeton University Press, Princeton, N.J., 1946.

[20]   G. Szegő,   *Orthogonal polynomials,*   American Math. Soc. Colloquium publications, New York, 1959.

J. Blum

University of California, Davis, USA.

V. Susarla – G. Walter

University of Wisconsin, Milwaukee, USA.

COLLOQUIA MATHEMATICA SOCIETATIS JÁNOS BOLYAI

32. NONPARAMETRIC STATISTICAL INFERENCE,

BUDAPEST (HUNGARY), 1980.

AN APPROXIMATION OF EMPIRICAL FIELDS

I.S. BORISOV

Let $\{X_i;\ i \geqslant 1\}$ be independent identically distributed random variables (i.i.d. r.v.) with values in R^k, and let $F(t)$, $t \equiv (t^{(1)}, \ldots, t^{(k)}) \in$ $\in R^k$, be the distribution function of r.v. X_1. Denote by $F_n(t)$ the empirical distribution function defined as

$$F_n(t) = \frac{1}{n} \sum_{i=1}^{n} I_{X_i}(t)$$

where

$$(1) \qquad I_{X_i}(t) = \begin{cases} 1 & \text{if} \quad X_i < t, \\ 0 & \text{if} \quad X_i \geqslant t. \end{cases}$$

(The relation $Z < t$ (or $Z \leqslant t$) for arbitrary $Z, t \in R^k$ will be understood as $Z^{(j)} < t^{(j)}$ (or $Z^{(j)} \leqslant t^{(j)}$) for all $j = 1, 2, \ldots, k$).

Consider the sequence of the so-called k-dimensional empirical processes (or fields)

$$S_n(t) = \sqrt{n}(F_n(t) - F(t)) \qquad (n = 1, 2, \ldots).$$

It is well known that the finite dimensional distributions of $S_n(\cdot)$ converge weakly to that of a Gaussian field $W_F(\cdot)$ with zero expectation and covariance function

$$\mathsf{E}\,W_F(t)W_F(s) = F(\min(t,s)) - F(t)F(s)$$

where $\min(t,s) = (\min(t^{(1)}, s^{(1)}), \ldots, \min(t^{(k)}, s^{(k)}))$.

During the last five years strong approximation of these fields (or construction of $S_n(\cdot)$ and $W_F(\cdot)$ on certain probability space and investigation of a distance between trajectories of that) has been studied by many authors (cf. for example, [1]-[4]). But strong results were obtained only in the case when the coordinates of X_1 are independent. We formulate one result.

Theorem A (cf. [1]-[3]). *Let X_1 be a random vector in R^k with independent coordinates. Then one can construct $S_n(\cdot)$ and $W_F(\cdot)$ on some probability space such that*

$$\mathsf{P}(\Delta_F \geqslant C(k)\gamma_k(n)) \leqslant \frac{1}{n^2}$$

where $\Delta_F = \sup\limits_{t \in R^k} |S_n(t) - W_F(t)|$; *the constant $C(k)$ depends on k only;* $\gamma_k(n) = (\log n)^k n^{-\frac{1}{2}}$ *if $k = 1, 2$;* $\gamma_k(n) = (\log n)^{\frac{3}{2}} n^{-\frac{1}{2(k+1)}}$ *if $k \geqslant 3$.*

This theorem is proved at first for the case when X_1 is uniformly distributed on the k-dimensional unit cube. The general case follows from here by a simple transformation of the argument t. However if the coordinates of X_1 are not independent, then such transformation does not exist.

In the present article we investigate the case when the coordinates are not independent.

Let X_1 have a discrete distribution concentrated on some m points $a_1, \ldots, a_m \in R^k$. We shall say that the distribution of X_1 belongs to the class $\mathscr{D}(r)$ $(r = 1, 2, \ldots)$, if there exists a permutation $k_1, \ldots, k_m$

of the numbers $1, \ldots, m$ such that for arbitrary $t \in R^k$

$$(2) \qquad \{a_i; \ i \leqslant m\} \cap \{z \in R^k : z < t\} = \bigcup_{j=1}^{r(t)} A(N_j(t), L_j(t))$$

where $A(N_j(t), L_j(t)) = \{a_{k_s}; \ N_j(t) \leqslant s < N_j(t) + L_j(t)\}$, $r(\cdot), N_j(\cdot), L_j(\cdot)$ are natural numbers, $N_j(t) + L_j(t) < N_{j+1}(t)$, $r(t) \leqslant r$ for all $t \in R^k$, and $r(t_0) = r$ for some $t_0 \in R^k$. It is clear that $r \leqslant m$. But as a rule $r = o(m)$ as $m \to \infty$. In particular, if $a_1 \leqslant a_2 \leqslant \ldots \leqslant a_m$ then $r = 1$.

Theorem 1. *Suppose that the distribution of X_1 belongs to the class $\mathscr{D}(r)$. Then one can construct $S_n(\cdot)$ and $W_F(\cdot)$ on some probability space such that*

$$P\left(\Delta_F \geqslant C \ \frac{r \log n}{\sqrt{n}}\right) \leqslant \frac{r}{n^2}$$

where C is an absolute constant.

In the case of a general distribution of X_1 we prove

Theorem 2. *One can construct $S_n(\cdot)$ and $W_F(\cdot)$ on some probability space such that*

$$\Delta_F = O(n^{-\frac{1}{2(2k-1)}} \log n) \qquad (n \to \infty) \quad \text{with probability } 1$$

where the constant $O(\cdot)$ depends only on k.

Theorem 2 improves a result of W. Philipp and L. Pinzur (cf. [5]).

Proof of Theorem 1. Let the distribution of X_1 belong to the class $\mathscr{D}(r)$. Then for some $m \geqslant 1$

$$F(t) = \sum_{i=1}^{m} p_i I_{a_i}(t)$$

where $p_i = P(X_1 = a_i)$, $I_{a_i}(t)$ defined in (1).

Denote by $n(a_i)$ the number of elements of the sample $X_1, \ldots, X_n$ which are equal to a_i. Then

$$F_n(t) = \sum_{i=1}^{m} \frac{n(a_i)}{n} I_{a_i}(t)$$

and the empirical field can be written as

$$(3) \qquad S_n(t) = \sum_{i=1}^{m} I_{a_i}(t) \frac{n(a_i) - np_i}{\sqrt{n}}.$$

Consider now the sequence of i.i.d. r.v.-s $\{v_i; \ i \geqslant 1\}$ with distribution function $\tilde{F}(u) = \sum_{i < u} p_i$, and denote by $\tilde{S}_n(u)$, $u \in R$, the empirical process based on the sample $v_1, \ldots, v_n$. It is easy to see that the r.v.'-s

$$\frac{n(a_i) - np_i}{\sqrt{n}} \qquad (i \leqslant m)$$

have the same common distribution as that of the r.v.'-s

$$\tilde{S}_n(i+1) - \tilde{S}_n(i) \qquad (i \leqslant m).$$

Hence by (3)

$$(4) \qquad S_n(t) \overset{d}{=} \sum_{i=1}^{m} I_{a_i}(t)[\tilde{S}_n(i+1) - \tilde{S}_n(i)]$$

It follows from here that

$$(5) \qquad W_F(t) \overset{d}{=} \sum_{i=1}^{m} I_{a_i}(t)[W_{\tilde{F}}(i+1) - W_{\tilde{F}}(i)].$$

We suppose that the identical permutation $1, 2, \ldots, m$ satisfies the condition (2). Let $\{N_j(t), L_j(t); \ j = 1, \ldots, r(t)\}$ be the sets of natural numbers defined in (2) for each $t \in R^k$. Then form (4) and (5)

$$(6) \qquad \begin{aligned} S_n(t) &\overset{d}{=} \sum_{j=1}^{r(t)} [\tilde{S}_n(N_j(t) + L_j(t)) - \tilde{S}_n(N_j(t))] \\ W_F(t) &\overset{d}{=} \sum_{j=1}^{r(t)} [W_{\tilde{F}}(N_j(t) + L_j(t)) - W_{\tilde{F}}(N_j(t))]. \end{aligned}$$

Using now Theorem A for $k = 1$ we can construct the processes $\tilde{S}_n(\cdot)$, $W_{\tilde{F}}(\cdot)$ on a suitable probability space such that

(7) $$P\left(\Delta_{\widetilde{F}} \geqslant C\frac{\log n}{\sqrt{n}}\right) \leqslant \frac{1}{n^2}$$

where C is absolute constant. Relations (6) and (7) together prove our statement.

For the proof of Theorem 2 we will need the following

Lemma. *There exists a sequence of i.i.d. r.v.'-s* $\{\hat{X}_i; i \geqslant 1\}$ *defined on the same probability space as that of* $\{X_i; i \geqslant 1\}$ *such that*

(1) *the distribution of* $\hat{X}_1$ *belongs to the class* $\mathscr{D}(m^{k-1})$,

(2) *there are constants* $C_1(k), C_2(k)$ *such that for all* $m \leqslant n$

$$P\left(\sup_{t \in R^k} |S_n(t) - \hat{S}_n(t)| > C_1(k)\frac{\log m}{\sqrt{m}}\right) \leqslant \frac{C_2(k)}{m^{2(2k-1)}}$$

where $\hat{S}_n(\cdot)$ *is the empirical process based on the sample* $\hat{X}_1, \ldots, \hat{X}_n$.

Proof. Let $X_1 = (X_1^{(1)}, \ldots, X_1^{(k)})$. Introduce the notation

$$t_i(u) = \inf\{y \in R: F^{(i)}(y) \geqslant u\} \qquad (u \in [0, 1], \ i = 1, \ldots, k),$$

where $F^{(i)}(\cdot)$ is the distribution function of the coordinate $X_1^{(i)}$. Consider for some natural m the set of lattice points

$$A_m = \left\{\left(t_1\left(\frac{i_1}{m}\right), t_2\left(\frac{i_2}{m}\right), \ldots, t_k\left(\frac{i_k}{m}\right)\right);\right.$$

$$i_s = 0, 1, \ldots, m-1, \ s = 1, \ldots, k\Big\}.$$

For the sake of simplicity we suppose that for all $s = 1, \ldots, k$, $t_s(0) > > -\infty$, $t_s(1) < \infty$ (that is $|X_1^{(s)}| \leqslant N$ with probability 1 for some $N > 0$) and $t_s\left(\frac{i_s}{m}\right) < t_s\left(\frac{i_s + 1}{m}\right)$ $(i_s = 0, 1, \ldots, m-1)$. (It is possible in the general case that $t_s\left(\frac{i_s^*}{m}\right) = t_s\left(\frac{i_s^* + 1}{m}\right)$ for some i_s^*; so that it is necessary in this case to consider only such numbers $i_s^{(1)}, \ldots, i_s^{(m_1)}$, $m_1 < m$, for which $t_s\left(\frac{i_s^{(l)}}{m}\right) < t_s\left(\frac{i_s^{(l+1)}}{m}\right)$, $s = 1, \ldots, k$, $l = 1, \ldots, m_1 - 1$.)

So introduce the following notation

$$b_m(i_1, \ldots, i_k) = \left(t_1\left(\frac{i_1}{m}\right), \ldots, t_k\left(\frac{i_k}{m}\right) \right)$$

$$(i_s = 0, 1, \ldots, m, \; s = 1, \ldots, k).$$

Now define the r.v.'s $\hat{X}_i$ $(i = 1, 2, \ldots)$, as follows. If

$$b_m(i_1, \ldots, i_k) < X_i \leqslant b_m(i_1 + 1, \ldots, i_k + 1)$$

for some $i_1, \ldots, i_k$ then we put

$$\hat{X}_i = b_m(i_1, \ldots, i_k).$$

It is clear that $\hat{X}_1, \hat{X}_2, \ldots$, are i.i.d. r.v.'s with discrete distribution concentrated on the set A_m. We number the points of A_m in the following way:

$$a_s \quad = b_m(0, \ldots, 0, s) \qquad (s = 0, 1, \ldots, m-1),$$

$$a_{m+s} = b_m(0, \ldots, 0, 1, s) \qquad (s = 0, 1, \ldots, m-1),$$

$$\cdots \cdots \cdots \cdots \cdots \cdots \cdots$$

$$a_{lm+s} = b_m(0, \ldots, 0, l, s) \qquad (s = 0, 1, \ldots, m-1),$$

$$\cdots \cdots \cdots \cdots \cdots \cdots \cdots$$

and so on. It is easy to see that the distribution of $\hat{X}_1$ belongs to the class $\mathscr{D}(m^{k-1})$.

Further note that it follows from definition of $\hat{X}_i$ that

$$I_{\hat{X}_i}(t) = I_{X_i}(b_m(i_1, \ldots, i_k))$$

if $b_m(i_1 - 1, \ldots, i_k - 1) < t \leqslant b_m(i_1, \ldots, i_k)$. Then we have for such a t

$$(8) \qquad \hat{S}_n(t) = S_n(b_m(i_1, \ldots, i_k))$$

and we can estimate

(9) $\quad P(\sup_{t \in R^k} |S_n(t) - \hat{S}_n(t)| > 2z) =$

$$= P\left(\bigcup_{i_1,\ldots,i_k} \left\{ \sup_{\substack{b_m(i_1-1,\ldots,i_k-1)< \\ <t \leqslant b_m(i_1,\ldots,i_k)}} |S_n(t) - S_n(b_m(i_1,\ldots,i_k))| > 2z \right\} \right) \leqslant$$

$$\leqslant \sum_{i_1,\ldots,i_k} P\left(\sup_{\substack{b_m(i_1-1,\ldots,i_k-1)< \\ <t \leqslant b_m(i_1,\ldots,i_k)}} |S_n(t) - \right.$$

$$\left. - S_n(b_m(i_1-1,\ldots,i_k-1)+0)| > z \right) +$$

$$+ \sum_{i_1,\ldots,i_k} P(|S_n(b_m(i_1,\ldots,i_k)) -$$

$$- S_n(b_m(i_1-1,\ldots,i_k-1)+0)| > z)$$

where

$$S_n(b_m(\cdot)+0) = \lim_{t \to b_m(\cdot),\, t > b_m(\cdot)} S_n(t).$$

Note that each of the sums in the right-hand side of (9) has m^k summands.

Introduce the notation

(10) $\quad p_m = F(b_m(i_1,\ldots,i_k)) - F(b_m(i_1-1,\ldots,i_k-1)+0).$

It follows from the results of [6] that if there exists $\gamma \in (0,1)$ such that

(11)
$$p_m \leqslant \min\left\{ \frac{\gamma}{\gamma+3}, \frac{1}{3}\gamma(1-\gamma)^k z\sqrt{n} \right\},$$

$$9p_m(1-p_m) < [z\gamma(1-\gamma)^k]^2$$

then for $n \geqslant 3$

(12) $\quad P\left(\sup_{\substack{b_m(i_1-1,\ldots,i_k-1)< \\ <t \leqslant b_m(i_1,\ldots,i_k)}} |S_n(t) - \right.$

$$\left. - S_n(b_m(i_1-1,\ldots,i_k-1)+0)| > z \right) \leqslant$$

$$\leqslant 2\left[1 - \frac{9p_m(1 - p_m)}{[z\gamma(1 - \gamma)^k]^2}\right]^{-k} P(|S_n(b_m(i_1, \ldots, i_k)) -$$

$$- S_n(b_m(i_1 - 1, \ldots, i_k - 1) + 0)| > (1 - \gamma)^k z).$$

Further note that it follows from (10) and from the construction of points $b_m(\cdot), t_s(\cdot)$

$$(13) \qquad p_m \leqslant \sum_{s=1}^{k} \left[F^{(s)}\left(t_s\left(\frac{i_s}{m}\right)\right) - F^{(s)}\left(t_s\left(\frac{i_s - 1}{m}\right) + 0\right)\right] \leqslant \frac{k}{m}.$$

Now put $\gamma = \frac{1}{2}$, $z = C\sqrt{\frac{k}{m}} \log m$. It is clear that there is $m_0 = m_0(k, C)$ such that for all $m \geqslant m_0$ the conditions (11) will be satisfied. Hence

$$(14) \qquad P\left(\sup_{\substack{b_m(i_1 - 1, \ldots, i_k - 1) < \\ < t \leqslant b_m(i_1, \ldots, i_k)}} |S_n(t) - \right.$$

$$\left. - S_n(b_m(i_1 - 1, \ldots, i_k - 1) + 0)| > z\right) \leqslant$$

$$\leqslant \tilde{C}_1(k) P(|S_n(b_m(i_1, \ldots, i_k)) -$$

$$- S_n(b_m(i_1 - 1, \ldots, i_k - 1) + 0)| > \tilde{C}_2(k)z).$$

Since

$$S_n(b_m(i_1, \ldots, i_k)) - S_n(b_m(i_1 - 1, \ldots, i_k - 1) + 0) =$$

$$= \frac{1}{\sqrt{n}} \sum_{j=1}^{n} (\xi_j - E\xi_j)$$

where $\xi_j = I_{X_j}(b_m(i_1, \ldots, i_k)) - I_{X_j}(b_m(i_1 - 1, \ldots, i_k - 1) + 0)$ with $D\xi_j = p_m(1 - p_m)$, we can use the exponential estimators for the right-hand side of the inequality (14). We have (cf. [7], p. 358)

$$J \equiv P\left(|S_n(b_m(i_1, \ldots, i_k)) - S_n(b_m(i_1 - 1, \ldots, i_k - 1) + 0)| > \right.$$

$$\left. > C\tilde{C}_2(k)\sqrt{\frac{k}{m}} \log m\right) \leqslant$$

$$\leqslant 2 \exp\left\{-\frac{[C\tilde{C}_2(k) \log m]^2 k}{4p_m(1 - p_m)m}\right\} \leqslant$$

$$\leqslant 2 \exp\left\{-\frac{1}{4}[C\widetilde{C}_2(k)\log m]^2\right\}$$

if $C\widetilde{C}_2(k)\sqrt{\dfrac{k}{m}}\log m \leqslant \sqrt{n}\, p_m(1-p_m)$; and

$$J \leqslant 2 \exp\left\{-\frac{1}{4}C\widetilde{C}_2(k)\sqrt{\frac{kn}{m}}\log m\right\}$$

if $C\widetilde{C}_2(k)\sqrt{\dfrac{k}{m}}\log m > \sqrt{n}\, p_m(1-p_m)$. Now setting

$$C = \frac{4(k+2(2k-1))}{\widetilde{C}_2(k)\sqrt{k}},$$

we obtain for $m \leqslant n$,

(15) $$J \leqslant \frac{2}{m^{k+2(2k-1)}}.$$

Our statement follows from (9), (14) and (15).

Now we return to the

Proof of Theorem 2. Let

$$m \equiv m(n) = [n^{\frac{1}{2k-1}}].$$

Let further $\{\hat{X}_i;\ i \geqslant 1\}$ be the sequence of i.i.d. r.v.'s defined in the Lemma. It follows from Theorem 1 and the Lemma that there exists a probability space $(\Omega, \mathscr{F}, \mathrm{P})$ such that as $n \to \infty$

(16)
$$\Delta_{\hat{F}} = O\left(\frac{m^{k-1}}{\sqrt{n}}\log n\right) = O(n^{-\frac{1}{2(2k-1)}}\log n)$$

with probability   1

where constant $O(\cdot)$ is absolute, and $\hat{F}(\cdot)$ is the distribution function of $\hat{X}_1$. From (8) we obtain that the common distribution of the family of r.v.'s

$$\{\hat{S}_n(b_m(i_1,\ldots,i_k));\ i_s \leqslant m,\ s = 1,\ldots,k\}$$

coincides with that of

$$\{S_n(b_m(i_1,\ldots,i_k));\ i_s \leqslant m,\ s = 1,\ldots,k\}$$

and analogous statement is true for the families

$$\{W_{\hat{F}}(b_m (i_1, \dots, i_k)); \ i_s \leqslant m, \ s = 1, \dots, k\}$$

and

$$\{W_F(b_m (i_1, \dots, i_k)); \ i_s \leqslant m, \ s = 1, \dots, k\}.$$

Therefore, we construct on the probability space $(\Omega, \mathscr{F}, P)$ the values of the random fields $S_n(\cdot)$ and $W_F(\cdot)$ in the lattice points

$$\{b_m (i_1, \dots, i_k); \ i_s \leqslant m, \ s = 1, \dots, k\}.$$

It is clear that by Kolmogorov's theorem one can define on the extended probability space the values of our random fields in other points of R^k.

Now we must estimate the increments of the fields $S_n(\cdot)$ and $W_F(\cdot)$ on each cube

$$\{t \in R^k: b_m (i_1 - 1, \dots, i_k - 1) < t \leqslant b_m (i_1, \dots, i_k)\}.$$

It follows from the Lemma and the Borel — Cantelli theorem that as $n \to \infty$

(17)
$$\max_{i_1, \dots, i_k} \ \sup_{\substack{b_m(i_1 - 1, \dots, i_k - 1) < \\ < t \leqslant b_m(i_1, \dots, i_k)}} |S_n(t) -$$

$$- S_n(b_m (i_1 - 1, \dots, i_k - 1) + 0)| =$$

$$= O(n^{-\frac{1}{2(2k-1)}} \log n) \quad \text{with probability} \ 1$$

where the constant $O(\cdot)$ depends only on k.

An analogous statement is true for the random field $W_F(\cdot)$ since one can obtain from (13), (14), by taking limits the analogous inequality for

$$P(\sup_{b_m(\cdot) < t \leqslant b_m(\cdot)} |W_F(t) - W_F(b_m (\cdot) + 0)| > z)$$

(cf. [5] or [8]).

Theorem 2 follows from (16) and (17).

REFERENCES

[1] M. Csörgő − P. Révész, A new method to prove Strassen type laws of invariance principle. II, *Z. Wahrscheinlichkeitstheorie und verw. Gebiete,* 31 (1975), 261-269.

[2] J. Komlós − P. Major − G. Tusnády, An approximation of partial sums of independent r.v.'s and the sample d.f.I., *Z. Wahrscheinlichkeitstheorie und verw. Gebiete,* 32 (1975), 111-131.

[3] G. Tusnády, A remark on the approximation of the sample DF in the multidimensional case, *Periodica Math. Hung.,* 8 (1977), 53-55.

[4] I.S. Borisov, On the rate of approximation of empirical fields, (in Russian), to appear in *Theor. Probability Appl.,* 1979.

[5] M.D. Burke − M. Csörgő − S. Csörgő − P. Révész, Approximation of the empirical process when parameters are estimated, *Ann. Probab.,* 7 (1979), 790-810.

[6] V.V. Yurinskiĭ, Exponential estimators for sums of independent random vectors, (in Russian), Doctors dissertation, Mathematical Institute of the USSR Academy of Sciences, Moscow, 1973.

[7] V.V. Petrov, *Sums of independent random variables,* (in Russian), Nauka, Moscow, 1972.

[8] R.M. Dudley, Weak convergence of probabilities on non-separable metric spaces and empirical measures on euclidean spaces, *Illinois J. Math.,* 10 (1966), 109-126.

I.S. Borisov

Institute of Mathematics, SU-Novosibirsk-90.

COLLOQUIA MATHEMATICA SOCIETATIS JÁNOS BOLYAI

32. NONPARAMETRIC STATISTICAL INFERENCE,

BUDAPEST (HUNGARY), 1980.

TESTS FOR EXPONENTIALITY BASED ON RANDOMLY CENSORED DATA

M.D. BURKE

1. INTRODUCTION

The following experimental situation often arises in life-testing and medical follow-up studies: let $X_1^0, X_2^0, \ldots$ be a sequence of independent and identically distributed random variables (i.i.d. r.v.) having continuous distribution function F^0. We wish to estimate F^0. However, the random variables $X_1^0, X_2^0, \ldots$ are known to be censored on the right by another sequence of i.i.d. r.v. $Y_1, Y_2, \ldots$ having continuous distribution function H. That is, one observes only the sequence $(X_1, \delta_1), (X_2, \delta_2), \ldots$, where $X_i = \min(X_i^0, Y_i)$ and $\delta_i = I\{X_i^0 \leqslant Y_i\}$ is the indicator function of the event $\{X_i^0 \leqslant Y_i\}$. For example X_i^0 might be the survival time of cancer patient number i undergoing a certain treatment. Censoring occurs when the patient, for example, dies from another cause, changes treatments or leaves the study. In this instance, $X_i = Y_i$ and $\delta_i = 0$. We will assume that $\{X_i^0\}$ and $\{Y_i\}$ are independent sequences.

To estimate F^0 in the above model, the K a p l a n $-$ M e i e r [12] estimator, $\hat{F}_n^0$, has been widely used. It is defined as

$$1 - \hat{F}_n^0(t) = \begin{cases} \prod_{i:X_i \leqslant t} \left[\dfrac{n - R_i}{n - R_i + 1} \right]^{\delta_i}, & \text{if} \quad t < X_{(n)} \\ 0, & \text{if} \quad t \geqslant X_{(n)}, \end{cases}$$

where $X_{(n)} = \max(X_1, X_2, \ldots, X_n)$ and R_i is the rank of $(X_i, 1 - \delta_i)$ in the lexicographic ordering of the sequence $(X_1, 1 - \delta_1), (X_2, 1 - \delta_2), \ldots, (X_n, 1 - \delta_n)$.

Consider the process Z_n defined by

(1) $\qquad Z_n(t) = \sqrt{n}[\hat{F}_n^0(t) - F^0(t)], \qquad t \geqslant 0.$

The problem of testing whether the underlying distribution function of $X_1^0, X_2^0, \ldots, X_n^0$ is a specific distribution function F^0, when n is large, can be handled by obtaining the limiting distribution of Z_n. Then, various statistics based on Z_n would have limiting distributions equal to the corresponding functionals of the limit, (cf. Corollary 1).

However, in many cases, the underlying distribution function cannot be completely specified under the null hypothesis. Instead, one is able only to specify that the underlying distribution belongs to a certain family $\{F^0(\cdot; \theta)\}$ of distribution functions indexed by a vector of parameters $\theta \in R^p$. One method of handling this situation is to estimate the nuisance parameters θ by $\hat{\theta}_n$, a sequence of estimators based on the observations $(X_1, \delta_1), \ldots, (X_n, \delta_n)$, and to study the asymptotic behaviour of the process

(2) $\qquad \hat{Z}_n(t) = \sqrt{n}[\hat{F}_n^0(t) - F^0(t; \hat{\theta}_n)], \qquad t \geqslant 0.$

This program is carried out in Section 2 under certain regularity conditions on the estimator sequence $\{\hat{\theta}_n\}$. These conditions are satisfied, for example, in the case when F^0 is a negative exponential distribution with unknown scale parameter, H is an arbitrary continuous distribution function and the estimation method of Section 3 is used. In the case of other families of distributions, the statistician would be advised to follow the suggestion of Durbin [7] concerning the empirical process with parameters estimated, namely to check in each case whether the putative estimator sequence satisfies the regularity conditions. As with the usual empirical process when parameters are estimated (with no censoring in the model), the asymptotic distribution of various statistics based on $\hat{Z}_n$ in the present model would depend on the underlying family of distribution functions and, in certain instances, on the true unknown value of θ. In Section 3, the negative exponential family is studied and tests for exponentiality are obtained.

2. STRONG APPROXIMATIONS

Closely related to the Kaplan – Meier estimator $\hat{F}_n^0$ are the empirical distribution functions F_n and $\tilde{F}_n$, where

$$F_n(t) = \frac{1}{n} \sum_{j=1}^{n} I_{\{X_i < t\}}$$

and

$$\tilde{F}_n(t) = \frac{1}{n} \sum_{j=1}^{n} I_{\{X_i < t, \, \delta_i = 1\}}$$

are based on all the observations $X_1, X_2, \ldots, X_n$, and the uncensored observations from $X_1, X_2, \ldots, X_n$, respectively. The $X_1, X_2, \ldots, X_n$ are i.i.d. r.v. with continuous distribution function F, where $(1 - F) = = (1 - F^0)(1 - H)$, while the uncensored r.v. from $X_1, X_2, \ldots, X_n$ have sub-distribution function $\tilde{F}$ given by $\tilde{F}(t) = \int_{-\infty}^{t} (1 - H) \, dF^0$.

The limiting behaviour of Z_n has been studied by several authors, including E f r o n [9], B r e s l o w and C r o w l e y [3], M e i e r [14] and B u r k e , C s ö r g ő and H o r v á t h [5].

In the latter paper, strong approximations of a more general Z_n is obtained for the competing risks model. In the sequel, we will use the following consequence of a result by B u r k e , C s ö r g ő and H o r v á t h [5]:

Theorem 1. *One can construct two Brownian Bridge sequences, B_n and $\tilde{B}_n$, such that*

$$\sup_{(s,t)\in R^2} \|(\xi_n(s), \widetilde{\xi}_n(t)) - (B_n[F(s)], \widetilde{B}_n[\widetilde{F}(t)])\| \overset{\text{a.s.}}{=} O\left(\frac{\log n}{\sqrt{n}}\right),$$

where $\xi_n(s) = \sqrt{n}[F_n(s) - F(s)]$, $\widetilde{\xi}_n(t) = \overline{n}[\widetilde{F}_n(t) - \widetilde{F}(t)]$ and $(B_n, \widetilde{B}_n)$ are jointly Gaussian with mean zero and covariance structure given by

$$\mathbb{E}B_n[F(s)]B_n[F(t)] = F(s) \wedge F(t) - F(s)F(t)$$

$$\mathbb{E}\widetilde{B}_n[\widetilde{F}(s)]\widetilde{B}_n[\widetilde{F}(t)] = \widetilde{F}(s) \wedge \widetilde{F}(t) - \widetilde{F}(s)\widetilde{F}(t)$$

$$\mathbb{E}B_n[F(s)]\widetilde{B}_n[\widetilde{F}(t)] = \widetilde{F}(s) \wedge \widetilde{F}(t) - F(s)\widetilde{F}(t).$$

Moreover, a sequence G_n of Gaussian processes can be constructed such that

$$\sup_{t \leqslant T_n} |Z_n(t) - G_n(t)| \overset{\text{a.s.}}{=} O\{r(n)\},$$

where $r(n) = \max\left\{\dfrac{b_n^2 (\log n)^{\frac{3}{2}}}{n^{\frac{1}{3}}}, \dfrac{b_n^4 \log n}{n^{\frac{1}{2}}}, \dfrac{b_n^6 (\log n)^2}{n^{\frac{3}{2}}}\right\}$, $b_n = \dfrac{1}{1 - F(T_n)}$

and T_n satisfies the condition

$$T_n < T = \inf\{t\colon F(t) = 1\}$$

and

$$1 - F(T_n) \geqslant \sqrt{\frac{2(1 + \delta) \log n}{n}}$$

with $\delta > -1$. The process G_n is defined by

$$G_n(t) = -[1 - F^0(t)] \int_{-\infty}^t \frac{B_n[F(s)]}{(1 - F(s))^2} d\widetilde{F}(s) -$$

$$-\frac{\widetilde{B}_n[\widetilde{F}(t)]}{1 - H(t)} + [1 - F^0(t)] \int_{-\infty}^t \frac{\widetilde{B}_n[\widetilde{F}(s)]}{(1 - F(s))^2} dF(s).$$

Remarks. We will, of course, assume that our underlying probability space is rich enough to make the above construction (cf. B u r k e et. al. [5]). If, for example, $1 - F(T_n) = \dfrac{1}{n^\delta}$, $0 < \delta < \dfrac{1}{8}$, then $r(n) = \dfrac{1}{n^\epsilon}$, for some $\epsilon > 0$. As a consequence of Theorem 1, $Z_n(t)$ converges weakly to the Gaussian process G_1 over the interval $(-\infty, T^*]$, where $F(T^*) < 1$. The process G_n has mean equal to zero and covariance

$$(3) \qquad \mathbf{E} G_n(s) G_n(t) = (1 - F^0(s))(1 - F^0(t)) \int_{-\infty}^{s \wedge t} \frac{dF^0}{(1-F)(1-F^0)}.$$

On letting $a(t) = \int_{-\infty}^{t} (1-F)^{-1}(1-F^0)^{-1} \, dF^0$, the Gaussian process

$$W_n(x) = \frac{G_n(a^{-1}(x))}{1 - F^0(a^{-1}(x))}$$

has mean zero and covariance $\mathbf{E} W_n(x) W_n(y) = x \wedge y$, that is, W_n is a Wiener process. Consequently, we obtain

$$\int_{-\infty}^{T_n} Z_n^*(t) \, d\left\{\frac{1}{1 + a(t)}\right\} \xrightarrow{\mathscr{D}} \int_0^1 [B(y)]^2 \, dy$$

$$(4) \qquad \sup_{t \leqslant T_n} |Z_n^*(t)| \xrightarrow{\mathscr{D}} \sup_{0 \leqslant y \leqslant 1} |B(y)|$$

and

$$\sup_{t \leqslant T_n} Z_n^*(t) \xrightarrow{\mathscr{D}} \sup_{0 \leqslant y \leqslant 1} B(y),$$

where

$$Z_n^*(t) = \frac{Z_n(t)}{(1 - F^0(t))(1 + a(t))},$$

B is a Brownian Bridge on $[0, 1]$ and $1 - F(T_n) = \frac{1}{n^\delta}$, $0 < \delta < \frac{1}{8}$. The proof of (4) is based on the fact that on letting $y = \frac{1}{1 + a(t)}$,

$$B_n(y) = \frac{G_n(t)}{(1 - F^0(t))(1 + a(t))}, \qquad 0 \leqslant y < 1,$$

is a Brownian Bridge for each n.

The limiting distributions in (4) do not depend on F^0 or H and they are well-known, (cf. A n d e r s o n and D a r l i n g [1] and B i l l i n g s l e y [2], p. 85). If the function $x = a(t)$ is not specified, then under mild regularity conditions $a(t)$ may be replaced by a uniformly consistent estimator in the definition of Z_n^* and the results of (4) will continue to hold. Thus, the results of (4) can be used for asymptotically distribution-free goodness of fit tests when F^0 is completely specified under the null hypothesis.

The idea of using the transformation $x = a(t)$ is from E f r o n [9] who, in turn, got it from D o o b [16]. A detailed account of this transmation can be found in B u r k e et al. [5] and C s ö r g ő and H o r v á t h [6]. The papers of N a i r (this volume), G i l l e s p i e and F i s h e r [10], and H a l l and W e l l n e r [11], use similar approaches to obtain confidence bands for F^0. Their results are obtained for the interval $[0, T]$, where $F(T) < 1$.

Consider, now, the situation where F^0 is not completely specified, that is, suppose that only the distributional form of $F^0(\cdot; \theta)$ can be specified while some parameters θ, a p-dimensional vector, are unknown. Our approach is to estimate θ by $\hat{\theta}_n$, a sequence of estimators based on the observations $(X_1, \delta_1), \dots, (X_n, \delta_n)$, and to study statistics based on $\hat{Z}_n$ defined by (2).

We will assume that $\{\hat{\theta}_n\}$ satisfies

$$(5) \qquad \sqrt{n}(\hat{\theta}_n - \theta_0) = \frac{1}{\sqrt{n}} \sum_{j=1}^{n} \{l(X_j) + A(\delta_j - \tilde{F}(\infty))\} + \epsilon_n,$$

where A is a p-dimensional constant vector depending possibly on the true unknown value θ_0 of θ, $\tilde{F}(\infty) = \lim_{t \to \infty} \tilde{F}(t)$ is equal to $E\delta_i$, and $\epsilon_n \xrightarrow{\mathscr{P}} 0$ as $n \to \infty$. We have

Theorem 2. *One can construct a Gaussian process* $\hat{G}_n$ *such that*

$$(6) \qquad \sup_{t \le T_n} |\hat{Z}_n(t) - G_n(t)| \xrightarrow{\mathscr{P}} 0, \quad \text{as} \quad n \to \infty,$$

where $1 - F(T_n) = \dfrac{1}{n^\delta}$, $0 < \delta < \dfrac{1}{8}$ *and*

$$\hat{G}_n(t) = G_n(t) - \left\{ \int l(x)\,d_x \frac{B_n[F(x)]}{\sqrt{n}} + A\tilde{B}_n(\tilde{F}(\infty)) \right\} \times$$

(7)

$$\times \nabla_\theta F^0(t;\theta_0)'$$

with $G_n(t)$ defined as in Theorem 1 to approximate $Z_n(t) = \sqrt{n}[\hat{F}_n^0(t) - F^0(t;\theta_0)]$, if (5) as well as the following conditions are satisfied:

(i) l is a measurable p-dimensional vector-valued function, (possibly depending on θ_0), and $\mathsf{E}\,l(X_j) = 0$.

(ii) $\mathsf{E}\,l(X_i)'l(X_j)$ is a finite nonnegative definite matrix.

(iii) The vector $\nabla_\theta F^0(t;\theta)$ of partial derivatives of F^0 with respect to the components of θ is uniformly continuous in t and $\theta \in \Lambda$, where Λ is the closure of a given neighborhood of θ_0. $\nabla_\theta F^0(t;\theta_0)$ is the above vector evaluated at $\theta = \theta_0$.

(iv) Each component of $l(x)$ is of bounded variation on each finite interval.

Proof. The proof of Theorem 2 is similar to that of Theorem 3.1 in Burke — M. Csörgő — S. Csörgő — Révész [4] and so we will only outline it and point out the differences. We have

(8)
$$\hat{Z}_n(t) = \sqrt{n}[\hat{F}_n^0(t) - F^0(t;\hat{\theta}_n)] =$$
$$= Z_n(t) - \sqrt{n}[F^0(t;\hat{\theta}_n) - F^0(t;\theta_0)],$$

where $Z_n(t) = \sqrt{n}[\hat{F}_n^0(t) - F^0(t;\theta_0)]$ and θ_0 is the true unknown value of θ. By Theorem 1,

$$\sup_{t \leqslant T_n} |Z_n(t) - G_n(t)| \overset{\mathscr{P}}{\to} 0 \quad \text{as} \quad n \to \infty,$$

where $T_n = F^{-1}\left(1 - \dfrac{1}{n^\delta};\theta_0\right)$. Now, under the conditions (5) and (i)-(iv)

$$\sqrt{n}[F^0(t;\hat{\theta}_n) - F^0(t;\theta_0)] =$$
$$= \sqrt{n}(\hat{\theta}_n - \theta_0)\nabla_\theta F^0(t;\theta_0) + R_n(t) =$$

$$= \frac{1}{\sqrt{n}} \sum_{j=1}^{n} \{l(X_j) + A(\delta_j - \widetilde{F}(\infty))\} \nabla_\theta F^0(t; \theta_0) + R_n'(t),$$

where $\sup_t |R_n(t)| \overset{\mathscr{P}}{\to} 0$ and $\sup_t |R_n'(t)| \overset{\mathscr{P}}{\to} 0$. As in B u r k e et al. [4],

$$\frac{1}{\sqrt{n}} \sum_{j=1}^{n} l(X_j) = \int l(x) d_x B_n[F(x)] + R_n'',$$

where $R_n'' \overset{\mathscr{P}}{\to} 0$ as $n \to \infty$. We also have

$$\frac{1}{\sqrt{n}} \sum_{j=1}^{n} (\delta_j - \widetilde{F}(\infty)) = \widetilde{\xi}_n(\infty) = \widetilde{B}_n(\widetilde{F}(\infty), n) + R_n''',$$

where $R_n''' \overset{\text{a.s.}}{=} O\{\frac{\log n}{\sqrt{n}}\}$ by Theorem 1. Combining these results, we obtain (6). ∎

Remarks. As in B u r k e et al. [4], if ϵ_n of (5) converges almost surely and further conditions are imposed on $l(x)$, then (6) would converge almost surely at a rate dependent upon the rate of convergence of ϵ_n to 0.

In the general estimation problem, one may or may not know the distribution function H of the censoring variables completely. One might be faced with the situation where both F^0 and H cannot be completely specified: $F^0 = F^0(\cdot; \theta)$ and $H = H(\cdot; \lambda)$ where θ and λ are unknown. When writing $l(x), A, \widetilde{F}(x)$ and $F(x)$ in Theorem 2, it is understood that $l, A, \widetilde{F}$ and F would, in general, depend on θ_0 and λ_0 the true unknown values of θ and λ. For certain families of distribution functions, (cf. Section 3), this difficulty in trying to obtain goodness of fit tests independent of θ_0 and λ_0 can be overcome.

The supremum in (6) can be taken over the range $\{t: t \leq \hat{T}_n\}$, where $\hat{T}_n = F^{-1}\left(1 - \frac{1}{n^\delta}; \hat{\theta}_n, \hat{\lambda}_n\right)$ if the following additional assumption is made:

(9) $\hat{\lambda}_n - \lambda_0 \overset{\text{a.s.}}{=} o\left(\frac{1}{n^\gamma}\right)$ for some $\gamma > 0$, if λ is an

unknown parameter of F, estimated by $\hat{\lambda}_n$,

and the vector $\nabla_{(\theta,\lambda)} F(t; \theta, \lambda)$ is bounded
uniformly in $t \in R$ and (θ, λ) belonging to
a compact neighbourhood of (θ_0, λ_0),
the true theoretical value of (θ, λ).

Lastly, the representation (5) is not the only form which can be handled by the methodology of Theorem 2. The method can accommodate similar sum representations based on, for example, the uncensored observations from among $X_1, X_2, \ldots, X_n$ as well as some other variants. The particular form of (5) is proposed since it is satisfied by the estimation method of our next section.

3. THE NEGATIVE EXPONENTIAL FAMILY

In this section, we will assume that $X_1^0, X_2^0, \ldots$ have a negative exponential distribution function $F^0(t; \theta) = 1 - e^{-\frac{t}{\theta}}$, $t \geq 0$, where $\theta > 0$ is unknown. Consequently, the observations $X_i = \min(X_i^0, Y_i)$ have distribution function F where $1 - F(t) = e^{-\frac{t}{\theta}}(1 - H(t))$ and we will assume that $H(0) = 0$.

We have

$$(10) \qquad \mathrm{E}X_i = \int_0^\infty [1 - F(u)]\, du = \xi,$$

say. Also

$$\mathrm{E}\delta_i = \int_0^\infty d\tilde{F}(u) = \frac{\xi}{\theta}.$$

We use $\bar{X}_n = \frac{1}{n} \sum_{i=1}^n X_i$ to estimate ξ and $\bar{\delta}_n = \frac{1}{n} \sum_{i=1}^n \delta_i$ to estimate $\frac{\xi}{\theta}$ and arrive at

$$(11) \qquad \hat{\theta}_n = \frac{\bar{X}_n}{\bar{\delta}_n}$$

as an estimator for θ.

It can easily be shown that $\hat{\theta}_n$ satisfies relation (5) with $l(x) =$

$$= \frac{\theta_0 (x - \xi)}{\xi} \quad \text{and} \quad A = -\frac{\theta_0^2}{\xi}. \quad \text{An application of Theorem 2 yields}$$

$$(12) \qquad \sup_{t \leq T_n} |\hat{Z}_n(t) - \hat{G}_n(t)| \xrightarrow{\mathcal{P}} 0,$$

where $\hat{Z}_n(t) = \sqrt{n}[\hat{F}_n^0(t) - F^0(t; \hat{\theta}_n)]$ and $\hat{G}_n$ is defined by (7) with $l(x)$ and A as above.

The calculation of the covariance of $\hat{G}_n$ is fairly complicated. If we assume that the censoring distribution function H is also negative exponential with scale parameter λ, that is,

$$(13) \qquad H(t; \lambda) = 1 - e^{-\frac{t}{\lambda}},$$

for some $\lambda > 0$, then the covariance of $\hat{G}_n$ reduces to

$$\mathsf{E}\hat{G}_n(s)\hat{G}_n(t) = [1 - F^0(s)][1 - F^0(t)] \times$$

$$(14) \qquad \times \left\{ \int_0^{s \wedge t} \frac{dF^0(u; \theta_0)}{(1 - F(u; \theta_0, \lambda))(1 - F^0(u; \theta_0))} - \right.$$

$$\left. - \frac{\lambda}{\theta_0 + \lambda} \log[1 - F(s; \theta_0, \lambda)] \log[1 - F(t; \theta_0, \lambda)] \right\}$$

with

$$\int_0^{s \wedge t} \frac{dF^0}{(1 - F)(1 - F^0)} = \frac{\lambda}{(\theta_0 + \lambda)[1 - F(s \wedge t; \theta_0, \lambda)]}.$$

If we base our goodness of fit tests on

$$(15) \qquad \hat{U}_n(t) = \frac{1}{\bar{\delta}_n} \exp\left(-\frac{t}{\hat{\lambda}_n}\right) \hat{Z}_n(t)$$

where $\hat{\lambda}_n = \bar{X}_n \dfrac{\bar{\delta}_n}{1 - \bar{\delta}_n}$ is a consistent estimator for λ, then it can easily

be shown that

$$\sup_{t \leqslant T_n} |\hat{U}_n(t) - U_n^*(t)| \overset{\mathscr{P}}{\rightarrow} 0, \quad \text{as} \quad n \rightarrow \infty,$$

where $U_n^*(t) = \dfrac{\theta_0 + \lambda_0}{\lambda_0} \exp\left(-\dfrac{t}{\lambda_0}\right) \hat{Z}_n(t)$ and λ_0 is true theoretical value of λ. Since, by (12),

$$\sup_{t \leqslant T_n} |U_n^*(t) - G_n^*(t)| \overset{\mathscr{P}}{\rightarrow} 0, \quad \text{as} \quad n \rightarrow \infty,$$

where $G_n^*(t) = \dfrac{\theta_0 + \lambda_0}{\lambda_0} \exp\left(-\dfrac{t}{\lambda_0}\right) \hat{G}_n(t)$, we obtain

$$\sup_{t \leqslant T_n} |\hat{U}_n(t) - G_n^*(t)| \overset{\mathscr{P}}{\rightarrow} 0, \quad \text{as} \quad n \rightarrow \infty.$$

Since condition (9) is satisfied, the supremum can be taken over the set $\{t: t \leqslant \hat{T}_n\}$, where $\hat{T}_n = \bar{X}_n \gamma \log n$, for some $\gamma > 0$. Thus, we have

Theorem 3. *If* F^0 *and* H *are negative exponential distribution functions with scale parameters* θ *and* λ, *respectively, then*

$$\int_0^{\hat{T}_n} [\hat{U}_n(t)]^2 \, dF(t; \hat{\theta}_n, \hat{\lambda}_n) \overset{\mathscr{D}}{\rightarrow} \int_0^{\infty} [G_1^*(t)]^2 \, dF(t),$$

$$\sup_{0 \leqslant t \leqslant \hat{T}_n} |\hat{U}_n(t)| \overset{\mathscr{D}}{\rightarrow} \sup_{0 \leqslant t < \infty} |G_1^*(t)|$$

and

$$\sup_{0 \leqslant t < \hat{T}_n} \hat{U}_n(t) \overset{\mathscr{D}}{\rightarrow} \sup_{0 \leqslant t < \infty} G_1^*(t),$$

where $\hat{U}_n$ *is defined by (15) and* G_1^* *is a Gaussian process with mean zero and covariance given by*

(16) $\quad \mathsf{E} G_1^*(s) G_1^*(t) = F(s) \wedge F(t) - F(s)F(t) - \varphi(s)\varphi(t),$

where $\varphi(s) = [1 - F(s)] \log [1 - F(s)]$ *and* $F(s) = F(s; \theta_0, \lambda_0)$.

Remarks. That the range of integration and the supremum for the limiting distributions in Theorem 3 is over all of $\mathscr{R}$ follows from the fact that

$$\sup_{\hat{T}_n < t < \infty} |G_1^*(t)| \xrightarrow{\mathscr{P}} 0, \quad \text{as } n \to \infty.$$

The covariance function (16) is the same as that of the asymptotic distribution of the usual empirical process, (with no censorship), when the scale parameter of the underlying negative exponential distribution function is estimated. Thus, asymptotic significance points of the statistics in Theorem 3 can be found in S t e p h e n s [15] and in D u r b i n [8]. As in these latter papers, weight functions such as $\{F(t; \hat{\theta}_n, \hat{\lambda}_n)(1 - F(t; \hat{\theta}_n, \hat{\lambda}_n))\}^{-1}$ can be incorporated into our results.

As in B u r k e et al. ([4], [5]), the approximations can be obtained in terms of the two-parameter Kiefer processes as well as the Brownian Bridge sequences.

Acknowledgement. I wish to thank H.L. K o u l for pointing out that the estimator defined by (11) is consistent when H is an arbitrary continuous distribution function.

REFERENCES

[1] T.W. A n d e r s o n – D.A. D a r l i n g, Asymptotic theory of certain 'goodness of fit' criteria based on stochastic processes, *Ann. Math. Statist.*, 23 (1952), 193-212.

[2] P. B i l l i n g s l e y, *Convergence of Probability Measures,* J. Wiley and Sons, San Francisco, 1968.

[3] N. B r e s l o w – J. C r o w l e y, A large sample study of the life table and product limit estimates under random censorship, *Ann. Statist.*, 2 (1974), 437-453.

[4] M.D. B u r k e – M. C s ö r g ő – S. C s ö r g ő – P. R é v é s z, Approximations of the empirical process when parameters are estimated, *Ann. Probability*, 7 (1979), 790-810.

[5] M.D. B u r k e – S. C s ö r g ő – L. H o r v á t h, Strong approximations of some biometric estimates under random censorship, *Z. Wahrscheinlichkeitstheorie verw. Geb.*, (1981), to appear.

[6] S. Csörgő – L. Horváth, On the Koziol – Green model of random censorship, to appear.

[7] J. Durbin, Weak convergence of the sample distribution function when parameters are estimated, *Ann. Statist.,* 1 (1973), 279-290.

[8] J. Durbin, Kolmogorov – Smirnov tests when parameters are estimated with applications to tests of exponentiality and tests on spacings, *Biometrika,* 62 (1975), 5-22.

[9] B. Efron, The two sample problem with censored data, *Proc. 5th Berkeley Symp.,* 4 (1967), 831-853.

[10] M.J. Gillespie – L. Fisher, Confidence bands for Kaplan – Meier, *Ann. Statist.,* 7 (1979), 920-924.

[11] W.J. Hall – J.A. Wellner, Confidence bands for a survival curve from censored data, *Biometrika,* 67 (1980), 133-143.

[12] E.L. Kaplan – P. Meier, Nonparametric estimation from incomplete observations, *J. Amer. Statist. Assoc.,* 53 (1958), 457-481.

[13] J.A. Koziol – S.B. Green, A Cramér – van Mises statistic for randomly censored data, *Biometrika,* 63 (1976), 465-474.

[14] P. Meier, Estimation of a distribution function from incomplete observations, in *Perspectives in Probability and Statistics,* Ed. J. Gani, pp. 67-87, Academic Press, London, 1975.

[15] M.A. Stephens, Asymptotic results and percentage points for goodness-of-fit statistics with unknown parameters, *Ann. Statist.,* 4 (1976), 357-369.

[16] J.L. Doob, Heuristic approach to the Kolmogorov – Smirnov theorems, *Ann. Math. Stat.,* 20 (1949), 393-403.

M.D. Burke
Department of Mathematics and Statistics, University of Calgary, Calgary, Alberta, T2N 1N4 Canada.

COLLOQUIA MATHEMATICA SOCIETATIS JÁNOS BOLYAI

32. NONPARAMETRIC STATISTICAL INFERENCE,

BUDAPEST (HUNGARY), 1980.

LARGE-SAMPLE PROPERTIES OF NONPARAMETRIC BIVARIATE ESTIMATORS WITH CENSORED DATA

G. CAMPBELL — A. FÖLDES[*]

1. INTRODUCTION AND SUMMARY

The estimation of a bivariate distribution function under random censoring is considered. The problem is to estimate the distribution of the life times under random censoring in which one knows whether the observations are losses (censored) or deaths (uncensored). There are numerous examples to demonstrate the importance of this bivariate problem. In some experiments the data are naturally paired such as observations on eyes, lungs, twins, married couples, or matched pairs. Often-times there are two sequential observations on the same individual (pre-test, post-test). In a reliability setting, a pair of components in a system can be observed.

[*]Mathematical Institute of the Hungarian Academy of Sciences. This research was done while the author was visiting the Departments of Statistics and Mathematics at Purdue University. The author would like to acknowledge the financial assistance of the two departments.

Examples of (possibly bivariate) censoring mechanisms are plentiful. There is, of course, the censoring due to patient drop-out or non-compliance. The competing risk framework can be thought of as censoring. The censoring may be essentially univariate, such as the random entry of subjects into the study with fixed cutoff time for evaluation.

The one-dimensional random censoring model has been treated in great detail in the recent literature, beginning with the landmark papers of K a p l a n and M e i e r [7] and E f r o n [3]. The asymptotic normality and weak convergence of the product-limit estimator of Kaplan and Meier was treated by B r e s l o w and C r o w l e y [1]. Strong uniform consistency was treated by W i n t e r, F ö l d e s and R e j t ő [11], F ö l d e s, R e j t ő and W i n t e r [5], and F ö l d e s and R e j t ő [4].

The bivariate estimation problem with discrete times of deaths or losses has been considered by C a m p b e l l [2] using an extension of the self-consistent approach of E f r o n [3]. A self-consistent approach for the continuous case has been suggested by K o r w a r and D a h i y a [9]. H a n l e y and P a r n e s [6] have treated maximum likelihood approaches to bivariate estimation. In contrast to the iterative estimators of these researchers this paper considers several new closed-form estimators for the bivariate model and proves strong uniform consistency to the true bivariate distribution of the lifetimes.

Two path-dependent estimators are introduced in Section 2. Each estimator is the product of two one-dimensional Kaplan − Meier product limit estimators.

A hazard function approach is employed in Section 3 to estimate $-\ln F(s, t)$ and hence $F(s, t)$. Two path-dependent estimators of $-\ln F(s, t)$ are proposed and these lead to estimators of the bivariate distribution function.

Section 4 explores the relationship of the estimators of Sections 2 and 3. In Section 5 the pointwise consistency of the estimators that are products of Kaplan − Meier estimators is considered under mild conditions on F. Under stronger conditions, all the estimators of Sections 2 and 3 are proved to be uniformly almost sure consistent for F with rate

$O\left(\sqrt{\dfrac{\log \log n}{n}}\right)$ on the rectangle $[0, S] \times [0, T]$. The final section presents an example and some discussion.

2. TWO PATH-DEPENDENT PRODUCT-LIMIT ESTIMATORS

Let $\{X_i^0, Y_i^0\}_{i=1}^{\infty}$ be independent identically distributed pairs of nonnegative random variables with continuous bivariate survival function $F(s, t) = P(X^0 > s,\ Y^0 > t)$. Let $\{C_i, D_i\}_{i=1}^{\infty}$ denote another sequence of nonnegative i.i.d. pairs of random (censoring) variables with continuous survival function $G(s, t) = P(C > s,\ D > t)$. Define

$$X_i = \min\{X_i^0, C_i\}, \qquad Y_i = \min\{Y_i^0, D_i\} \qquad (i = 1, 2, \ldots, n),$$

$$\epsilon_{1i} = \begin{cases} 1 & X_i = X_i^0 \ \text{(uncensored)} \\ 0 & X_i < X_i^0 \ \text{(censored)} \end{cases}$$

$$\epsilon_{2i} = \begin{cases} 1 & Y_i = Y_i^0 \ \text{(uncensored)} \\ 0 & Y_i < Y_i^0 \ \text{(censored)}. \end{cases}$$

It is assumed that the two sequences $\{X_i^0, Y_i^0\}_{i=1}^{\infty}$ and $\{C_i, D_i\}_{i=1}^{\infty}$ are mutually independent. Let $H(s, t) = P(X_i > s,\ Y_i > t)$ denote the survival function of (X_i, Y_i). By independence,

(2.1) $\qquad H(s, t) = F(s, t)G(s, t).$

Based on the elementary observation

(2.2) $\qquad F(s, t) = F(s, 0)F_1(t \mid s),$

where $F_1(t \mid s) = P(Y^0 > t \mid X^0 > s)$, the survival function $F(s, t)$ is estimated by separately estimating each of the two terms on the right of (2.2). This leads to an estimator $\hat{F}_1(s, t)$ based on the path from $(0, 0)$ to (s, t) which is linear from $(0, 0)$ to $(s, 0)$ and linear from $(s, 0)$ to (s, t).

The following notation is established: Let

(2.3) $\qquad N_n(s, t) = N(s, t) = \sum_{i=1}^{n} I_{\{X_i > s,\, Y_i > t\}};$

- 105 -

(2.4) $\alpha_i(s, t) = I_{\{X_i \leqslant s, Y_i > t, \epsilon_{1i} = 1\}}$ $(i = 1, 2, \ldots, n)$,

(2.5) $\beta_j(s, t) = I_{\{X_j > s, Y_j \leqslant t, \epsilon_{2j} = 1\}}$ $(j = 1, 2, \ldots, n)$.

To estimate $F(s, 0)$, project all points vertically onto the line $y = 0$, and, ignoring the (Y_i, ϵ_{2i}) values, calculate the Kaplan – Meier product-limit estimator of $F(s, 0)$ using the one-dimensional censored sample $\{X_i, \epsilon_{1i}\}_{i=1}^n$. This produces the estimator

$$\hat{F}_{1n}(s, 0) = \begin{cases} \prod\limits_{i=1}^n \left(\dfrac{N(X_i, 0)}{N(X_i, 0) + 1}\right)^{\alpha_i(s, 0)} & \text{if } s \leqslant \tau_{1n} \\ 0 & \text{otherwise,} \end{cases}$$

where $\tau_{1n} = \max\limits_{1 \leqslant i \leqslant n} \{X_i\}$.

(The one-dimensional convention that the last observation is converted to a death (if it is censored) is adhered to here.) To estimate $F_1(t \mid s)$, the second term of (2.2), project all points for which $X_i > s$ horizontally to the line $X = s$, and ignoring the (X_i, ϵ_{1i}) values calculate the Kaplan – Meier product-limit estimator based on the data $\{Y_j, \epsilon_{2j}\}_{j=1}^n$ for which $X_j > s$ (see Figure 1). Observe that this method estimates the probability $P(Y^0 > t \mid X > s)$ but

(2.6)
$$P(Y^0 > t \mid X > s) = P(Y^0 > t \mid X^0 > s, C > s) =$$
$$= P(Y^0 > t \mid X^0 > s)$$

since C is independent of the pair (X^0, Y^0). Thus the following estimator of $F_1(t \mid s)$ is obtained:

(2.7) $$\hat{F}_{1n}(t \mid s) = \begin{cases} \prod\limits_{j=1}^n \left(\dfrac{N(s, Y_j)}{N(s, Y_j) + 1}\right)^{\beta_j(s, t)} & \text{if } t \leqslant \tau_{2n}(s) \\ 0 & \text{otherwise,} \end{cases}$$

where $\tau_{2n}(s) = \max\limits_{1 \leqslant i \leqslant n} \{Y_i : X_i > s\}$.

Consequently the estimator for the $F(s, t)$ is

$$
(2.8) \quad \hat{F}_{1n}(s, t) = \begin{cases} \prod_{i=1}^{n} \left(\dfrac{N(X_i, 0)}{N(X_i, 0) + 1} \right)^{\alpha_i(s,0)} \prod_{j=1}^{n} \left(\dfrac{N(s, Y_j)}{N(s, Y_j) + 1} \right)^{\beta_j(s,t)} \\ \qquad\qquad\qquad\qquad\qquad\qquad \text{if } N(s, t) > 0, \\ 0 \qquad\qquad\qquad\qquad\qquad\qquad\quad \text{otherwise.} \end{cases}
$$

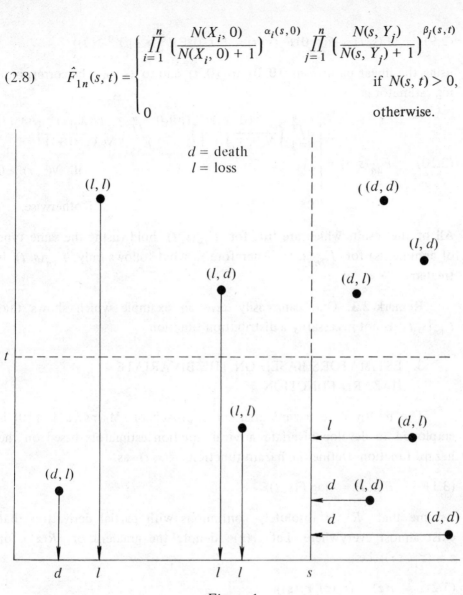

Figure 1

Remark 2.1. In the event of no censoring in either coordinate this estimator reduces to the ordinary empirical survival function $\frac{1}{n} \sum I_{\{X_i^0 > s, Y_i^0 > t\}}$.

Remark 2.2. By changing the role of s and t it is possible to develop our estimator $\hat{F}_{2n}(s, t)$ (based on the relation

(2.9) $F(s, t) = F(t, 0)F_2(s \mid t) = F(t, 0)PX^0 > s \mid Y^0 > t))$

using the linear path from $(0, 0)$ to $(0, t)$ and to (s, t). The corresponding estimator is

$$(2.10) \quad \hat{F}_{2n}(s, t) = \begin{cases} \prod_{j=1}^{n} \left(\dfrac{N(0, Y_j)}{N(0, Y_j) + 1} \right)^{\beta_j(0, t)} \prod_{j=1}^{n} \left(\dfrac{N(X_i, t)}{N(X_i, t) + 1} \right)^{\alpha_i(s, t)} & \text{if } N(s, t) > 0 \\ \\ 0 & \text{otherwise.} \end{cases}$$

All of the results which are true for $\hat{F}_{1n}(s, t)$ hold (using the same type of arguments) for $\hat{F}_{2n}(s, t)$. Therefore in what follows only $\hat{F}_{1n}(s, t)$ is treated.

Remark 2.3. One can easily give an example which shows that $\hat{F}_{1n}(s, t)$ is not necessarily a distribution function.

3. ESTIMATORS BASED ON THE BIVARIATE HAZARD FUNCTION

The multivariate hazard gradient approach of Marshall [10] is employed to develop bivariate survival function estimators based on the hazard function. Define the hazard function $R(s, t)$ as

(3.1) $R(s, t) = - \log F(s, t)$.

Assume that R is absolutely continuous with partial derivatives that exist almost everywhere. Let $r(z)$ denote the gradient of $R(z)$ for $z = (z_1, z_2)$; i.e.,

(3.2) $r(z) = (r_1(z), r_2(z))$,

where

$$(3.3) \quad r_1(z) = \frac{\partial R(z)}{\partial z_1} \quad \text{and} \quad r_2(z) = \frac{\partial R(z)}{\partial z_2}.$$

Then $R(s, t)$ can be reconstructed as the path intgeral of $r(z)$ from $(0, 0)$ to (s, t). By path independence one can write

$$(3.4) \qquad R(s, t) = \int\limits_{(0,0)}^{(s,t)} r(z)\, dz.$$

In particular, consider the path linear from $(0, 0)$ to $(0, s)$ and linear from $(s, 0)$ to (s, t). Then

$$(3.5) \qquad R(s, t) = \int\limits_0^s r_1(u, 0)\, du + \int\limits_0^t r_2(s, v)\, dv.$$

By (3.1) and (3.3) from (3.5)

$$(3.6) \qquad -\log F(s, t) = -\int\limits_0^s \frac{1}{F(u, 0)}\, d_u F(u, 0) - \int\limits_0^t \frac{1}{F(s, v)}\, d_v F(s, v)$$

where $d_u F(u, t)$ denotes Lebesgue $-$ Stieltjes integration over u for t fixed. Using (2.1),

$$(3.7) \qquad -\log F(s, t) = -\int\limits_0^s \frac{G(u, 0)}{H(u, 0)}\, \frac{\partial F(u, 0)}{\partial u}\, du - \int\limits_0^t \frac{G(s, v)}{H(s, v)}\, \frac{\partial F(s, v)}{\partial v}\, dv.$$

Introduce the following functions:

$$(3.8) \qquad \tilde{K}(s, t) = \int\limits_0^s G(u, t)\, \frac{\partial P(X^0 \leqslant u, Y^0 > t)}{\partial u}\, du;$$

$$(3.9) \qquad \tilde{L}(s, t) = \int\limits_0^t G(s, v)\, \frac{\partial P(X^0 > s, Y^0 \leqslant v)}{\partial v}\, dv.$$

Applying the trivial

$$(3.10) \qquad \frac{\partial P(X^0 \leqslant u, Y^0 > t)}{\partial u} = -\frac{\partial P(X^0 > u, Y^0 > t)}{\partial u}$$

and

$$(3.11) \qquad \frac{\partial P(X^0 > s, Y^0 \leqslant v)}{\partial v} = -\frac{\partial P(X^0 > s, Y^0 > v)}{\partial v}$$

relations yields

$$(3.12) \qquad -\log F(s, t) = \int\limits_0^s \frac{1}{H(u, 0)}\, d_u \tilde{K}(u, 0) + \int\limits_0^t \frac{1}{H(s, t)}\, d_v \tilde{L}(s, v).$$

Equation (3.12) suggests that H, $\tilde{K}$ and $\tilde{L}$ be estimated first. The natural estimator of $H(u, v)$ is the empirical survival function:

$$(3.13) \qquad H_n(s, t) = \frac{1}{n} \Sigma I_{\{X_i > s, Y_i > t\}} = \frac{N(s, t)}{n}.$$

The basic idea of estimating $\tilde{K}(u, v)$ and $\tilde{L}(u, v)$ is the following observation:

$$\tilde{K}(s, t) = \int_0^s G(u, t) \, d_u P(X^0 \leqslant u, Y^0 > t) =$$

$$= \int_0^s P(C > u, D > t) \, d_u P(X^0 \leqslant u, Y^0 > t) =$$

$$(3.14) \qquad = \int_0^s P(C > u, D > t \mid X^0 = u, Y^0 > t) \, d_u P(X^0 \leqslant u, Y^0 > t) =$$

$$= \int_0^s P(C > u, Y > t \mid X^0 = u, Y^0 > t) \, d_u P(X^0 \leqslant u, Y^0 > t) =$$

$$= P(C \geqslant X^0, Y > t, X^0 < s) = P(X^0 \leqslant C, Y > t, X \leqslant s) =$$

$$= E(\alpha_i(s, t)),$$

the last equality following from (2.4). And similarly

$$(3.15) \qquad \tilde{L}(s, t) = E(\beta_j(s, t)).$$

Hence the natural estimators of $\tilde{K}(s, t)$ and $\tilde{L}(s, t)$ are

$$(3.16) \qquad \tilde{K}_n(s, t) = \frac{1}{n} \sum_{i=1}^n \alpha_i(s, t)$$

and

$$(3.17) \qquad \tilde{L}_n(s, t) = \frac{1}{n} \sum_{j=1}^n \beta_j(s, t).$$

Consequently by (3.12)-(3.17) estimate $R(s, t) = -\log F(s, t)$ by

$$\tilde{R}_{1n}(s, t) = \int_0^s \frac{1}{H_n(u, 0)} \, d_u \tilde{K}_n(u, 0) + \int_0^t \frac{1}{H_n(s, v)} \, d_v \tilde{L}_n(s, v) =$$

$$(3.18)$$

$$= \frac{1}{n} \sum_{i=1}^n \frac{\alpha_i(s, 0)}{H_n(X_i, 0)} + \frac{1}{n} \sum_{j=1}^n \frac{\beta_j(s, t)}{H_n(s, Y_j)}$$

if $N(s, t) > 0$ and let $\tilde{R}_1(s, t) = + \infty$ otherwise. Moreover, let

(3.19) $\tilde{F}_{1n}(s, t) = \exp(-\tilde{R}_{1n}(s, t))$.

4. RELATIONSHIP OF THE PRODUCT-LIMIT AND THE HAZARD FUNCTION ESTIMATORS

Lemma 4.1.

(4.1) $\displaystyle\sup_{\substack{0 \leqslant s < \infty \\ 0 \leqslant t < \infty}} |H_n(s, t) - H(s, t)| = O\left(\sqrt{\frac{\log \log n}{n}}\right)$ *a.s.*

(4.2) $\displaystyle\sup_{\substack{0 \leqslant s < \infty \\ 0 \leqslant t < \infty}} |\tilde{K}_n(s, t) - \tilde{K}(s, t)| = O\left(\sqrt{\frac{\log \log n}{n}}\right)$ *a.s.*

(4.3) $\displaystyle\sup_{\substack{0 \leqslant s < \infty \\ 0 \leqslant t < \infty}} |\tilde{L}_n(s, t) - \tilde{L}(s, t)| = O\left(\sqrt{\frac{\log \log n}{n}}\right)$ *a.s.*

Proof. Result (4.1) simply follows from the multi-dimensional law of the iterated logarithm for empirical distributions of K i e f e r [8]. To prove (4.2) it is enough to observe that

(4.4)
$$\tilde{K}(s, t) = P(X^0 \leqslant s, X^0 - C \leqslant 0) -$$
$$- P(X^0 \leqslant s, X^0 - C \leqslant 0, Y \leqslant t).$$

Therefore $\tilde{K}_n(s, t)$ can be considered as the difference of two empirical distributions:

(4.5)
$$\tilde{K}_n(s, t) = \frac{1}{n} \sum_{i=1}^{n} I_{\{X_i^0 \leqslant s, X_i^0 - C_i \leqslant 0\}} -$$
$$- \frac{1}{n} \sum_{j=1}^{n} I_{\{X_j^0 \leqslant s, X_j^0 - C_j \leqslant 0, Y_j \leqslant t\}}.$$

That means that applying again Kiefer's result (once in two-, once in three-dimensions) (4.2) is obtained. A similar argument proves (4.3). ∎

Remark 4.2. Suppose that $H(S, T) > 0$ for $S, T < \infty$. Then from the above mentioned Kiefer theorem for almost all ω there exist an

$n_0(\omega)$ such that if $n > n_0(\omega)$ then

(4.6) $H_n(s, t) > \frac{1}{2} H(s, t)$ for all $0 \leqslant s \leqslant S,\ 0 \leqslant t \leqslant T.$

For technical reasons introduce F_{1n}, the modified Kaplan – Meier-type estimator (a similar idea was used in F ö l d e s and R e j t ő [4]):

(4.7) $\check{F}_{1n}(s, t) = \begin{cases} \displaystyle\prod_{i=1}^{n} \left(\frac{N(X_i, 0) + 1}{N(X_i, 0) + 2} \right)^{\alpha_i(s,0)} \prod_{j=1}^{n} \left(\frac{N(s, Y_j) + 1}{N(s, Y_j) + 2} \right)^{\beta_j(s,t)} & \\ & \text{if } N(s, t) > 0 \\[2mm] 0 & \text{otherwise.} \end{cases}$

Lemma 4.3. *If $H(S, T) > 0$ $(0 < S, T < \infty)$ then*

(4.8) $\displaystyle\sup_{\substack{0 \leqslant s \leqslant S \\ 0 \leqslant t \leqslant T}} |\hat{F}_{1n}(s, t) - \check{F}_{1n}(s, t)| = O\left(\frac{1}{n}\right) \quad a.s.$

Proof. By the well-known inequality

(4.9)
$$\left| \prod_{k=1}^{m} a_k - \prod_{k=1}^{m} b_k \right| \leqslant \sum_{k=1}^{m} |a_k - b_k|$$

$$\text{if } |a_k| \leqslant 1,\ |b_k| \leqslant 1 \qquad (k = 1, 2, \ldots, m),$$

estimate the difference of the original and the modified estimators as follows:

(4.10)
$$|\hat{F}_{1n}(s, t) - \check{F}_{1n}(s, t)| \leqslant$$
$$\leqslant \sum_{i=1}^{n} \frac{\alpha_i(s, 0)}{(N(X_i, 0) + 1)^2} + \sum_{i=1}^{n} \frac{\beta_j(s, t)}{(N(s, Y_j) + 1)^2}.$$

Hence applying Remark 4.2

(4.11)
$$\sup_{\substack{0 \leqslant s \leqslant S \\ 0 \leqslant t \leqslant T}} |\hat{F}_{1n}(s, t) - \check{F}_{1n}(s, t)| \leqslant$$

$$\leqslant \frac{2n}{N^2(S, T)} \leqslant \frac{2n}{n^2 H_n^2(S, T)} \leqslant \frac{8}{n H^2(S, T)} = O\left(\frac{1}{n}\right) \quad a.s. \qquad \blacksquare$$

Lemma 4.4. *If* $H(S, T) > 0$ $(0 < S, T < \infty)$ *then*

$$(4.12) \qquad \sup_{\substack{0 \leqslant s \leqslant S \\ 0 \leqslant t \leqslant T}} |\hat{F}_{1n}(s, t) - \tilde{F}_{1n}(s, t)| = O\left(\frac{1}{n}\right) \quad a.s.$$

Proof. First observe that by the elementary inequality

$$(4.13) \qquad |x - y| < |\log x - \log y| \quad \text{(for } 0 < x, y \leqslant 1\text{)}$$

and by Lemma 4.3:

$$\sup_{\substack{0 \leqslant s \leqslant S \\ 0 \leqslant t \leqslant T}} |\hat{F}_{1n}(s, t) - \tilde{F}_{1n}(s, t)| \leqslant \sup_{\substack{0 \leqslant s \leqslant S \\ 0 \leqslant t \leqslant T}} |\hat{F}_{1n}(s, t) - \check{F}_{1n}(s, t)| +$$

$$(4.14) \qquad + \sup_{\substack{0 \leqslant s \leqslant S \\ 0 \leqslant t \leqslant T}} |\check{F}_{1n}(s, t) - \tilde{F}_{1n}(s, t)| =$$

$$= O\left(\frac{1}{n}\right) + \sup_{\substack{0 \leqslant s \leqslant S \\ 0 \leqslant t \leqslant T}} |\log \check{F}_{1n}(s, t) + \tilde{R}_{1n}(s, t)| \quad \text{a.s.}$$

Using logarithmic expansion

$$
\begin{aligned}
\log \check{F}_{1n}(s, t) &= - \sum_{i=1}^{n} \frac{\alpha_i(s, 0)}{N(X_i, 0) + 2} - \\
&\quad - \sum_{i=1}^{n} \alpha_i(s, 0) \sum_{k=2}^{\infty} \frac{1}{k(N(X_i, 0) + 2)^k} - \\
(4.15) \\
&\quad - \sum_{j=1}^{n} \frac{\beta_j(s, t)}{N(s, Y_j) + 2} - \sum_{j=1}^{n} \beta_j(s, t) \sum_{k=2}^{\infty} \frac{1}{k(N(s, Y_j) + 2)^k} = \\
&= - \sum_{i=1}^{n} \frac{\alpha_i(s, 0)}{N(X_i, 0) + 2} - \sum_{j=1}^{n} \frac{\beta_j(s, t)}{N(s, Y_j) + 2} + D_n(s, t)
\end{aligned}
$$

where $D_n(s, t)$ denotes the remainder term. Now

$$(4.16) \qquad \sup_{\substack{0 \leqslant s \leqslant S \\ 0 \leqslant t \leqslant T}} |D_n(s, t)| \leqslant \frac{2n}{(N(S, T) + 2)^2} \leqslant \frac{8n}{n^2 H^2(S, T)} = O\left(\frac{1}{n}\right) \quad \text{a.s.}$$

by estimating the infinite sums by geometric series, and using Remark 4.2.

– 113 –

Hence from (4.13)-(4.16) and by (3.18)

$$\sup_{\substack{0 \leqslant s \leqslant S \\ 0 \leqslant t \leqslant T}} |\hat{F}_{1n}(s, t) - \tilde{F}_{1n}(s, t)| \leqslant$$

$$\leqslant \left| \frac{1}{n} \sum_{i=1}^{n} \alpha_i(s, 0) \left(\frac{1}{H_n(X_i, 0)} - \frac{1}{H_n(X_i, 0) + \frac{2}{n}} \right) \right| +$$

(4.17) $$+ \left| \frac{1}{n} \sum_{j=1}^{n} \beta_j(s, t) \left(\frac{1}{H_n(s, Y_j)} - \frac{1}{H_n(s, Y_j) + \frac{2}{n}} \right) \right| + O\left(\frac{1}{n}\right) \leqslant$$

$$\leqslant \frac{2\frac{2}{n}}{(H_n(S, T))\left(H_n(S, T) + \frac{2}{n}\right)} + O\left(\frac{1}{n}\right) \leqslant$$

$$\leqslant \frac{16}{nH^2(S, T)} + O\left(\frac{1}{n}\right) = O\left(\frac{1}{n}\right) \quad \text{a.s.}$$

again using Remark 4.2. ∎

Remark 4.5. Lemma 4.4 gives a large sample result for the proximity of $\hat{F}_{1n}$ and $\tilde{F}_{1n}$. In fact, an absolute bound can be obtained by repeated application of a one-dimensional result of B r e s l o w and C r o w l e y [1]; namely,

$$0 < - \ln \hat{F}_{1n}(s, t) + \ln \tilde{F}_{1n}(s, t) \leqslant$$

$$\leqslant \frac{n - N_n(s, 0)}{nN_n(s, 0)} + \frac{N_n(s, 0) - N_n(s, t)}{N_n(s, 0)N_n(s, t)} = \frac{n - N_n(s, t)}{nN_n(s, t)}.$$

In particular, $\tilde{F}_{1n}(s, t) \geqslant \hat{F}_{1n}(s, t)$.

5. CONSISTENCY

The pointwise consistency of the estimator $\hat{F}_{1n}$ follows from the corresponding one-dimensional results, in that the estimator was constructed as a product of two one-dimensional Kaplan − Meier estimators. The pointwise consistency remains true in case of not necessarily continuous functions F and G, as one can develop, using the same projecting argument the corresponding bivariate estimator as the product of two one-dimensional Kaplan – Meier estimators. Observe that for this point-wise consistency neither the continuity condition on F nor G is required in that the Kaplan – Meier estimator is consistent (see W i n t e r, F ö l d e s and R e j t ő [11] and F ö l d e s, R e j t ő and W i n t e r [5]). Under some smoothness conditions the following much stronger theorem is now proved.

Theorem 5.1. *If F and G are continuous and if F is such that $- \ln F$ is absolutely continuous with partial derivatives that exist almost everywhere and if, for $0 < S, T < \infty$, $H(S, T) > 0$, then*

$$
(5.1) \qquad \sup_{\substack{0 \leqslant s \leqslant S \\ 0 \leqslant t \leqslant T}} |\widetilde{F}_{1n}(s, t) - F(s, t)| = O\left(\sqrt{\frac{\log \log n}{n}} \right) \quad a.s.
$$

Proof. Applying again (4.13), (3.12), (3.18) and (3.19) it is possible to estimate the left-hand side of (5.1) as follows:

$$
|\widetilde{F}_{1n}(s, t) - F(s, t)| \leqslant
$$

$$
(5.2) \qquad \leqslant \left| \int_0^s \frac{1}{H_n(u, 0)} \, d_u \widetilde{K}_n(u, 0) - \int_0^s \frac{1}{H(u, 0)} \, d_u \widetilde{K}(u, 0) \right| +
$$

$$
+ \left| \int_0^t \frac{1}{H_n(s, v)} \, d_v \widetilde{L}_n(s, v) - \int_0^t \frac{1}{H(s, v)} \, d_v L(s, v) \right|.
$$

Both of the terms of (5.2) can be estimated using Lemma 4.1, Remark 4.2 and *partial integration* as follows (we perform only their estimation for the first terms). (Observe that $\widetilde{K}_n(u, 0)$ and $\widetilde{K}(u, 0)$ are nondecreasing in u, $\widetilde{L}_n(s, t)$ and $\widetilde{L}(s, t)$ are nondecreasing in t for fixed s, and $H_n(s, t)$ and $H(s, t)$ nonincreasing in both arguments.)

$$
\sup_{\substack{0 \leqslant s \leqslant S \\ 0 \leqslant t \leqslant T}} \left| \int_0^s \frac{1}{H_n(u, 0)} \, d_u \widetilde{K}_u(u, 0) - \int_0^s \frac{1}{H(u, 0)} \, d_u \widetilde{K}_n(u, 0) \right| \leqslant
$$

$$
\leqslant \sup_{\substack{0 \leqslant s \leqslant S \\ 0 \leqslant t \leqslant T}} \int_0^s \left| \frac{1}{H_u(u, 0)} - \frac{1}{H(u, 0)} \right| d_u \widetilde{K}_n(u, 0) +
$$

$$+ \sup_{\substack{0 \leqslant s \leqslant S \\ 0 \leqslant t \leqslant T}} \left| \int \frac{1}{H(u, 0)} d_u (\tilde{K}_n(u, 0) - \tilde{K}(u, 0)) \right| \leqslant$$

$$\leqslant \sup_{\substack{0 \leqslant s \leqslant S \\ 0 \leqslant t \leqslant T}} |H_n(s, t) - H(s, t)| \int_0^s \frac{2}{H^2(u, 0)} d_u \tilde{K}_n(u, 0) +$$

$$+ \frac{1}{H(S, T)} 2 \sup_{\substack{0 \leqslant s \leqslant S \\ 0 \leqslant t \leqslant T}} |\tilde{K}_n(s, t) - \tilde{K}(s, t)| \leqslant$$

$$= O\left(\sqrt{\frac{\log \log n}{n}} \right) \left(\frac{2}{H^2(S, T)} + \frac{2}{H(S, T)} \right) =$$

$$= O\left(\sqrt{\frac{\log \log n}{n}} \right) \quad \text{a.s.} \qquad \blacksquare$$

Corollary 5.2. *Under the conditions of Theorem 5.1,*

$$(5.3) \qquad \sup_{\substack{0 \leqslant s \leqslant S \\ 0 \leqslant t \leqslant T}} |\hat{F}_{1n}(s, t) - F(s, t)| = O\left(\sqrt{\frac{\log \log n}{n}} \right) \quad \text{a.s.}$$

Proof. Apply Lemmas 4.3 and 4.4 in conjunction with Theorem 5.1.

6. AN EXAMPLE

Consider the example of Figure 2 consisting of four points. Here l denotes loss, and d death, so that the (d, l) at (x_3, y_3) denotes a point which is a death in the first coordinate and censored in the second. At each point in the rectangle $[0, x_4] \times [0, y_4]$ the estimator $\hat{F}_{14}$ can be calculated. Note that in the calculation of $\hat{F}_{14}(x, 0)$ the final loss in the first coordinate is converted to a death (as is the convention in one-dimensional Kaplan – Meier estimation). Suppose we wish to compute $\hat{F}_1(s, t)$ where $s \in (x_2, x_3)$ and $t \in (y_3, y_2)$. Then

$$\hat{F}_{14}(s, 0) = \left(\frac{3}{4} \right)^0 \left(\frac{2}{3} \right)^1.$$

To compute $\hat{F}_{14}(t \mid s)$, note that there are two points such that $x_i > s$, the point (x_3, y_3) projected back to the line $x = s$ is a loss, the point

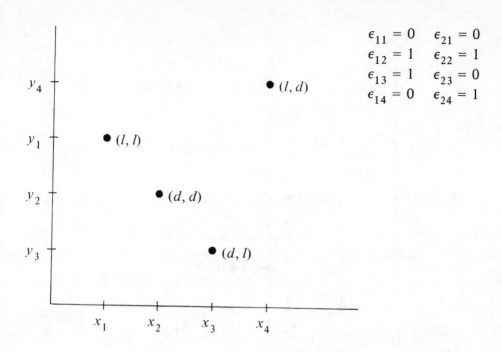

$$
\begin{array}{ll}
\epsilon_{11} = 0 & \epsilon_{21} = 0 \\
\epsilon_{12} = 1 & \epsilon_{22} = 1 \\
\epsilon_{13} = 1 & \epsilon_{23} = 0 \\
\epsilon_{14} = 0 & \epsilon_{24} = 1
\end{array}
$$

Figure 2. A four-point example

(x_4, y_4) projected to $x = s$ is a death. Thus $\hat{F}_{14}(t \mid s) = \left(\frac{1}{2}\right)^0 = 1$.

Therefore $\hat{F}_{14}(s, t) = \hat{F}_{14}(s, 0)\hat{F}_{14}(t \mid s) = \frac{2}{3} 1 = \frac{2}{3}$. Another way to arrive to the same conclusion is to follow (2.8) and then we get (see Figure 3)

$$
\hat{F}_{14}(s, t) =
$$

$$
= \left(\frac{N(x_1, 0)}{N(x_1, 0) + 1}\right)^{\epsilon_{11}} \left(\frac{N(x_2, 0)}{N(x_2, 0) + 1}\right)^{\epsilon_{12}} \left(\frac{N(s, y_3)}{N(s, y_3) + 1}\right)^{\epsilon_{23}} =
$$

$$
= \left(\frac{3}{4}\right)^0 \left(\frac{2}{3}\right)^1 \left(\frac{1}{2}\right)^0 = \frac{2}{3}.
$$

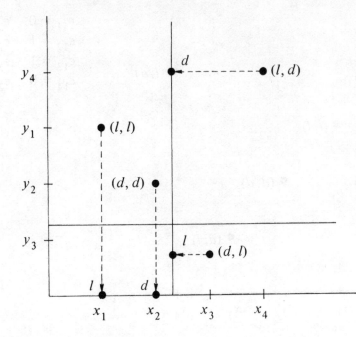

Figure 3. The projection rule

In this way the entire estimator $\hat{F}_{14}$ can be calculated. It is displayed in Figure 4 where the function is constant on the smaller rectangles. It was remarked earlier that $\hat{F}_{14}$ need not be a bivariate survival function and it is not one for this example. Figure 5 presents the estimator $\hat{F}_{24}$ based on the alternate path

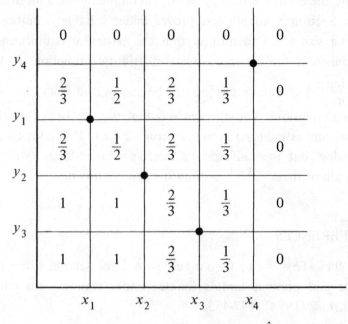

Figure 4. The survival estimate $\hat{F}_1$

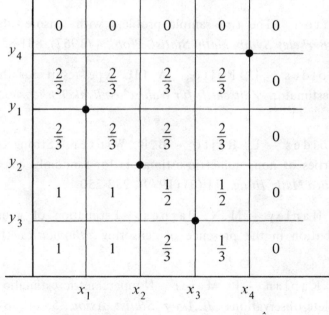

Figure 5. The survival estimate $\hat{F}_2$

While the estimator in $\hat{F}_{14}$ is not guaranteed to be a bivariate survival function, Section 5 nonetheless proves under suitable conditions that as the sample size tends to infinity that the estimator is uniformly almost surely consistent for the true survival distribution function F (with rate $O\left(\sqrt{\frac{\log \log n}{n}}\right)$). This technique can be generalized from two dimensions to higher dimensions. The difference is that the number of possible paths (and hence the estimators) increases from 2 to 2^{k-1} for the k-dimensional analog, but the uniformly almost sure consistency (with the same rate) of each of these 2^{k-1} estimators can be shown.

REFERENCES

[1] N. Breslow – J. Crowley, A large sample study of the life table and product limit estimates under random censorship, *Ann. Statist.*, 2 (1974), 437-453.

[2] G. Campbell, Nonparametric bivariate estimation with randomly censored data, *Biometrika*, 68 (2) (1981), 417-423.

[3] B. Efron, The two sample problem with censored data, *Proc. Fifth Berkeley Symp. Math. Statist. Prob.*, 4 (1967), 831-853.

[4] A. Földes – L. Rejtő, A LIL type result for the product limit estimator, *Zeitschrift für Wahrscheinlichkeitstheorie*, 56 (1981), 75-86.

[5] A. Földes – L. Rejtő – B.B. Winter, Strong consistency properties of nonparametric estimators for randomly censored data, *Periodica Math. Hung.*, 11 (3) (1979), 233-250.

[6] J.A. Hanley – M.N. Parnes, Estimation of a multivariate distribution in the presence of censoring, *Biometrics*, (to appear), 1980.

[7] E.L. Kaplan – P. Meier, Nonparametric estimation from incomplete observations, *J. Amer. Statist. Assoc.*, 53 (1958), 457-481.

[8] J. Kiefer, On large deviations of the empiric d.f. of vector chance variables and a law of the iterated logarithm, *Pacific J. of Math.*, 11 (1961), 649-660.

[9] R. Korwar – R. Dahiya, Estimation of a bivariate distribution function from incomplete observations, Air Force Office of Scientific Research Technical Report #2, Department of Mathematics, University of Massachusetts, 1979.

[10] A.W. Marshall, Some comments on the hazard gradient, *Stochastic Processes and their Applications*, 3 (1975), 293-300.

[11] B.B. Winter – A. Földes – L. Rejtő, Glivenko – Cantelli theorems for the PL estimate, *Problems of Control and Information Theory*, 7 (1978), 213-225.

G. Campbell

Dept. of Stat., Purdue University, West Lafayette, IN 47907, USA.

A. Földes

Mathematical Institute of the Hungarian Academy of Sciences, Reáltanoda u. 13-15, Budapest, Hungary.

COLLOQUIA MATHEMATICA SOCIETATIS JÁNOS BOLYAI

32. NONPARAMETRIC STATISTICAL INFERENCE,

BUDAPEST (HUNGARY), 1980.

ON THE STANDARDIZED EMPIRICAL DISTRIBUTION FUNCTION

E. CSÁKI

1. INTRODUCTION

Let $X_1, X_2, \ldots$ be a sequence of i.i.d. random variables, each uniformly distributed over the interval $(0, 1)$. Denote by $F_n(x)$ the empirical distribution function of the variables $X_1, \ldots, X_n$. We investigate the standardized empirical process defined by

$$(1.1) \qquad \alpha_n(x) = \sqrt{n} \, \frac{F_n(x) - x}{\sqrt{x(1 - x)}} \qquad (0 < x < 1)$$

and its usual sup and L_2 functionals:

$$(1.2) \qquad \sqrt{n} \, T_n = \sup_{0 < x < 1} |\alpha_n(x)|$$

$$(1.3) \qquad n A_n = \int_0^1 \alpha_n^2(x) \, dx.$$

We define also some related variables as

$$(1.4) \qquad U_n = \sup_{0 < x \leqslant \frac{1}{2}} \left(\frac{F_n(x) - x}{\sqrt{x}} \right)$$

$$(1.5) \qquad V_n = \sup_{0 < x \leqslant \frac{1}{2}} \left(\frac{x - F_n(x)}{\sqrt{x}} \right).$$

The limiting distributions of both T_n and A_n are known. Jaeschke [10] (see also Eicker [7]) has shown that

$$(1.6) \qquad \lim_{n \to \infty} P\left(\sqrt{n}\, T_n < \frac{t + b_n}{a_n} \right) = e^{-2e^{-t}},$$

where $a_n = \sqrt{2 \log \log n}$ and $b_n = 2 \log \log n + \frac{1}{2} \log \log \log n - \frac{1}{2} \log \pi$.

In the sequel we need the following corollary, given also in [10]:

$$(1.7) \qquad \frac{\sqrt{n}\, T_n}{\sqrt{2 \log \log n}} \xrightarrow{\mathscr{P}} 1 \quad \text{as} \quad n \to \infty,$$

where $\xrightarrow{\mathscr{P}}$ denotes convergence in probability.

Anderson and Darling [1] have determined the limiting distribution of A_n. They have shown that

$$(1.8) \qquad nA_n \xrightarrow{\mathscr{D}} \int_0^1 \frac{B^2(t)}{t(1 - t)}\, dt \quad \text{as} \quad n \to \infty,$$

where $\xrightarrow{\mathscr{D}}$ denotes convergence in distribution and $B(t)$ is a Brownian Bridge. Anderson and Darling [1] give also an explicit expression for the limiting distribution, based on the identity

$$(1.9) \qquad \int_0^1 \frac{B^2(t)}{t(1 - t)}\, dt = \sum_{j=1}^{\infty} \frac{Y_j^2}{j(j + 1)},$$

where $Y_1, Y_2, \ldots$ are i.i.d. standard normal random variables.

Our concern in this paper is to give lim inf and lim sup results for T_n and A_n. It is well known that

$$(1.10) \qquad \limsup_{n \to \infty} \frac{\sqrt{n}\, T_n}{\sqrt{\log \log n}} = \infty \quad \text{a.s.}$$

and the same result holds for T_n replaced by U_n. On the other hand we have shown [3] that

$$(1.11) \qquad \limsup_{n \to \infty} \frac{\sqrt{n}\, V_n}{\sqrt{\log \log n}} = 2 \quad \text{a.s.,}$$

pointing out the unsymmetrical behavior between $\alpha_n(x)$ and $-\alpha_n(x)$ near 0.

Concerning the lower limit of T_n, in Section 2 we show that

$$(1.12) \qquad \liminf_{n \to \infty} \frac{\sqrt{n}\, T_n}{\sqrt{2 \log \log n}} = 1 \quad \text{a.s.}$$

This is worth comparing with the result of Mogul'skiĭ [12], for $D_n = \sup_{0 \leqslant x \leqslant 1} |F_n(x) - x|$;

$$(1.13) \qquad \liminf_{n \to \infty} D_n \sqrt{n \log \log n} = \frac{\pi}{\sqrt{8}} \quad \text{a.s.}$$

The above mentioned results for T_n show that the standardized empirical process $\alpha_n(x)$ is too sensitive on the tail (near 0 and 1) and in many respects T_n is much larger in order of magnitude than its unweighted counterpart D_n.

On the other hand, the tail behavior of $\alpha_n(x)$ has much less influence on A_n and it is seen from (1.8) and the results given in Section 3, that the norming factors are the same as for $\omega_n^2 = \int_0^1 (F_n(x) - x)^2 \, dx$.

For the sake of comparison, the corresponding results for ω_n^2 are as follows.

Smirnov [13]:

$$(1.14) \qquad n\omega_n^2 \overset{\mathscr{D}}{\to} \int_0^1 B^2(t)\, dt \overset{\mathscr{D}}{=} \sum_{j=1}^{\infty} \frac{Y_j^2}{j^2 \pi^2} \quad \text{as } n \to \infty,$$

Finkelstein [9]:

$$(1.15) \qquad \limsup_{n \to \infty} \frac{n\omega_n^2}{\log \log n} = \frac{2}{\pi^2} \quad \text{a.s.,}$$

Mogul'skiǐ [12]:

$$(1.16) \qquad \liminf_{n \to \infty} n\omega_n^2 \log \log n = \frac{1}{8} \quad \text{a.s..}$$

Most of our results will be based on the strong approximation theorem of K o m l ó s, M a j o r and T u s n á d y [11]:

Theorem A ([11]). *If the random variables are defined on a rich enough probability space then one can construct a Kiefer process* $K(x, y)$ *on the same space such that*

$$(1.17) \qquad \sup_{0 \leqslant x \leqslant 1} |n(F_n(x) - x) - K(x, n)| = O(\log^2 n) \quad a.s.$$

as $n \to \infty$.

The Kiefer process $K(x, y)$ is defined by

$$(1.18) \qquad K(x, y) = W(x, y) - xW(1, y) \qquad (0 \leqslant x \leqslant 1, \ 0 \leqslant y < \infty),$$

where $W(x, y)$ is a two-parameter Wiener process. We shall refer to $K(x, y)$ satisfying (1.17) as KMT construction.

We shall apply the following inequalities given in M o g u l' s k i ǐ ([12], Lemma 1):

Lemma B ([12]). *Let* S_i $(i = 1, 2, \ldots)$ *be partial sums of independent Banach space-valued random variables. Let* $v > 0$, $y > 0$, $2 \leqslant m \leqslant r$, $c = \min_{m \leqslant n \leqslant r} P(\|S_{r-n}\| \leqslant y)$. *Then*

$$(1.19) \qquad P(\min_{m \leqslant n \leqslant r} \|S_n\| \leqslant v) \leqslant \frac{1}{c} P(\|S_r\| \leqslant v + y)$$

and

$$(1.20) \qquad P(\max_{m \leqslant n \leqslant r} \|S_n\| > v) \leqslant \frac{1}{c} P(\|S_r\| > v - y).$$

2. STRONG LIMIT THEOREMS FOR T_n

First we give a slight correction to Theorem 3.1 in [2].

Theorem 2.1.

(i) *If* $\sum\limits_{n=1}^{\infty} a_n = \infty$, *then*

$$(2.1) \qquad \limsup_{n \to \infty} (n\sqrt{a_n}\, T_n) = \infty \quad a.s.$$

(ii) *If* $\sum\limits_{n=1}^{\infty} b_n < \infty$ *and* nb_n *is nonincreasing, then*

$$(2.2) \qquad \lim_{n \to \infty} (n\sqrt{b_n}\, T_n) = 0 \quad a.s..$$

The condition that nb_n is nonincreasing was missing in [2]. It is not hard to see that the condition $\sum\limits_{n=1}^{\infty} b_n < \infty$ itself does not imply (2.2). We have to assume somewhat more, the above given condition is a simple and natural one. We also give a short proof here.

Proof. Part (i) is easy. For a simple proof we may refer to G a e n s s l e r and S t u t e [8].

For part (ii), observe that it suffices to prove (2.2) for U_n and V_n defined by (1.4) and (1.5), respectively. Observe furthermore that the conditions of part (ii) imply that b_n is nonincreasing and $\lim\limits_{n \to \infty} (nb_n \log n) = 0$. Hence from (1.11) it follows that

$$(2.3) \qquad \lim_{n \to \infty} (n\sqrt{b_n}\, V_n) = 0 \quad a.s..$$

Split U_n into two parts by defining

$$(2.4) \qquad U_n^{(1)} = \sup_{0 < x < x_n'} \left(\frac{F_n(x) - x}{\sqrt{x}} \right),$$

$$(2.5) \qquad U_n^{(2)} = \sup_{x_n' \le x \le \frac{1}{2}} \left(\frac{F_n(x) - x}{\sqrt{x}} \right),$$

where $x'_n = \dfrac{\alpha}{n^2 b_n}$ and the constant α will be specified later. Since

$x'_n \geqslant \dfrac{\alpha \log n}{n}$ for large n, it follows from [3] that

$$(2.6) \qquad \lim_{n \to \infty} (n\sqrt{b_n}\, U_n^{(2)}) = 0 \quad \text{a.s..}$$

It suffices therefore to show that $\displaystyle\sum_{n=1}^{\infty} b_n < \infty$ and $nb_n \downarrow$ imply

$$(2.7) \qquad P\left(U_n^{(1)} \geqslant \frac{1}{n\sqrt{b_n}} \text{ infinitely often}\right) = 0.$$

Define the events C_n and B_{nk} by

$$(2.8) \qquad C_n = \left\{ U_{n-1}^{(1)} < \frac{1}{(n-1)\sqrt{b_{n-1}}}, \ U_n^{(1)} \geqslant \frac{1}{n\sqrt{b_n}} \right\},$$

$$(2.9) \qquad \begin{aligned} B_{nk} &= \{nF_n(x_k) = k, \ (n-1)F_{n-1}(x_k) < k\} = \\ &= \{(n-1)F_{n-1}(x_k) = k-1, \ X_n \leqslant x_k\}, \end{aligned}$$

where x_k is the positive solution of the equation

$$(2.10) \qquad nx + \frac{1}{\sqrt{b_n}}\sqrt{x} = k.$$

We may assume without loss of generality that x'_n is decreasing. Then it can be easily seen that

$$(2.11) \qquad C_n \subseteq \bigcup_{k=1}^{k_0} B_{nk},$$

where $k_0 = \max\,(k: x_k \leqslant x'_n)$ and hence

$$(2.12) \qquad P(C_n) \leqslant \sum_{k=1}^{k_0} P(B_{nk}) = \sum_{k=1}^{k_0} \frac{k}{n} \binom{n}{k} x_k^k (1 - x_k)^{n-k}.$$

From (2.10) it follows that $x_k \leqslant k^2 b_n$, thus

$$(2.13) \qquad P(C_n) \leqslant \sum_{k=1}^{k_0} \frac{k}{n} \frac{(nx_k)^k}{k!} \leqslant \sum_{k=1}^{k_0} b_n (nb_n)^{k-1} \frac{k^{2k}}{(k-1)!}.$$

- 128 -

Put $\alpha_k = \dfrac{(nb_n)^{k-1}k^{2k}}{(k-1)!}$. Then

$$\frac{\alpha_{k+1}}{\alpha_k} = nb_n k\left(\frac{k+1}{k}\right)^{2k+2} \leqslant 16k_0 nb_n.$$

Since $k_0 \leqslant nx'_n + \sqrt{\dfrac{x'_n}{b_n}} \leqslant \dfrac{\alpha + \sqrt{\alpha}}{nb_n}$, we can choose α such that

$\dfrac{\alpha_{k+1}}{\alpha_k} \leqslant \dfrac{1}{2}$ for $1 \leqslant k \leqslant k_0$. Therefore $\alpha_{k+1} \leqslant \left(\dfrac{1}{2}\right)^k$ and

$$(2.14) \qquad P(C_n) \leqslant b_n \sum_{k=1}^{\infty} \left(\frac{1}{2}\right)^{k-1} = 2b_n.$$

Hence $\displaystyle\sum_{n=1}^{\infty} P(C_n) < \infty$ and by Borel – Cantelli lemma

$$(2.15) \qquad P(C_n \text{ infinitely often}) = 0.$$

Similar argument shows that

$$(2.16) \qquad \lim_{n \to \infty} P\left(U_n^{(1)} \geqslant \frac{1}{n\sqrt{b_n}}\right) = 0,$$

which together with (2.15) implies (2.7), and this completes the proof of Theorem 2.1.

We note that (2.3) and (2.6) can be proved also directly, without reference to [3], by using the normal approximation of binomial probabilities.

A corrected form of Corollary 3.1 in [2] is given by

Corollary 2.1. *There is no numerical sequence* a_n *such that* $na_n \downarrow$ *and*

$$(2.17) \qquad \limsup_{n \to \infty} n\sqrt{a_n}\, T_n = 1 \quad a.s..$$

Our next result concerns the lower limit of T_n.

Theorem 2.2.

$$(2.18) \qquad \liminf_{n \to \infty} \frac{\sqrt{n}\, T_n}{\sqrt{2 \log \log n}} = 1 \quad a.s.$$

Proof. The $\leqslant$ part follows from (1.7). We prove the $\geqslant$ part only.

As shown by C s ö r g ő and R é v é s z [4], for KMT construction we have

$$(2.19) \qquad \sup_{\delta_n \leqslant x \leqslant 1 - \delta_n} \left| \alpha_n(x) - \frac{K(x, n)}{\sqrt{nx(1 - x)}} \right| = o(\sqrt{\log \log n}) \quad a.s.,$$

where $\delta_n = \dfrac{(\log n)^4}{n}$. Therefore our statement will follow from

$$(2.20) \qquad \liminf_{n \to \infty} \left(\frac{1}{\sqrt{2n \log \log n}} \sup_{\delta_n \leqslant x \leqslant 1 - \delta_n} \frac{|K(x, n)|}{\sqrt{x(1 - x)}} \right) \geqslant 1 \quad a.s..$$

Let $\epsilon > 0$ small enough. Choose a subsequence $n_k = \left(1 - \dfrac{\epsilon^2}{8}\right)^{-k}$. and let the events B_k be defined by

$$(2.21) \qquad B_k = \left\{ \inf_{n_{k-1} \leqslant n < n_k} \sup_{\delta_{n_{k-1}} \leqslant x \leqslant 1 - \delta_{n_{k-1}}} \frac{|K(x, n)|}{\sqrt{x(1 - x)}} < \right.$$

$$\left. < (1 - \epsilon)\sqrt{2n_k \log \log n_k} \right\}.$$

We show that $\displaystyle\sum_{k=1}^{\infty} P(B_k) < \infty$.

To apply M o g u l' s k i ǐ's inequality (1.19), let $m = n_{k-1}$, $r = n_k$, $v = (1 - \epsilon)\sqrt{2n_k \log \log n_k}$, $y = 2\sqrt{(n_k - n_{k-1}) \log \log n_k}$. Since $K(x, n)$ is the sum of n independent Brownian Bridges, we can define

$$S_n = \frac{|K(x, n)|}{\sqrt{x(1 - x)}}$$

and

$$(2.22) \qquad \| S_n \| = \sup_{\delta_{n_{k-1}} \leqslant x \leqslant 1 - \delta_{n_{k-1}}} \frac{|K(x, n)|}{\sqrt{x(1 - x)}}.$$

Then using the fact that $\dfrac{K(x,n)}{\sqrt{n}}$ is also a Brownian Bridge, we have

$$c = \min_{n_{k-1} \leqslant n < n_k} \mathbf{P}\left(\sup_{\delta_{n_{k-1}} \leqslant x \leqslant 1 - \delta_{n_{k-1}}} \left| \frac{K(x, n_k - n)}{\sqrt{x(1-x)}} \right| \leqslant \right.$$

$$\left. \leqslant 2\sqrt{(n_k - n_{k-1}) \log \log n_k} \right) =$$

$$(2.23) \qquad = \mathbf{P}\left(\sup_{\delta_{n_{k-1}} \leqslant x \leqslant 1 - \delta_{n_{k-1}}} \left| \frac{K(x, n_k - n_{k-1})}{\sqrt{x(1-x)(n_k - n_{k-1})}} \right| \leqslant \right.$$

$$\left. \leqslant 2\sqrt{\log \log n_k} \right) =$$

$$= \mathbf{P}\left(\sup_{0 \leqslant t \leqslant \log \frac{1 - \delta_{n_{k-1}}}{\delta_{n_{k-1}}}} |U(t)| \leqslant 2\sqrt{\log \log n_k} \right),$$

where $U(t)$ is the Ornstein – Uhlenbeck process, i.e. stationary Gaussian process with zero mean and covariance $\mathbf{E}(U(t)U(0)) = e^{-|t|}$.

From the results of D a r l i n g and E r d ő s [6] concerning the limiting distribution of $\sup |U(t)|$, it is easily seen that $c = c_k \to 1$ as $k \to \infty$, hence $c \geqslant \dfrac{1}{2}$ for k large enough. Thus by applying (1.19),

$$\mathbf{P}(B_k) \leqslant$$

$$\leqslant 2\mathbf{P}\left(\sup_{\delta_{n_{k-1}} \leqslant x \leqslant 1 - \delta_{n_{k-1}}} \frac{|K(x, n_k)|}{\sqrt{x(1-x)}} \leqslant \right.$$

$$\left. \leqslant (1 - \epsilon)\sqrt{2n_k \log \log n_k} + 2\sqrt{(n_k - n_{k-1}) \log \log n_k} \right) =$$

$$= 2\mathbf{P}\left(\sup_{\delta_{n_{k-1}} \leqslant x \leqslant 1 - \delta_{n_{k-1}}} \frac{|K(x, n_k)|}{\sqrt{n_k x(1-x)}} \leqslant \right.$$

$$\left. \leqslant \sqrt{2 \log \log n_k} \left(1 - \epsilon + \sqrt{\frac{2(n_k - n_{k-1})}{n_k}} \right) \right) =$$

$$= 2\mathbf{P}\left(\sup_{0 \leqslant t \leqslant \log \frac{1-\delta_{n_k-1}}{\delta_{n_k-1}}} |U(t)| \leqslant \left(1 - \frac{\epsilon}{2}\right)\sqrt{2 \log \log n_k} \right).$$

Put $\alpha_0 = \log \dfrac{8 + 4\epsilon + \epsilon^2}{4\epsilon + \epsilon^2}$ and $U_i = U(i\alpha_0)$, $i = 1, \ldots, N_k$, where N_k is the greatest integer such that $N_k \alpha_0 \leqslant \log \dfrac{1 - \delta_{n_k-1}}{\delta_{n_k-1}}$. Then $U_1, \ldots, U_{N_k}$ is a stationary Gauss – Markov sequence with mean zero and $\mathbf{E}(U_i U_j) = e^{-\alpha_0|i-j|}$. We show that

$$(2.24) \qquad \mathbf{P}(\max_{1 \leqslant i \leqslant N} |U_i| < u) \leqslant \left(\Phi\left(u\left(1 + \frac{\epsilon}{2}\right)\right)\right)^N.$$

$$\mathbf{P}(\max_{1 \leqslant i \leqslant N} |U_i| < u) =$$

$$= \int_{-u}^{u} \mathbf{P}(|U_1| < u, \ldots, |U_{N-2}| < u \mid U_{N-1} = y)\varphi(y)\, dy =$$

$$= \int_{-u}^{u} \mathbf{P}(|U_1| < u, \ldots, |U_{N-2}| < u \mid U_{N-1} = y) \times$$

$$\times \mathbf{P}(|U_N| < u \mid U_{N-1} = y)\varphi(y)\, dy \leqslant$$

$$\leqslant \int_{-u}^{u} \mathbf{P}(|U_1| < u, \ldots, |U_{N-2}| < u \mid U_{N-1} = y) \times$$

$$\times \Phi\left(\frac{u - ry}{\sqrt{1 - r^2}}\right)\varphi(y)\, dy \leqslant$$

$$\leqslant \Phi\left(\frac{u + ru}{\sqrt{1 - r^2}}\right) \times$$

$$\times \int_{-u}^{u} \mathbf{P}(|U_1| < u, \ldots, |U_{N-2}| < u \mid U_{N-1} = y)\varphi(y)\, dy =$$

$$= \Phi\left(u\sqrt{\frac{1+r}{1-r}}\right)P(\max_{1\leqslant i\leqslant N-1}|U_i|<u),$$

where $r=e^{-\alpha_0}$. Since $\sqrt{\frac{1+r}{1-r}}=1+\frac{\epsilon}{2}$, by repeating the same procedure several times, we get (2.24). Furthermore

$$P(B_k)\leqslant 2P\left(\max_{1\leqslant i\leqslant N_k}|U_i|\leqslant\left(1-\frac{\epsilon}{2}\right)\sqrt{2\log\log n_k}\right)\leqslant$$

(2.25)

$$\leqslant 2\left(\Phi\left(\left(1-\frac{\epsilon^2}{4}\right)\sqrt{2\log\log n_k}\right)\right)^{N_k}.$$

By using the inequalities

(2.26) $1-u\leqslant e^{-u}\qquad(u\geqslant 0)$

and

(2.27) $\left(1-\frac{1}{u^2}\right)\frac{1}{u\sqrt{2\pi}}e^{-\frac{u^2}{2}}\leqslant 1-\Phi(u)\qquad(u>0)$

one can see from (2.25) that for k large

$$P(B_k)\leqslant 2\exp\left\{-N_k\left(1-\Phi\left(\left(1-\frac{\epsilon^2}{4}\right)\sqrt{2\log\log n_k}\right)\right)\right\}\leqslant$$

(2.28)

$$\leqslant 2\exp\left\{-c_1\frac{N_k}{\sqrt{\log\log n_k}}(\log n_k)^{-\left(1-\frac{\epsilon^2}{4}\right)^2}\right\},$$

with some constant c_1.

It can be seen further that both N_k and $\log n_k$ are of order k, therefore with some constant c_2 (which may depend on ϵ), we have

(2.29) $P(B_k)\leqslant 2\exp\left(-c_2\frac{k^{\frac{\epsilon^2}{2}}}{\sqrt{\log k}}\right),$

showing that $\sum_{k=1}^{\infty}P(B_k)<\infty$, which in turn proves the theorem.

Remark 1. It would also be interesting to obtain both the limiting distribution (1.6) of T_n and (2.18) by direct methods and to characterize the sequences a_n such that $P(T_n < a_n$ infinitely often$) = 0$ or 1.

Remark 2. In [3] we investigated the lim sup of

$$\sup_{\delta_n \leqslant x \leqslant 1 - \delta_n} |\alpha_n(x)|$$

suitably normalized, for different sequences δ_n. It would also be interesting to determine the corresponding lim inf and investigate how the factor $\sqrt{\log \log n}$ gets down from the numerator (when $\delta_n = \delta > 0$ constant) to the denominator (when $\delta_n = 0$).

3. STRONG LIMIT THEOREMS FOR A_n

First we establish an a.s. invariance principle for A_n:

Lemma 3.1. *For KMT construction there holds*

$$(3.1) \qquad \left| nA_n - \int_0^1 \frac{K^2(x, n)}{nx(1 - x)} \, dx \right| = O\left(\left(\frac{\log \log n}{n} \right)^{\frac{1}{3}} \log^2 n \right) \quad a.s.$$

Proof. Let $\delta_n = \left(\dfrac{\log \log n}{n} \right)^{\frac{1}{3}}$, then

$$\left| nA_n - \int_0^1 \frac{K^2(x, n)}{nx(1 - x)} \, dx \right| \leqslant$$

$$\leqslant \int_0^{\delta_n} \frac{n(F_n(x) - x)^2}{x(1 - x)} \, dx + \int_{1 - \delta_n}^1 \frac{n(F_n(x) - x)^2}{x(1 - x)} \, dx +$$

$$(3.2) \qquad + \int_0^{\delta_n} \frac{K^2(x, n)}{nx(1 - x)} \, dx + \int_{1 - \delta_n}^1 \frac{K^2(x, n)}{nx(1 - x)} \, dx +$$

$$+ \int_{\delta_n}^{1 - \delta_n} \left| \frac{n^2(F_n(x) - x)^2 - K^2(x, n)}{nx(1 - x)} \right| dx.$$

Now

$$(3.3) \qquad \int_0^{\delta_n} \frac{n(F_n(x) - x)^2}{x(1-x)} \, dx \leqslant n\delta_n T_n^2 = O(\delta_n \log^2 n) \quad \text{a.s.}$$

by Theorem 2.1. The same holds for the second term. To estimate the next two terms we use the following result of Csörgő and Révész [5]:

$$(3.4) \qquad \sup_{0 < x < 1} \frac{K^2(x, n)}{nx(1-x) \log \log \frac{n}{x(1-x)}} = O(1) \quad \text{a.s..}$$

Hence

$$(3.5) \qquad \int_0^{\delta_n} \frac{K^2(x, n)}{nx(1-x)} \, dx \leqslant O(1) \int_0^{\delta_n} \log \log \frac{n}{x(1-x)} \, dx =$$
$$= O(\delta_n \log \log n)$$

and the same holds for the fourth term.

Finally

$$\int_{\delta_n}^{1-\delta_n} \left| \frac{n(F_n(x) - x)^2}{x(1-x)} - \frac{K^2(x, n)}{nx(1-x)} \right| dx \leqslant$$

$$\leqslant \frac{1}{\sqrt{\delta_n(1-\delta_n)}} \sup_{0 < x < 1} \left| \sqrt{n} (F_n(x) - x) - \frac{K(x, n)}{\sqrt{n}} \right| \times$$

$$(3.6) \qquad \times \left(\sup_{\delta_n \leqslant x \leqslant 1 - \delta_n} \frac{\sqrt{n} |F_n(x) - x|}{\sqrt{x(1-x)}} + \right.$$

$$+ \sup_{\delta_n \leqslant x \leqslant 1 - \delta_n} \left. \frac{|K(x, n)|}{\sqrt{nx(1-x)}} \right) =$$

$$= O\left(\frac{1}{\sqrt{\delta_n}} \frac{\log^2 n}{\sqrt{n}} \sqrt{\log \log n} \right) = O(\delta_n \log^2 n) \quad \text{a.s.}$$

by Theorem A and by the law of the iterated logarithm (Csáki [3], Csörgő and Révész [5]). Lemma 3.1 is proved.

Theorem 3.1.

$$(3.7) \qquad \limsup_{n \to \infty} \frac{nA_n}{\log \log n} = 1 \quad \text{a.s..}$$

Proof. By Lemma 3.1 it suffices to show that

$$(3.8) \qquad \limsup_{n \to \infty} \frac{Z_n}{\sqrt{n \log \log n}} = 1 \quad a.s.,$$

where

$$(3.9) \qquad Z_n^2 = \int_0^1 \frac{K^2(x, n)}{x(1 - x)} \, dx,$$

i.e. Z_n is the L_2-norm of $\dfrac{K(x, n)}{\sqrt{x(1 - x)}}$. Since $\dfrac{K(x, n)}{\sqrt{n}}$ $(0 \leqslant x \leqslant 1)$ is a Brownian Bridge for all fixed n, we may use the representation (1.9) for $\dfrac{Z_n^2}{n}$, i.e.

$$(3.10) \qquad \frac{Z_n^2}{n} \overset{\mathscr{D}}{=} \sum_{j=1}^{\infty} \frac{Y_j^2}{j(j + 1)}.$$

It follows from Zolotarev [15] that

$$(3.11) \qquad \log \mathsf{P}\Big(\sum_{j=1}^{\infty} \frac{Y_j^2}{j(j + 1)} > u^2 \Big) \sim - u^2 \quad \text{as} \quad u \to \infty.$$

The proof of (3.7) can be completed by the usual method to prove laws of the iterated logarithm, applying (3.11) and the inequality (1.20).

Theorem 3.2.

$$(3.12) \qquad \liminf_{n \to \infty} (nA_n \log \log n) = \frac{\pi^2}{8} \quad a.s..$$

Proof. Again, by Lemma 3.1 it suffices to deal with Z_n defined by (3.9) and show that

$$(3.13) \qquad \liminf_{n \to \infty} \frac{Z_n \sqrt{\log \log n}}{\sqrt{n}} = \frac{\pi}{\sqrt{8}} \quad a.s..$$

We need the lower tail of the distribution of (3.10). It follows from Sytaja [14] that

$$(3.14) \qquad \log \mathsf{P}\Big(\sum_{j=1}^{\infty} \frac{Y_j^2}{j(j+1)} < u^2 \Big) \sim -\frac{\pi^2}{8u^2} \quad \text{as} \quad u \to 0.$$

Our result (3.12) follows from Theorem 1 of Mogul'skiĭ [12].

REFERENCES

[1] T.W. Anderson – D.A. Darling,   Asymptotic theory of certain "goodness of fit" criteria based on stochastic processes, *Ann. Math. Statist.*, 23 (1952), 193-212.

[2] E. Csáki, Some notes on the law of the iterated logarithm for empirical distribution function,   *Colloquia Mathematica Societatis János Bolyai*, 11. *Limit theorems of probability theory*, Keszthely, (1974), 47-58.

[3] E. Csáki, The law of the iterated logarithm for normalized empirical distribution function,   *Z. Wahrscheinlichkeitstheorie verw. Geb.*, 38 (1977), 147-167.

[4] M. Csörgő – P. Révész,   Some notes on the empirical distribution function and the quantile process, *Colloquia Mathematica Societatis János Bolyai*, 11. *Limit theorems of probability theory*, Keszthely, (1974), 59-71.

[5] M. Csörgő – P. Révész,   How big are the increments of a multi-parameter Wiener process?, *Z. Wahrscheinlichkeitstheorie verw. Geb.*, 42 (1978), 1-12.

[6] D.A. Darling – P. Erdős, A limit theorem for the maximum of normalized sums of independent random variables, *Duke Math. J.*, 23 (1956), 143-155.

[7] F. Eicker, The asymptotic distribution of the suprema of the standardized empirical processes, *Ann. Statist.*, 7 (1979), 116-138.

[8] P. Gaenssler – W. Stute,   Empirical process: a survey of results for independent and identically distributed random variables, *Ann. Probab.*, 7 (1979), 193-243.

[9]   H. Finkelstein, The law of the iterated logarithm for empirical distributions, *Ann. Math. Statist.*, 42 (1971), 607-615.

[10] D. Jaeschke, The asymptotic distribution of the supremum of the standardized empirical distribution function on subintervals, *Ann. Statist.*, 7 (1979), 108-115.

[11] J. Komlós – P. Major – G. Tusnády, An approximation of partial sums of independent r.v.'s and the sample df. I., *Z. Wahrscheinlichkeitstheorie verw. Geb.*, 32 (1975), 111-131.

[12] A.A. Mogul'skiĭ, On the law of the iterated logarithm in Chung's form for functional spaces, *Teor. Verojatnost. i Primenen.*, 24 (1979), 399-407 (in Russian).

[13] N.V. Smirnov, Sur la distribution de ω^2, *C.R. Acad. Sci. Paris*, 202 (1936), 449.

[14] G.N. Sytaja, Certain asymptotic representations for a Gaussian measure in Hilbert space, *Theory of random processes*, No. 2, 93-104, 140., Izdat. "Naukova Dumka", Kiev, 1974 (in Russian).

[15] V.M. Zolotarev, Concerning a certain probability problem, *Theor. Probability Appl.*, 6 (1961), 201-204.

E. Csáki

Mathematical Institute of the Hungarian Academy of Sciences, 1053 Budapest, Reáltanoda u. 13-15, Hungary.

COLLOQUIA MATHEMATICA SOCIETATIS JÁNOS BOLYAI

32. NONPARAMETRIC STATISTICAL INFERENCE,

BUDAPEST (HUNGARY), 1980.

THE EMPIRICAL MOMENT GENERATING FUNCTION

S. CSÖRGŐ

1. INTRODUCTION AND SUMMARY

Consider a univariate non-degenerate distribution function F whose moment generating function

$$m(t) = \int_{-\infty}^{\infty} e^{tx} \, dD(x)$$

exists (finite) in a non-degenerate interval J. J has one of the forms $(c, d), [c, d], (c, d]$, or $[c, d)$, where $-\infty \leqslant c \leqslant 0 \leqslant d \leqslant \infty$ and $c < d$, and m has derivatives of all orders in the interior of J. Moreover, m is convex over J and if 0 is an inner point of J, then it uniquely determines F. Let $X_1, X_2, \ldots$ be a sequence of independent random variables with common distribution function F, and let F_n denote the empirical distribution function of the first n variables. The empirical moment generating function

$$m_n(t) = \frac{1}{n} \sum_{k=1}^{n} e^{tX_k} = \int_{-\infty}^{\infty} e^{tx} \, dF_n(x)$$

appears in a paper by Quandt and Ramsey [7] who used it for estimating the parameters in a mixture of two normal distributions and switching regressions. The only theoretical feature of $m_n(t)$ they needed was that it converges almost surely to $m(t)$ at each fixed $t \in J$. Of course, more is true. Breaking up the integral in

$$m_n(t) - m(t) = \int_{-\infty}^{\infty} e^{tx} \, d(F_n(x) - F(x))$$

into two parts $\int\limits_{|x|>A} + \int\limits_{|x| \leqslant A}$ and making use of the Glivenko – Cantelli theorem and of another classical result, Dini's theorem (Sz.-Nagy [8], p. 66), we arrive at the following

Proposition. *If S is any closed interval contained in J, then almost surely, as $n \to \infty$,*

$$\sup_{t \in S} |m_n(t) - m(t)| \to 0.$$

On a generally shorter interval I, this result will be accompanied in (4) below with the usual rate of convergence.

The aim of this note is to prove the weak convergence of the sequence of processes

$$M_n(t) = \sqrt{n} \, (m_n(t) - m(t))$$

in the Banach space $\mathscr{C}(I)$ of continuous functions, endowed with the supremum norm, over a suitable interval $I = [a, b] \subseteq J$. The choice of the endpoints of this interval are the following:

$$a = \begin{cases} \text{arbitrary negative number,} & \text{if } c = -\infty \\ \dfrac{c}{2} + \epsilon, & \text{if } -\infty < c < 0 \\ 0, & \text{if } c = 0, \end{cases}$$

$$b = \begin{cases} \text{arbitrary positive number,} & \text{if } d = \infty \\ \dfrac{d}{2} - \delta, & \text{if } 0 < d < \infty \\ 0, & \text{if } d = 0, \end{cases}$$

where $\epsilon = 0$ if $c \in J$ and ϵ is an arbitrarily small positive number if $c \notin J$, and $\delta = 0$ if $d \in J$ and δ is an arbitrarily small positive number if $d \notin J$. If at least one of ϵ and δ is positive, then of course they are chosen in such a way that $a < b$ be satisfied, i.e., $I \neq \phi$.

Since M_n has the form

$$(1) \qquad M_n(t) = \frac{1}{\sqrt{n}} \sum_{k=1}^{n} (e^{tX_k} - m(t)),$$

it follows from the multidimensional central limit theorem that the finite-dimensional distributions of $\{M_n\}$ over I converge to those of a Gaussian process M specified by

$$\mathsf{E}M(t) = 0, \quad \mathsf{E}M(t)M(s) = m(t + s) - m(t)m(s) \qquad (t, s \in I).$$

Since M_n has the alternative form

$$(2) \qquad M_n(t) = \int_{-\infty}^{\infty} e^{tx} \, d\beta_n(x),$$

where

$$\beta_n(x) = \sqrt{n} \, (F_n(x) - F(x))$$

is the empirical process distributed asymptotically as $B(F(\cdot))$, where B is the Brownian Bridge process on $[0, 1]$, it is not surprising that the Gaussian process M can be represented as

$$(3) \qquad M(t) = \int_{-\infty}^{\infty} e^{tx} \, dB(F(x)) \qquad (t \in I).$$

Note that this stochastic integral is meaningful again in the usual sense if and only if $t \in I$. It will be shown in the next section that M is sample-continuous (i.e., almost all sample paths of each separable version of it are continuous) over I. Thus the anomalies existing when treating the related notion of the empirical characteristic function (Csörgő [1], [2]) do not appear here once $m(t)$ exists. Indeed, we shall prove the following result which can be useful when treating the problems of Quandt and Ramsey [7] and similar ones.

Theorem. $\{M_n(\cdot)\}$ *converges weakly in* $\mathscr{C}(I)$ *to* $M(\cdot)$.

Introducing now the Finkelstein subset of the absolutely continuous functions

$$\mathscr{F} = \Big\{f \mid f\colon [0, 1] \to (-\infty, \infty),\ f(0) = f(1) = 0,$$

$$\int_0^1 \Big(\frac{df(x)}{dx}\Big)^2 dx \leqslant 1\Big\}$$

consider

$$\mathscr{K}_F = \Big\{g(t) = \int_{-\infty}^{\infty} e^{tx}\, df(F(x)),\ t \in I \mid f \in \mathscr{F}\Big\}.$$

Then, by the result of Pisier [6] (cf. also Heinkel [4]) and the argument in Csörgő [2], the Theorem implies the following

Corollary. $\Big\{\dfrac{M_n(\cdot)}{\sqrt{2 \log \log n}}\Big\}$ *is almost surely relatively compact in* $\mathscr{C}(I)$, *and the set of its limit points is* $\mathscr{K}_F$.

This corollary gives, in particular, the usual rate of convergence to the above Proposition, at least on I. That is, almost surely,

$$(4) \qquad \limsup_{n \to \infty} \Big(\frac{n}{2 \log \log n}\Big)^{\frac{1}{2}} \sup_{t \in I} |m_n(t) - m(t)| = \sup_{g \in \mathscr{K}_F} \sup_{t \in I} |g(t)|.$$

Let us replace the real variable t by the complex z in the definitions of m and m_n. Then $m(z)$ and $m_n(z)$ are analytic in the strip $T_J = \{z \mid \operatorname{Re} z \in J\}$. The Proposition also holds for $m_n(z)$ and $m(z)$ in any compact part of T_J. Also, the Theorem and its Corollary remain valid for the complex Laplace transform of the empirical process on any compact subset of the strip T_I. Of course, $M(z)$ is then a complex Gaussian process with

$$\mathsf{E}M(z) = 0, \quad \mathsf{E}M(z_1)\overline{M(z_2)} = m(z_1 + \bar{z}_2) - m(z_1)m(\bar{z}_2).$$

2. PROOFS OF THE THEOREM

First we show that M is sample-continuous over I. The reason of this fact will also play a role in the second proof of the Theorem.

Since B in (3) can itself be represented as $B(u) = W(u) - uW(1)$ with a standard Wiener process W, it follows that

$$M(t) = \int_{-\infty}^{\infty} e^{tx}\, dW(F(x)) - m(t)W(1) =$$

$$= Z(t) - m(t)W(1) \qquad (t \in I).$$

Thus M is sample-continuous if and only if Z is such. For Z we have $\mathsf{E}Z(t) = 0$, $\mathsf{E}Z(t)Z(s) = m(t+s)$, and hence

$$f(t,s) = \mathsf{E}(Z(t) - Z(s))^2 = m(2t) + m(2s) - 2m(t+s) \quad (s, t \in I).$$

Define the function

$$\varphi(h) = \sup\{\sqrt{f(t,s)}: t, s \in I, |s - t| \leqslant h\} \qquad (h \geqslant 0).$$

For each fixed $t \in I$ let $f_t(u) = f(t, t+u)$, $0 \leqslant u \leqslant b - t$, and for $k = 1, 2, \ldots$ let $m^{(k)}$ denote the k-th order derivative of m over I. Of course, $f_t(0) = 0$. Since $m^{(2)}$ is strictly positive, $m^{(1)}$ is strictly increasing in I. Consequently,

$$\frac{df_t(u)}{du} = 2\{m^{(1)}(2t + 2u) - m^{(1)}(2t + u)\} > 0,$$

if $0 < u \leqslant b - t$. This implies that

$$\varphi(h) = \begin{cases} \sup\{\sqrt{f(t, t+h)}: t \in [a, b - h]\}, & \text{if } h \leqslant b - a, \\ \sup\{\sqrt{f(t, s)}: t, s \in I\}, & \text{if } h > b - a. \end{cases}$$

Moreover

$$f(t, t+h) = \sum_{k=2}^{\infty} \frac{h^k}{k!} (2^k - 2) m^{(k)}(2t),$$

$$0 \leqslant h \leqslant b - a, \quad a \leqslant t \leqslant b - h,$$

and the last two equations imply that

$$\limsup_{h \downarrow 0} \frac{\varphi(h)}{h} \leqslant \sup_{2a \leqslant t \leqslant 2b} m^{(2)}(t) = K_{(2)}.$$

Therefore

$$(5) \qquad \int_1^\infty \varphi(e^{-x^2})\, dx = \frac{1}{2} \int_0^{\frac{1}{e}} \frac{\varphi(h)}{h\sqrt{\log \frac{1}{h}}}\, dh < \infty,$$

and this is F e r n i q u e ' s [3] sufficient condition for the sample-continuity of Z over I.

As the classical approach to weak convergence suggests, it would be enough to show that the sequence $\{M_n\}$ is tight. It is, indeed, because one easily shows that $\mathsf{E}\,|M_n(t) - M_n(s)|^2 \leqslant K_{(2)}\,|t - s|^2$ for all $t, s \in I$. This inequality proves the Theorem and, of course, the sample-continuity of M.

The much lengthier second proof below is presented mainly for didactic reasons. Such proofs work often when the classical tightness proofs do not (cf. [1], [2]), and a rate of convergence can also be deduced along these lines (à la [1], [2]). This rate now depends on how fast $\sup\left\{\int_{|x|>A} e^{tx}\, dF(x):\ 2a \leqslant t \leqslant 2b\right\}$ decreases as $A \uparrow \infty$.

We assume, without loss of generality, that our basic sequence $X_1, X_2, \ldots$ is defined on the probability space $(\Omega, \mathscr{A}, \mathsf{P})$ of Theorem 3 of K o m l ó s , M a j o r and T u s n á d y [5]. Then there exists a sequence $\{B_n\}$ of Brownian Bridge such that

$$(6) \qquad \limsup_{n \to \infty} \frac{\sqrt{n}}{\log n} \sup_{-\infty < x < \infty} |\Delta_n(x)| \leqslant L \qquad \text{almost surely,}$$

where L is some non-random positive constant and

$$\Delta_n(x) = \beta_n(x) - B_n(F(x)).$$

In order to prove the Theorem it is enough to show that

(7) $\qquad \sup_{t \in I} \left| \int_{-\infty}^{\infty} e^{tx} \, d\Delta_n(x) \right| \to 0 \qquad\qquad$ in probability

as $n \to \infty$.

Set $C = \max(|a|, |b|)$ and let K be a positive number such that $K < \dfrac{1}{2C}$. If $u(n) = K \log n$, then integration by parts and (6) give that

$$\sup_{t \in I} \left| \int_{|x| \leqslant u(n)} e^{tx} \, d\Delta_n(x) \right| \leqslant \sup_{-\infty < x < \infty} |\Delta_n(x)| e^{Cu(n)} \to 0$$

almost surely. Hence

(8) $\qquad \sup_{t \in I} \left| \int_{|x| > u(n)} e^{tx} \, dB_n(F(x)) \right| \to 0 \quad$ in probability

and

(9) $\qquad \sup_{t \in I} \left| \int_{|x| > u(n)} e^{tx} \, d\beta_n(x) \right| \to 0 \qquad$ in probability

will imply (7).

To prove (8) first of all we note that B_n there can be replaced by a single Brownian Bridge $B(u) = W(u) - uW(1)$. Because

$$|W(1)| \sup_{t \in I} \left| \int_{|x| > u(n)} e^{tx} \, dF(x) \right| \to 0$$

in probability (as $n \to \infty$) by Dini's theorem,

(10) $\qquad \sup_{t \in I} |G_n(t)| \to 0 \qquad\qquad$ in probability

implies (8), where the sequence $\{G_n\}$ of Gaussian processes is defined as

$$G_n(t) = \int_{|x| > u(n)} e^{tx} \, dW(F(x)) \qquad (t \in I, \ n = 1, 2, \ldots).$$

The proof of (10) is based on Fernique's inequality ([3], Lemma 4.1.3) which is the main tool for his sample-continuity result just applied. For the norm

$$\| \Gamma_n \| = \sup \{ |\Gamma_n(s, t)| : s, t \in I \}$$

of the covariance

$$\Gamma_n(s, t) = \mathsf{E}\, G_n(s) G_n(t) = \int\limits_{|x| > u(n)} e^{(t+s)x}\, dF(x)$$

we have

(11) $\qquad \|\Gamma_n\| \to 0$

again by Dini's theorem. By the same reason

$$\varphi_n(h) = \sup\{\sqrt{\mathsf{E}(G_n(s) - G_n(t))^2} : s, t \in I,\ |s - t| \leqslant h\} =$$

$$= \sup\Big\{\Big(\int\limits_{|x| > u(n)} (e^{sx} - e^{tx})^2\, dF(x)\Big)^{\frac{1}{2}} :$$

$$s, t \in I,\ |s - t| \leqslant h\Big\} \to 0, \qquad n \to \infty,$$

for each $h \geqslant 0$. But

$$\varphi_n(h) \leqslant \varphi(h) =$$

$$= \sup\Big\{\Big(\int\limits_{-\infty}^{\infty} (e^{sx} - e^{tx})^2\, dF(x)\Big)^{\frac{1}{2}} : s, t \in I,\ |s - t| \leqslant h\Big\}$$

where this function φ is the same which figures in (5), and hence by Lebesgue's dominated convergence theorem we obtain from (5) that for any integer $p \geqslant 2$ (say)

$$Q_n(p) = (2 + \sqrt{2}) \int\limits_1^\infty \varphi_n\Big(\frac{b-a}{2}\, p^{-x^2}\Big)\, dx =$$

(12)

$$= \frac{2 + \sqrt{2}}{2\sqrt{\log p}} \int\limits_0^{\frac{b-a}{2e}} \frac{\varphi_n(h)}{h\sqrt{\log \dfrac{2}{(b-a)h}}}\, dh \to 0, \qquad n \to \infty.$$

Now let ϵ and η be arbitrarily small positive numbers. Choose the integer $p \geqslant 2$ so large that $\dfrac{5}{2p} < \eta$. Let $x(p) = \sqrt{6 \log p}$ with this p, and finally, by (11) and (12), let n be so large that

$$x(p)[\| \Gamma_n \|^{\frac{1}{2}} + Q_n(p)] < \epsilon.$$

Then by the Fernique inequality we obtain

$$P(\sup_{t \in I} |G_n(t)| > \epsilon) \leqslant$$

$$\leqslant P(\sup_{t \in I} |G_n(t)| > x(p)[\| \Gamma_n \|^{\frac{1}{2}} + Q_n(p)]) \leqslant$$

$$\leqslant \frac{5}{2} p^2 \int_{x(p)} e^{-\frac{u^2}{2}} \, du \leqslant \frac{5}{2} p^2 e^{-\frac{x^2(p)}{2}} = \frac{5}{2p} < \eta,$$

and this is (10).

Now we proceed to the proof of (9). Let the process in question be denoted by H_n:

$$H_n(t) = \int_{|x| > u(n)} e^{tx} \, d\beta_n(x) =$$

$$= n^{-\frac{1}{2}} \sum_{k=1}^{n} \left[e^{tX_k} \chi(|X_k| > u(n)) - \int_{|x| > u(n)} e^{tx} \, dF(x) \right]$$

where, for an event A, $\chi(A)$ denotes the indicator of A. For an integer $r \geqslant 1$ introduce the division

$$I_i^r = \left[a + \frac{i}{r}(b-a), a + \frac{i+1}{r}(b-a) \right) \qquad (i = 0, \ldots, r-1),$$

of the interval I, and let

$$s_i^r = a + \frac{2i+1}{2r}(b-a) \qquad (i = 0, \ldots, r-1),$$

denote the centre of the subinterval I_i^r. Let

$$H_n^r(t) = H_n(s_i^r) \quad \text{whenever} \quad t \in I_i^r \qquad (i = 0, \ldots, r-1),$$

be the pure jump process approximating H_n. For any continuous function $A(t)$ defined on I let $\|A\| = \sup\{|A(t)|: t \in I\}$ denote its norm in $\mathscr{C}(I)$. Now let $1 \leqslant r_1 < r_2 < \ldots$ be an infinite sequence of integers such that for each $j = 1, 2, \ldots$ r_{j+1} is divisible by r_j, and consider the corresponding subintervals $I_i^{r_j}$ with their centres $s_i^{r_j}$ $(i = 0, \ldots, r_j - 1)$, and approximating jump processes $H_n^{r_j}$ $(j = 1, 2, \ldots)$. Since H_n is continuous and the set

$$S = \bigcup_{j=1}^{\infty} \{s_i^{r_j}: i = 0, \ldots, r_j - 1\}$$

is everywhere dense in I, the distributions of $\|H_n\|$ and $\sup\{|H_n(t)|: t \in S\}$ are the same and therefore

$$\|H_n\| \leqslant \|H_n^{r_1}\| + \sum_{j=1}^{\infty} \|H_n^{r_{j+1}} - H_n^{r_j}\|.$$

Up to now the setup is identical with that of F e r n i q u e [3] when proving his inequality just used for the estimation of $\|G_n\|$. As H_n is not Gaussian, we have to deviate from that proof in the sequel and use the presently available knowledge about the pointwise behaviour of H_n.

Let $\epsilon > 0$ be arbitrary and put $\epsilon_j = \dfrac{\epsilon}{2^{j+1}}$ $(j = 0, 1, 2, \ldots)$. For $i = 0, \ldots, r_j - 1$, $j = 1, 2, \ldots$ set

$$t_i^{r_j} = s^{r_j}_{\{i \frac{r_j}{r_{j+1}}\}}$$

where $\{x\}$ denotes the smallest integer not less than x. Then

$$P(\|H_n\| > \epsilon) \leqslant P\left(\|H_n^{r_1}\| + \sum_{j=1}^{\infty} \|H_n^{r_{j+1}} - H_n^{r_j}\| > \sum_{j=0}^{\infty} \epsilon_j\right) \leqslant$$

$$\leqslant \frac{1}{\epsilon_0^2} \sum_{i=0}^{r_1-1} E(H_n(s_i^{r_1}))^2 +$$

$$+ \sum_{j=1}^{\infty} \frac{1}{\epsilon_j^2} \sum_{i=0}^{r_{j+1}-1} E(H_n(s_i^{r_{j+1}}) - H_n(t_i^{r_j}))^2$$

by the Chebyshev inequality. Define

$$R_n = \sup_{t \in I} \int_{|x| > u(n)} e^{2tx} \, dF(x)$$

and

- 148 -

$$D_n = \sup_{t \in I} \int_{|x| > u(n)} x^2 e^{2tx}\, dF(x) \leqslant \sup_{t \in I} \frac{1}{4} m^{(2)}(2t).$$

We see from the second form of H_n that

$$\mathsf{E}(H_n(s_i^{r_1}))^2 \leqslant R_n \qquad (i = 0, \ldots, r_1 - 1),$$

and

$$\mathsf{E}(H_n(s_i^{r_{j+1}}) - H_n(t_i^{r_j}))^2 \leqslant \int_{|x| > u(n)} (e^{s_i^{r_{j+1}}x} - e^{t_i^{r_j}x})^2\, dF(x) \leqslant$$

$$\leqslant \left(\frac{b-a}{2r_j}\right)^2 D_n \qquad (i = 0, \ldots, r_{j+1} - 1),$$

where the last inequality was obtained by an application of the Lagrange theorem for the integrand and the fact that $|s_i^{r_{j+1}} - t_i^{r_j}| \leqslant \frac{b-a}{2r_j}$. Consequently,

$$\mathsf{P}(\|H_n\| > \epsilon) \leqslant \frac{4}{\epsilon^2} r_1 R_n + \frac{(b-a)^2}{\epsilon^2} D_n \sum_{j=1}^{\infty} 2^{2j} \frac{r_{j+1}}{r_j^2}.$$

Apart from its divisibility property, our monotone sequence $\{r_j\}$ is arbitrary. We now choose it in such a way that the series on the right be convergent. For example, if $r_j = 5^j$, then

$$\mathsf{P}(\|H_n\| > \epsilon) \leqslant \frac{20}{\epsilon^2} R_n + \frac{20(b-a)^2}{\epsilon^2} D_n.$$

Since $R_n \to 0$ and $D_n \to 0$ by Dini's theorem, the proof is completed.

REFERENCES

[1] S. Csörgő, Limit behaviour of the empirical characteristic functions, *Ann. Prob.*, 9 (1981), (to appear).

[2] S. Csörgő, Multivariate empirical characteristic functions, *Z. Wahrscheinlichkeitstheorie verw. Geb.*, 55 (1981), to appear.

[3] X. Fernique, Régularité des trajectoires des fonctions aléatoires gaussiennes, *École d'Été de Probabilités de Saint-Flour* IV-1974, *Lecture Notes in Mathematics* 480 (1975), 1-96, Springer, Berlin – Heidelberg – New York.

[4] B. Heinkel, Relation entre théorème central-limite et loi du logarithme iteré dans les espaces de Banach, *Z. Wahrscheinlichkeitstheorie verw. Geb.*, 49 (1979), 211-220.

[5] J. Komlós – P. Major – G. Tusnády, An approximation of partial sums of independent rv's and the sample df' I, *Z. Wahrscheinlichkeitstheorie verw. Geb.*, 32 (1975), 111-131.

[6] G. Pisier, Le théorème de la limite centrale et la loi du logarithme iteré dans les espaces de Banach, *Séminaire Maurey – Schwartz*, 1975-76, exposés nos. 3 et 4.

[7] R.E. Quandt – J.B. Ramsey, Estimating mixtures of normal distributions and switching regressions, *J. Amer. Statist. Assoc.*, 73 (1978), 730-752.

[8] B. Sz.-Nagy, *Introduction to real functions and orthogonal expansions,* Oxford University Press, New York, 1965.

S. Csörgő

Bolyai Institute, Szeged University, Aradi vértanuk tere 1, H-6720 Szeged, Hungary.

COLLOQUIA MATHEMATICA SOCIETATIS JÁNOS BOLYAI

32. NONPARAMETRIC STATISTICAL INFERENCE,

BUDAPEST (HUNGARY), 1980.

AN INVARIANCE PRINCIPLE FOR N. N. EMPIRICAL DENSITY FUNCTIONS

M. CSÖRGŐ* — P. RÉVÉSZ

1. INTRODUCTION

Let $X_1, X_2, \ldots$ be a sequence of i.i.d. r.v. with density function

$$f(x) = \frac{d}{dx} F(x) = \frac{d}{dx} P(X_i \leqslant x) \qquad (i = 1, 2, \ldots).$$

Further let $X_{1:n} \leqslant X_{2:n} \leqslant \ldots \leqslant X_{n:n}$ be the ordered sample based on the random sample $X_1, X_2, \ldots, X_n$, and let $\lambda(x)$ be an arbitrary density function. For any $0 < \alpha < \beta < 1$ define the sequences

$$k_n = [n^\alpha], \quad A_n = X_{[n^\beta]:n}, \quad B_n = X_{n-[n^\beta]:n}.$$

On the interval $[A_n, B_n]$ define the (k_n, λ)-N.N. empirical density function of the sample $X_1, X_2, \ldots, X_n$ (N.N. stands for nearest neighbour) by

*Research supported by a Canada Council Senior Research Fellowship and by a Canadian National Sciences and Engineering Research Council operating grant, both held at Carleton University.

$$(1) \qquad f_n(x) = \frac{1}{nR_n(x)} \sum_{k=1}^{n} \lambda\left(\frac{x - X_k}{R_n(x)}\right) = \frac{1}{R_n(x)} \int_{-\infty}^{+\infty} \lambda\left(\frac{x - y}{R_n(x)}\right) dF_n(y),$$

where

$$F_n(y) = \frac{1}{n} \sum_{k=1}^{n} I_{(-\infty, y]}(X_k)$$

is the empirical distribution function based on the sample $X_1, X_2, \ldots, X_n$ and $R_n(x)$ is the smallest positive number for which the interval $x - \dfrac{R_n(x)}{2}, x + \dfrac{R_n(x)}{2}$ contains k_n elements of the sample $X_1, X_2, \ldots, X_n$. Further let $j = j_n = j_n(x)$ be the smallest integer for which $X_{j:n} \geqslant x - \dfrac{R_n(x)}{2}$. Note that at least one of the relations

$$X_{j:n} = x - \frac{R_n(x)}{2}, \qquad X_{j+k_n:n} = x + \frac{R_n(x)}{2}$$

holds, and clearly we have

$$X_{j-1:n} < x - \frac{R_n(x)}{2}, \qquad X_{j+1+k_n:n} > x + \frac{I_n(x)}{2}.$$

In most studies of empirical density functions the following definition is used: let the (h_n, λ)-R empirical density function of the sample $X_1, X_2, \ldots, X_n$ (R stands for R o s e n b l a t t) be

$$\rho_n(x) = \frac{1}{nh_n} \sum_{k=1}^{n} \lambda\left(\frac{x - X_k}{h_n}\right) = \frac{1}{h_n} \int_{-\infty}^{+\infty} \lambda\left(\frac{x - y}{h_n}\right) dF_n(y),$$

where $\{h_n\}$ is a decreasing sequence of positive numbers with

$$h_n \searrow 0, \qquad nh_n \to +\infty,$$

When studying ρ_n (see especially B i c k e l — R o s e n b l a t t [1]), the investigation of the process

$$\beta_n(x) = \sqrt{nh_n} \; \frac{\rho_n(x) - f(x)}{\sqrt{f(x)}}$$

is of main interest. For example, under several regularity conditions the limit distribution of $\sup\limits_{x} \beta_n(x)$ is evaluated. Strong theorems for the limit behaviour of $\sup\limits_{x} \beta_n(x)$ can be obtained as well (cf. R é v é s z [8]).

In this paper we intend to study the limit behaviour of f_n (cf. (1)). In fact we are going to consider the process

$$\theta_n(x) = \sqrt{k_n} \, \frac{f_n(x) - f(x)}{f(x)}.$$

Let

$$\bar{f}_n(x) = \frac{1}{R_n(x)} \int\limits_{-\infty}^{+\infty} \lambda\left(\frac{x-y}{R_n(x)}\right) dF(y).$$

As a first step we will prove that $\bar{f}_n(x)$ is so close to $f(x)$ that instead of the process $\theta_n(x)$ one can study the process

$$\theta_n^{(1)}(x) = \sqrt{k_n} \, \frac{f_n(x) - \bar{f}_n(x)}{f(x)} =$$

$$= \frac{\sqrt{k_n}}{R_n(x)f(x)} \int\limits_{-\infty}^{+\infty} \lambda\left(\frac{x-y}{R_n(x)}\right) d(F_n(y) - F(y)).$$

The second step of our argumentation will be a Lemma, stating that $R_n(x)$ is so close to $\dfrac{k_n}{nf(x)}$ that $\theta_n^{(1)}$ can be replaced by

$$\theta_n^{(2)}(x) = \frac{n}{\sqrt{k_n}} \int\limits_{-\infty}^{+\infty} \lambda\left(\frac{(x-y)nf(x)}{k_n}\right) d(F_n(y) - F(y)) =$$

$$= \sqrt{\frac{n}{h_n}} \int\limits_{-\infty}^{+\infty} \lambda\left(\frac{(x-y)f(x)}{h_n}\right) d(F_n(y) - F(y)),$$

where $h_n = \dfrac{k_n'}{n}$. Applying then the formula

$$F^{-1}(z) - F^{-1}(u) \approx \frac{z-u}{f(F^{-1}(z))}$$

(if $|z - u|$ is small enough) and the substitution $x = F^{-1}(z)$, $y = F^{-1}(u)$, we get that the process $\theta_n^{(2)}(F^{-1}(z))$ is close enough to

$$\theta_n^{(3)}(z) = \sqrt{\frac{n}{h_n}} \int_0^1 \lambda\left(\frac{z-u}{h_n}\right) d(F_n(F^{-1}(u)) - u).$$

Here $F_n(F^{-1}(u)) = E_n(u)$ can be considered as the empirical distribution function based on the sequence

$$U_1 = F(X_1), U_2 = F(X_2), \ldots, U_n = F(X_n)$$

of independent uniform-$(0, 1)$ r.v.

Looking at $\theta_n^{(3)}$ one observes that

$$\theta_n^{(3)}(x) = \sqrt{nh_n}\, (\rho_n(x) - \bar{\rho}_n(x)),$$

where $\rho_n(x)$ is the (h_n, λ)-R empirical density function of the uniform random sample $U_1, U_2, \ldots, U_n$ and

$$\bar{\rho}_n(x) = \frac{1}{h_n} \int_0^1 \lambda\left(\frac{x-y}{h_n}\right) dy.$$

Elementary and well-known estimations show that $\bar{\rho}_n(x)$ $(0 \leqslant x \leqslant 1)$ is so close to 1 (the density of the uniform-$(0, 1)$ distribution) that $\theta_n^{(3)}(x)$ can be replaced by

$$\theta_n^{(4)}(x) = \sqrt{nh_n}\, (\rho_n(x) - 1) = \beta_n(x).$$

Hence the above given argumentation shows that the $\theta_n(x)$ process of the sample $X_1, X_2, \ldots, X_n$ should be close to $\beta_n(F(x))$, where the process $\beta_n(\cdot)$ is obtained from the uniform $U_1, U_2, \ldots, U_n$ sample.

The aim of the present exposition is to prove that, under suitable regularity conditions on f, λ and k_n, this heuristic approach is correct. As a consequence of this fact some properties of the process θ_n will be detailed.

2. THE CONDITIONS

In this section we list the regularity conditions which will ensure the correctness of the above approach. These conditions form three groups:

(a) regularity conditions of the underlying distribution function $F(x)$;

(b) regularity conditions of the window (i.e., those of the density function λ);

(c) regularity conditions of the sequences $\{k_n\}, \{A_n\}, \{B_n\}$, i.e., conditions on α and β.

(a.i) $F(x)$ is twice differentiable on (a, b), where $-\infty \leqslant a = \sup\{x : F(x) = 0\}$, $+\infty \geqslant b = \inf\{x : F(x) = 1\}$,

(a.ii) $F' = f > 0$ on (a, b),

(a.iii)   For some $\gamma > 0$ we have

$$\sup_{a < x < b} F(x)(1 - F(x)) \frac{|f'(x)|}{f^2(x)} \leqslant \gamma,$$

(a.iv) $A = \lim_{x \downarrow a} f(x) < \infty$, $B = \lim_{x \uparrow b} f(x) < \infty$,

(a.v)   one of the following conditions hold

(a.v.α) $\min(A, B) > 0$,

(a.v.β)   if $A = 0$ (resp. $B = 0$) then f is nondecreasing (resp. nonincreasing) on an interval to the right of a (resp. to the left of b),

(a.vi) $\displaystyle\sup_{a < x < b} \frac{(F(x)(1 - F(x)))^2}{f(x)} \leqslant C$ with some $C > 0$,

(a.vii) $\displaystyle\sup_{a < x < b} |f''(x)| \leqslant C$ with some $C > 0$,

(b.i) λ is vanishing outside of $(-1, +1)$,

(b.ii) $\lambda(-x) = \lambda(x)$ $[-1 < x < +1]$,

(b.iii) $|\lambda''(x)| \leqslant C$ $(-1 < x < +1)$ with some $C > 0$,

(c)   let $\dfrac{1}{2} < \alpha = \dfrac{1}{2} + \epsilon < \beta = 1 - \delta < 1$ with

$$0 < \delta < \min \left(\frac{3 - 10\epsilon}{24}, \frac{1 - 2\epsilon}{16} \right).$$

Remark. The conditions (a.i)-(a.v) were used in our previous paper (Csörgő – Révész [3]; cf. also Csörgő – Révész [4]), where we also alluded to how wide a class of distributions satisfies these conditions. Condition (a.vi) also seems to be a weak one. As to our (b.) conditions, we mention that the optimal window of Epanechnikov [5] and also the uniform window satisfy these conditions. On condition (c) we note that about α it says that $\frac{1}{2} < \alpha < \frac{4}{5}$. This restriction, which is common to all density estimation studies, does not allow the optimal $\frac{4}{5}$ value of α.

3. SOME LEMMAS

Lemma A (Csörgő – Révész [3], Lemma 1). *For every pair* $y_1, y_2 \in (0, 1)$ *we have*

$$\frac{f(F^{-1}(y_1))}{f(F^{-1}(y_2))} \leqslant \left(\frac{y_1 \vee y_2}{y_1 \wedge y_2} \frac{1 - (y_1 \wedge y_2)}{1 - (y_1 \vee y_2)} \right)^{\gamma}.$$

provided conditions (a.i)-(a.iii) *hold.*

Lemma B (Csörgő – Révész [3], Theorem 5). *Put*

$$J_n(j) = \sqrt{n} f \left(F^{-1} \left(\frac{j}{n} \right) \right) \left(X_{j:n} - F^{-1} \left(\frac{j}{n} \right) \right).$$

Then, under conditions (a.i)-(a.v) *we have*

$$\limsup_{n \to \infty} \sqrt{\frac{2}{\log \log n}} \sup_{1 \leqslant j \leqslant n-1} |J_n(j)| = 1 \quad \textit{almost surely.}$$

Lemma C (Smirnov [10], Chung [2]). *For a continuous distribution* F *we have*

$$\limsup_{n \to \infty} \sup_{1 \leqslant j \leqslant n} \sqrt{\frac{2n}{\log \log n}} \left| F(X_{j:n}) - \frac{j}{n} \right| = 1 \quad \textit{almost surely.}$$

Lemma D (Csörgő – Révész [3], Theorem 6; Csörgő – Révész [11], Theorem 1.15.2). *Given that* α *(of* $k_n) \in \left(\frac{1}{2}, \frac{4}{5} \right)$, *then* *under the conditions* (a.i)−(a.v) *we have*

$$\lim_{n \to \infty} \sup_{1 \leqslant j \leqslant n - k_n} \left(\frac{2k_n}{n} \log \frac{n}{k_n}\right)^{-\frac{1}{2}} |J_n(j + k_n) - J_n(j)| = 1 \quad a.s.$$

Lemma 1. *Let* $\rho < \min \left(\frac{\alpha}{2}, \beta - \alpha\right)$ *with* $\alpha \in \left(\frac{1}{2}, \frac{4}{5}\right)$. *Then, given the conditions* (a.i)-(a.v) *we have*

$$\lim_{n \to \infty} n^\rho \sup_{A_n \leqslant x \leqslant B_n} \left| \frac{nf(x)}{k_n} R_n(x) - 1 \right| = 0 \quad almost \ surely.$$

Proof. Let $j = j_n(x)$ and $k = k_n$. Then we have

$$\frac{nf(x)}{k} (X_{j+k:n} - X_{j:n}) \leqslant \frac{nf(x)}{k} R_n(x) \leqslant$$

$$\leqslant \frac{nf(x)}{k} (X_{j+k+1:n} - X_{j-1:n}).$$

Consider, for example, the lower estimation

$$\frac{nf(x)}{k} (X_{j+k:n} - X_{j:n}) =$$

$$= \frac{nf(x)}{k} \left[\left(X_{j+k:n} - F^{-1} \left(\frac{j+k}{n}\right)\right) - \right.$$

$$\left. - \left(X_{j:n} - F^{-1} \left(\frac{j}{n}\right)\right) + \left(F^{-1} \left(\frac{j+k}{n}\right) - F^{-1} \left(\frac{j}{n}\right)\right) \right] =$$

$$= \frac{1}{n^\epsilon} \left[\frac{f(x)}{f\left(F^{-1} \left(\frac{j+k}{n}\right)\right)} J_n(j + k) - \frac{f(x)}{f\left(F^{-1} \left(\frac{j}{n}\right)\right)} J_n(j) \right] +$$

$$+ \frac{f(x)}{f(F^{-1}(\theta_n))},$$

where $\frac{j}{n} \leqslant \theta_n \leqslant \frac{j+k}{n}$, and $\epsilon = \alpha - \frac{1}{2}$. By Lemma A the last term of this expression can be estimated as follows:

$$\frac{f(x)}{f(F^{-1}(\theta_n))} = \frac{f(F^{-1}(F(x)))}{f(F^{-1}(\theta_n))} \leqslant$$

$$\leqslant \left(\frac{F(x) \vee \theta_n}{F(x) \wedge \theta_n} \, \frac{1 - (F(x) \wedge \theta_n)}{1 - (F(x) \vee \theta_n)} \right)^{\gamma} \leqslant$$

$$\leqslant \left(\frac{F(X_{j+k:n}) \vee \frac{j+k}{n}}{F(X_{j:n}) \wedge \frac{j}{n}} \, \frac{1 - \left(F(X_{j:n}) \wedge \frac{j}{n} \right)}{1 - \left(F(X_{j+k:n}) \vee \frac{j+k}{n} \right)} \right)^{\gamma}.$$

Since by Lemma A a similar estimation can be also obtained for $\dfrac{f(F^{-1}(\theta_n))}{f(x)}$, by Lemma C one gets

$$\lim_{n \to \infty} n^{\tau} \sup_{A_n \leqslant x \leqslant B_n} \left| \frac{f(x)}{f(F^{-1}(\theta_n))} - 1 \right| = 0 \quad \text{almost surely,}$$

where $\tau < \dfrac{1}{2} - \epsilon - \delta = \beta - \alpha$.

The factors $\dfrac{f(x)}{f\left(F^{-1}\left(\frac{j+k}{n}\right)\right)}$, $\dfrac{f(x)}{f\left(F^{-1}\left(\frac{j}{n}\right)\right)}$ can be estimated similarly. Hence by Lemmas C and D we have Lemma 1.

Lemma 2. *Given conditions* (a.i)-(a.vi) *and* (c) *we have*

$$\lim_{n \to \infty} \sup_{A_n \leqslant x \leqslant B_n} \sqrt{k_n} \, \frac{R_n^2(x)}{f(x)} = 0 \quad \text{almost surely,}$$

$$\lim_{n \to \infty} \sup_{A_n \leqslant x \leqslant B_n} \frac{R_n(x)}{f^3(x)} = 0 \quad \text{almost surely}$$

and

$$\lim_{n \to \infty} \sup_{A_n \leqslant x \leqslant B_n} \frac{k_n}{n f^4(x)} = 0 \quad \text{almost surely.}$$

Proof. By Lemma C the inequality

$$A_n \leqslant x \leqslant B_n$$

implies

$$C n^{\beta - 1} \leqslant F(x) \leqslant 1 - C n^{\beta - 1}$$

with probability 1 for any $C > 1$ if n is big enough. Now, applying condition (a.vi), our statements follow from Lemma 1. In the proof of the first equation the condition $\delta < \dfrac{3 - 10\epsilon}{24}$, while in the proof of the second and third statements the condition $\delta < \dfrac{1 - 2\epsilon}{16}$ should be utilized.

Lemma 3. *Given conditions* (a.i)-(a.vii), (b.i)-(b.iii) *and* (c) *we have*

$$\lim_{n \to \infty} \sup_{A_n \leqslant x \leqslant B_n} |\theta_n(x) - \theta_n^{(1)}(x)| = 0 \quad \textit{almost surely.}$$

Proof. In order to prove our statement we have to prove that

$$\lim_{n \to \infty} \sqrt{k_n} \sup_{A_n \leqslant x \leqslant B_n} \left| \frac{1}{f(x) R_n(x)} \int_{-\infty}^{+\infty} \lambda\left(\frac{x - y}{R_n(x)}\right) dF(y) - 1 \right| = 0$$

$$\text{almost surely.}$$

Since $A_n \leqslant x \leqslant B_n$ implies $a < x - R_n(x) < x + R_n(x) < b$ (with probability 1 if n is big enough), we have

$$\frac{1}{R_n(x)} \int_{-\infty}^{+\infty} \lambda\left(\frac{x - y}{R_n(x)}\right) dF(y) = \int_{-1}^{+1} \lambda(u) f(x - u R_n(x)) \, du =$$

$$= f(x) + \frac{R_n^2(x)}{2} \int_{-1}^{+1} u^2 \lambda(u) f''(x + \theta_n(x, u) u) \, du,$$

where $|\theta_n(x, u)| \leqslant R_n(x)$. Now our statement follows from Lemma 2.

Lemma 4. *Given* (a.i)-(a.vii), (b.i)-(b.iii) *and* (c) *we have*

$$\lim_{n \to \infty} \sup_{A_n \leqslant x \leqslant B_n} |\theta_n^{(1)}(x) - \theta_n^{(2)}(x)| = 0 \quad \textit{almost surely.}$$

Proof. Let

$$\frac{R_n(x) n f(x)}{k_n} = \frac{1}{1 + \xi_n(x)} \quad \text{and} \quad \frac{k_n}{n} = h_n = h,$$

where by Lemma 1

$$\lim_{n \to \infty} n^\delta \sup_{A_n \leqslant x \leqslant B_n} |\xi_n(x)| = 0 \quad \text{almost surely.}$$

$$- 159 -$$

In addition to (b.i)-(b.iii) in this proof we also assume that $\lambda(x)$ is absolutely continuous on $(-\infty, +\infty)$, which is then equivalent to the condition $\lambda(-1) = \lambda(+1) = 0$. In the case when this extra condition does not hold we have to take special care of those y's (given x and $\xi_n(x)$) for which one of the numbers $(1 + \xi_n(x))f(x)\frac{x-y}{h}$ and $f(x)\frac{x-y}{h}$ belongs to the interval $(-1, +1)$ but the other one does not. The proof without this extra condition is much longer but it does not require any further ideas. All we have to realize is that the Lebesgue measure of the trouble making y's is small (uniformly in $x \in (A_n, B_n)$) with probability 1 if n is big enough.

Now consider

$$|\theta_n^{(1)}(x) - \theta_n^{(2)}(x)| = \left| (1 + \xi_n(x)) \times \right.$$

$$\times \sqrt{\frac{n}{h}} \int_{-\infty}^{+\infty} \lambda\left((1 + \xi_n(x))f(x)\frac{x-y}{h}\right) d(F_n(y) - F(y)) -$$

$$\left. - \sqrt{\frac{n}{h}} \int_{-\infty}^{+\infty} \lambda\left(f(x)\frac{x-y}{h}\right) d(F_n(y) - F(y)) \right| \leqslant$$

$$\leqslant \left| \xi_n(x) \sqrt{\frac{n}{h}} \int_{-\infty}^{+\infty} \lambda\left((1 + \xi_n(x))f(x)\frac{x-y}{h}\right) d(F_n(y) - F(y)) \right| +$$

$$+ \left| \sqrt{\frac{n}{h}} \int_{-\infty}^{+\infty} \left[\lambda\left((1 + \xi_n(x))f(x)\frac{x-y}{h}\right) - \right. \right.$$

$$\left. \left. - \lambda\left(f(x)\frac{x-y}{h}\right) \right] d(F_n(y) - F(y)) \right| = \mathcal{L}_1 + \mathcal{L}_2 .$$

As to $\mathcal{L}_2$, by the mean value theorem and by Lemma C we get

$$\mathcal{L}_2 = \left| \sqrt{\frac{n}{h}} \int_{-\infty}^{+\infty} (F_n(y) - F(y)) \times \right.$$

$$\times \left[-(1 + \xi_n(x))\frac{f(x)}{h} \lambda'\left((1 + \xi_n(x))f(x)\frac{x-y}{h}\right) + \right.$$

$$\left. \left. + \frac{f(x)}{h} \lambda'\left(f(x)\frac{x-y}{h}\right) \right] dy \right| \leqslant$$

$$\leqslant \left| \sqrt{\frac{n}{h}} \, \xi_n(x) \frac{f(x)}{h} \int\limits_{-\infty}^{+\infty} (F_n(y) - F(y)) \times \right.$$

$$\left. \times \lambda'\!\left((1 + \xi_n(x))f(x)\frac{x-y}{h}\right) dy \right| +$$

$$+ \left| \sqrt{\frac{n}{h}} \int\limits_{-\infty}^{+\infty} (F_n(y) - F(y)) \times \right.$$

$$\left. \times \left[\lambda'\!\left((1 + \xi_n(x))f(x)\frac{x-y}{h}\right) - \lambda'\!\left(f(x)\frac{x-y}{h}\right) \right] dy \right| \leqslant$$

$$\leqslant O\!\left(\frac{\xi_n(x)}{\sqrt{h_n}}\right) \to 0 \quad \text{almost surely} \qquad (n \to \infty).$$

The term $\mathscr{L}_1$ can be estimated similarly.

Lemma 5. *Given* (a.i)-(a.vii), (b.i)-(b.iii) *and* (c) *we have*

$$\lim_{n \to \infty} \sup_{A_n \leqslant x \leqslant B_n} |\theta_n^{(2)}(x) - \theta_n^{(3)}(F(x))| = 0 \quad \text{almost surely}.$$

Proof. Because of the very same reason as in Lemma 4 we give the proof only in the case when λ is absolutely continuous on $(-\infty, +\infty)$. Then

$$|\theta_n^{(2)}(x) - \theta_n^{(3)}(x)| = \left| \sqrt{\frac{n}{h}} \int\limits_{-\infty}^{+\infty} \left[\lambda\!\left(\frac{x-y}{h} f(x)\right) - \right. \right.$$

$$\left. \left. - \lambda\!\left(\frac{F(x) - F(y)}{h}\right) \right] d(F_n(y) - F(y)) \right| =$$

$$= \left| \sqrt{\frac{n}{h}} \int\limits_{-\infty}^{+\infty} (F_n(y) - F(y)) \times \right.$$

$$\left. \times \left[-\frac{f(x)}{h} \lambda'\!\left(\frac{x-y}{h} f(x)\right) + \frac{f(x)}{h} \lambda'\!\left(\frac{F(x) - F(y)}{h}\right) \right] dy \right| \leqslant$$

$$\leqslant \left| \sqrt{\frac{n}{h}} \int\limits_{x-\frac{h}{f(x)}}^{x+\frac{h}{f(x)}} (F_n(y) - F(y)) \lambda'\!\left(\frac{x-y}{h} f(x)\right) \frac{f(y) - f(x)}{h} \, dy \right| +$$

$$+ \left| \sqrt{\frac{n}{h}} \int_{x-\frac{h}{f(x)}}^{x+\frac{h}{f(x)}} (F_n(y) - F(y)) \frac{f(y)}{h} \times \right.$$

$$\left. \times \left[\lambda'\left(\frac{F(x) - F(y)}{h}\right) - \lambda'\left(\frac{x-y}{h} f(x)\right) \right] dy \right|.$$

Since

$$\left| \frac{f(y) - f(x)}{h} \right| \leqslant \frac{C}{f(x)} \qquad \left(x - \frac{h}{f(x)} \leqslant y \leqslant x + \frac{h}{f(x)} \right)$$

and

$$\left| \frac{F(x) - F(y)}{h} - \frac{x-y}{h} f(x) \right| \leqslant C \frac{h}{f^2(x)}$$

we have our statement by Lemma 2.

Lemma 6. *Assume only conditions* (b.i)-(b.iii) *and let* $\{h_n\}$ *be a sequence of positive numbers with* $h_n \to 0$. *Then*

$$\sup_{h_n < x < 1 - h_n} \left| \frac{1}{h_n} \int_{-\infty}^{+\infty} \lambda\left(\frac{x-y}{h_n}\right) e(y) \, dy - 1 \right| = O(h_n^2),$$

where

$$e(u) = \begin{cases} 1 & if \quad 0 \leqslant u \leqslant 1, \\ 0 & otherwise. \end{cases}$$

The proof of this lemma is trivial and well-known and it will be omitted.

4. THE MAIN RESULTS

The above proved lemmas immediately imply the following

Theorem 1. *Let* $X_1, X_2, \ldots$ *be a sequence of i.i.d. r.v. with distribution function* $F(x)$ *satisfying conditions* (a.i)-(a.vii). *Further define the processes*

$$\theta_n(x) = \sqrt{k_n} \frac{f_n(x) - f(x)}{f(x)} \qquad (a < x < b)$$

and

$$\beta_n(u) = \sqrt{nh_n}\,(\rho_n(u) - 1) \qquad (0 < u < 1)$$

where

$$f_n(x) = \frac{1}{nR_n(x)} \sum_{k=1}^{n} \lambda\left(\frac{x - X_k}{R_n(x)}\right),$$

$$\rho_n(u) = \frac{1}{nh_n} \sum_{k=1}^{n} \lambda\left(\frac{u - F(X_k)}{h_n}\right),$$

$$h_n = \frac{k_n}{n},$$

$R_n(x)$ *is the "random window width" defined in the introduction and* λ *is a "window" satisfying conditions* (b.i)-(b.iii).

Then, with (A_n, B_n) *as in the introduction, we have*

$$\lim_{n \to \infty} \sup_{A_n \leqslant x \leqslant B_n} |\theta_n(x) - \beta_n(F(x))| = 0 \quad \textit{almost surely,}$$

provided that condition (c) *is satisfied.*

Remark. Since

$$A_n \to a, \quad B_n \to b \quad \text{almost surely} \qquad (n \to \infty),$$

we can say that our Theorem 1 states that the limit behaviour of $\theta_n(x)$ *on its whole support* (even if it is the whole real line) is the same as that of $\beta_n(u)$ $(0 < u < 1)$.

Paying more attention to rate of convergence in the above formulated lemmas one can prove the following more useful

Theorem 2. *Under the conditions of Theorem 1 we have*

$$\lim_{n \to \infty} n^{\nu} \sup_{A_n \leqslant x \leqslant B_n} |\theta_n(x) - \beta_n(F(x))| = 0 \quad \textit{almost surely}$$

where $\nu = \nu(\alpha, \beta)$ *is a positive number depending only on* α *and* β.

In our next theorem we investigate how the process $\beta_n(u)$ can be approximated by a Gaussian process, namely by a Kiefer process. For convenience we give here the

Definition. The separable Gaussian process $\{K(x, y);\ 0 \leqslant x \leqslant 1,\ 0 \leqslant y\}$ is called a Kiefer process if

$$\mathsf{E} K(x, y) = 0 \qquad (0 \leqslant x \leqslant 1,\ 0 \leqslant y)$$

and

$$\mathsf{E} K(x_1, y_1) K(x_2, y_2) = \min(y_1, y_2)[\min(x_1, x_2) - x_1 x_2].$$

Now we prove our

Theorem 3. *Let* $Y_1, Y_2, \ldots$ *be a sequence of i.i.d. uniform-$(0, 1)$ r.v. and let* $E_n(x)$ $(0 \leqslant x \leqslant 1;\ n = 1, 2, \ldots)$ *be the empirical distribution function based on the sample* $Y_1, Y_2, \ldots, Y_n$. *Assume that* $\lambda(x)$ *is a density function satisfying conditions* (b.i)-(b.iii). *Let*

$$\Delta_n(x) = n \int_0^1 \lambda\left(\frac{x-y}{h_n}\right) d(E_n(y) - y) - \int_0^1 \lambda\left(\frac{x-y}{h_n}\right) dK(y, n)$$

where $h_n = \dfrac{1}{n^\gamma}$ $(0 < \gamma < 1)$.

Then

$$\lim_{n \to \infty} \frac{1}{\sqrt{nh_n}} \sup_{0 < x < 1} |\Delta_n(x)| = 0 \quad \text{almost surely.}$$

Proof. By the Komlós – Major – Tusnády [6] approximation theorem we have

$$\sup_{0 < x < 1} |\Delta_n(x)| =$$

$$= \sup_{0 < x < 1} \frac{1}{h_n} \left| \int_{\max(0, x-h)}^{\min(1, x+h)} [n(E_n(y) - y) - K(y, n)] \times \right.$$

$$\left. \times \lambda'\left(\frac{x-y}{h_n}\right) dy \right| = O(\log^2 n),$$

and hence the theorem.

This theorem together with Lemma 6 clearly implies

Theorem 3*. *Let* $Y_1, Y_2, \ldots$ *be a sequence of i.i.d. uniform-$(0, 1)$ r.v. and let* $\rho_n(x)$ *be the* (h_n, λ)-*R empirical density function of* $Y_1, Y_2, \ldots, Y_n$. *Assume that* λ *satisfies conditions* (b.i)-(b.iii) *and* $h_n = \dfrac{1}{n^\epsilon}$ *with* $\dfrac{1}{5} < \epsilon < 1$. *Then*

$$\lim_{n \to \infty} \sup_{h_n < x < 1 - h_n} \left| \sqrt{nh_n} \, (\rho_n(x) - 1) - \frac{1}{\sqrt{nh_n}} \int_0^1 \lambda \left(\frac{x - y}{h_n} \right) dK(y, n) \right| = 0.$$

Theorems 3 and 3* are also true with a rate n^ν $(0 < \nu = \nu(\epsilon))$ (cf. Theorem 2). Theorems 2 and 3* and Lemma C imply:

Theorem 4. *Under the conditions of Theorem 1 we have*

$$\lim_{n \to \infty} n^\nu \sup_{A_n \leqslant x \leqslant B_n} \left| \theta_n(x) - \frac{1}{\sqrt{nh_n}} \int_0^1 \lambda \left(\frac{F(x) - y}{h_n} \right) dK(y, n) \right| = 0$$

almost surely.

It will be convenient in the sequel to view the Kiefer process in the form

$$K(y, n) = W(y, n) - yW(1, n)$$

where $W(\cdot, \cdot)$ is a Wiener sheet. It is immediately clear that

$$W(1, n) \frac{1}{\sqrt{nh_n}} \int_0^1 \lambda \left(\frac{F(x) - y}{h_n} \right) dy \to 0 \quad \text{almost surely} \quad (n \to \infty).$$

This fact implies

Theorem 4*. *Under the conditions of Theorem 1 we have*

$$\lim_{n \to \infty} n^\nu \sup_{A_n \leqslant x \leqslant B_n} \left| \theta_n(x) - \frac{1}{\sqrt{nh_n}} \int_0^1 \lambda \left(\frac{F(x) - y}{h_n} \right) dW(y, n) \right| = 0.$$

5. A CONSEQUENCE

At first we intend to investigate the Gaussian process

$$\Gamma_n(u) = \frac{1}{\sqrt{nh_n}} \int_0^1 \lambda\left(\frac{u-y}{h_n}\right) dW(y,n).$$

Since

$$\{W(Ty,n); \ y \geqslant 0, \ n = 1, 2, \ldots\} \overset{\mathscr{D}}{=}$$

$$\overset{\mathscr{D}}{=} \{\sqrt{T}\, W(y,n); \ y \geqslant 0, \ n = 1, 2, \ldots\} \qquad (T \geqslant 0)$$

and

$$\left\{\frac{1}{\sqrt{n}}\, W(y,n); \ y \geqslant 0\right\} \overset{\mathscr{D}}{=} \{W(y); \ y \geqslant 0\} \qquad (n = 1, 2, \ldots),$$

one can see that

$$\{\Gamma_n(u); \ h_n < u < 1 - h_n, \ n = 1, 2, \ldots\} \overset{\mathscr{D}}{=}$$

$$\overset{\mathscr{D}}{=} \left\{\frac{1}{\sqrt{n}} \int_{v-1}^{v+1} \lambda(v-z)\, dW(z,n); \ 1 < v = \frac{1}{h_n} - 1, \ n = 1, 2, \ldots\right\}$$

and for any $n = 1, 2, \ldots$

$$\left\{\frac{1}{\sqrt{n}} \int_{v-1}^{v+1} \lambda(v-z)\, dW(z,n); \ v > 1\right\} \overset{\mathscr{D}}{=}$$

$$\overset{\mathscr{D}}{=} \left\{\int_{v-1}^{v+1} \lambda(v-z)\, dW(z); \ v > 1\right\}.$$

These relationships show that instead of the properties of the process $\Gamma_n(u)$ one can study the properties of the Gaussian processes

$$\{g_n^{(1)}(v); \ v > 1, \ n = 1, 2, \ldots\} =$$

$$= \left\{\frac{1}{\sqrt{n}} \int_{v-1}^{v+1} \lambda(v-z)\, dW(z,n); \ v > 1, \ n = 1, 2, \ldots\right\}$$

and

$$\{g^{(2)}(v);\ v>1\} = \Big\{ \int_{v-1}^{v+1} \lambda(v-z)\, dW(z);\ v>1 \Big\}.$$

Theorem 4* implies that many of the properties of the processes $g_n^{(1)}(v)$ and $g^{(2)}(v)$ are inherited by the process $\{\theta_n(x);\ A_n \leqslant x \leqslant B_n,\ n=1,2,\ldots\}$.

Here we formulate a simple lemma and two known results for $g_n^{(1)}$ and $g^{(2)}$.

Lemma 7. *Let*

$$\Lambda^2 = \int_{-1}^{+1} \lambda^2(u)\, du.$$

Then the covariance function of the stationary Gaussian process $g^{(2)}$ *around* 0 *can be given as follows*

$$\mathsf{E}g^{(2)}(v)g^{(2)}(v+\Delta v) \approx$$

$$\approx \begin{cases} \Lambda^2 - \Delta v \lambda^2(1) & \text{if} \quad \lambda^2(1)>0, \\[2mm] \Lambda^2 + (\Delta v)^2 \dfrac{1}{2}\displaystyle\int_{-1}^{+1} \lambda(u)\lambda''(u)\, du & \text{if} \quad \lambda^2(1)=0, \end{cases}$$

for any $v>1$ *as* $\Delta v \to 0$.

Proof is trivial.

Theorem A (Qualls – Watanabe [7], Bickel – Rosenblatt [1]). *We have*

$$\lim_{T\to\infty} \mathsf{P}\Big\{\sqrt{2\log T}\Big[\sup_{1<v<T} \frac{1}{\Lambda}|g^{(2)}(v)| - b(T)\Big] < z\Big\} =$$

$$= \exp(-2e^{-z}) \qquad (-\infty < z < +\infty)$$

where

$$b(T) = \sqrt{2\log T} + \frac{1}{\sqrt{2\log T}} \times$$

$$\times \Big\{\Big(\frac{1}{\alpha} - \frac{1}{2}\Big)\log\log T + \Big(\log \frac{1}{\sqrt{2\pi}}\Big) C_\alpha^{\frac{1}{\alpha}} H_\alpha 2^{\frac{2-\alpha}{2\alpha}}\Big\},$$

$$\alpha = \begin{cases} 1 & if \quad \lambda(1) > 0, \\ 2 & if \quad \lambda(1) = 0, \end{cases}$$

$$C_\alpha = \begin{cases} \dfrac{\lambda^2(1)}{\Lambda^2} & if \quad \alpha = 1, \\ \dfrac{1}{2\Lambda^2} \int\limits_{-1}^{+1} \lambda(u)\lambda''(u)\,du & if \quad \alpha = 2, \end{cases}$$

and

$$H_\alpha = \begin{cases} 1 & if \quad \alpha = 1, \\ \dfrac{1}{\sqrt{\pi}} & if \quad \alpha = 2. \end{cases}$$

Theorem B (R é v é s z [9]). *We have*

$$\lim_{n\to\infty} \left(\sup_{1 < v < T} \frac{g_n^{(1)}(v)}{\Lambda\sqrt{n}} - \sqrt{2\log T} \right) = 0 \quad almost\ surely.$$

Now we give the main result of this section.

Consequence. *Assume that all the conditions of Theorem 1 are satisfied. Then*

$$\lim_{n\to\infty} P\left\{ \sqrt{2\log\frac{1}{h_n}} \left[\sup_{A_n \leqslant x \leqslant B_n} \frac{1}{\Lambda} |\theta_n(x)| - b\left(\frac{1}{h_n}\right) \right] < z \right\} =$$
$$= \exp(-2e^{-z})$$

and

$$\lim_{n\to\infty} \left(\sup_{A_n \leqslant x \leqslant B_n} \frac{1}{\Lambda} |\theta_n(x)| - \sqrt{2\log\frac{1}{h_n}} \right) = 0 \quad almost\ surely,$$

where $b(\cdot)$ *is as in Theorem* A.

REFERENCES

[1]   P.J. Bickel – M. Rosenblatt, On some global measures of the deviations of density function estimates, *Ann. Statist.*, 1 (1973), 1071-1095.

[2]   K.L. Chung, An estimate concerning the Kolmogorov limit distribution, *Trans. Amer. Math. Soc.*, 67 (1949), 36-50.

[3]   M. Csörgő – P. Révész, Strong approximations of the quantile process, *Ann. Statist.*, 6 (1978), 882-894.

[4]   M. Csörgő – P. Révész, Quantile processes and sums of weighted spacings for composite goodness-of-fit, I and II, to appear.

[5]   V.A. Epanechnikov, Nonparametric estimates of a multivariate probability density, *Theor. Probability Appl.*, 14 (1969), 153-158.

[6]   J. Komlós – P. Major – G. Tusnády, An approximation of partial sums of independent RV's and the sample DF, I. *Z. Wahrscheinlichkeitstheorie und verw. Gebiete*, 32 (1975), 111-131.

[7]   C. Qualls – H. Watanabe, Asymptotic properties of Gaussian processes, *Ann. Math. Stat.*, 43 (1972), 580-596.

[8]   P. Révész, A strong law of the empirical density function, *Periodica Math. Hung.*, 9 (1978), 317-324.

[9]   P. Révész, How to characterize the asymptotic properties of a stochastic process by four classes of deterministic curves? *Carleton Mathematical Lecture Note.* 1980.

[10]   N.V. Smirnov, Approximate laws of distribution of random variables from empirical data, *Uspehi Mat. Nauk.*, 10 (1944), 179-206, (in Russian).

[11]   M. Csörgő – P. Révész, *Strong approximations in probability and statistics,* Akadémiai Kiadó, Budapest, 1981.

M. Csörgő

Department of Mathematics and Statistics, Carleton University, Ottawa, Canada K1S 5B6.

P. Révész

Mathematical Institute of the Hungarian Academy of Sciences, Reáltanoda u. 13-15, Budapest, Hungary.

COLLOQUIA MATHEMATICA SOCIETATIS JÁNOS BOLYAI

32. NONPARAMETRIC STATISTICAL INFERENCE,

BUDAPEST (HUNGARY), 1980.

PARAMETRIC AND NONPARAMETRIC MODELS WITH SPECIAL SCHEMES OF STOCHASTIC DEPENDENCE

D. DABROWSKA

1. INTRODUCTION

General postulates concerning measurement of such aspects of univariate distributions as location, scale, skewness and kurtosis have been formulated by B i c k e l and L e h m a n n [1] in their study on descriptive statistics for nonparametric models. According to B i c k e l and L e h m a n n [1], particular attention should be paid to models with natural parameters, more precisely to parametric and nonparametric models with natural parameters and their nonparametric neighbourhoods. Typically, for any given model we should start the description of the chosen aspect with introducing a quasi-ordering, which indicates cases, when one distribution possesses the considered aspect more strongly than another one. A measure of this aspect should be then chosen among those which preserve this ordering and satisfy appropriate invariance conditions. Moreover, if the examined model contains a smaller one, with a natural parameter of the given aspect, then the measure in question should reduce to this parameter on the restricted model. Further requirements are those of robustness and/or effectivity.

It is not so easy to follow Bickel and Lehmann's postulates when dealing with measurement of stochastic dependence of bivariate (multivariate) distributions. In many cases it is not possible to find orderings and measures which provide both qualitative and quantitaive information about the specific scheme of stochastic dependence. In particular numerical i.e. real valued measures of stochastic dependence seem not to be satisfactory and it is therefore advisable to admit measures taking on values from sets having richer structure. Moreover, it is not clear what is meant by a model with a natural parameter of stochastic dependence.

Let us point out that it is necessary to make a distinction among three general types of stochastic dependence, namely the global, monotone and linear stochastic dependence of a random variable X on a random variable Y. Each of them might be considered as a generalization of an appropriate functional relationship between X and Y. Intuitively, global dependence of X and Y may be stronger or weaker in the sense that realizations of possible values of Y may have a more or a less significant influence on realizations of possible values of X. In the case of the so-called monotone or signed dependence, we usually say that in some sense large values of Y tend to associate either with large or with small values of X and this tendency is described by some inequalities taking into the account the joint distribution of (X, Y) or some less informative characteristics (see Yanagimoto [8]). Surprisingly enough, no meaningful formalizations of linear stochastic dependence have been proposed apart from the one corresponding to models with linear regression function $E(X \mid Y = y)$.

In Sections 3 and 4 of this paper we indicate some nonparametric models with natural real and function valued parameters of stochastic dependence. Parameters is question take into the account such characteristics of bivariate random variables (X, Y) as the marginal distributions and the distribution of the conditional expectation $E(X \mid Y)$ or the form of the regression function $E(X \mid Y = y)$. In Section 2 we recall some well known results on Frechet bounds.

2. FRECHET BOUNDS

Assume that X and Y are two univariate random variables with distribution functions F_X and F_Y respectively. Then

$$F^+_{X,Y}(x, y) = \min\{F_X(x), F_Y(y)\}$$

and

$$F^-_{X,Y}(x, y) = \max\{F_X(x) + F_Y(y) - 1, 0\}$$

are called the upper and the lower Frechet bounds of the family of all bivariate random variables (X, Y) with marginals distributed as X and Y (see Mardia [7]). They are bounds in the sense that for any bivariate random variable (X, Y) with marginal distributions equal to those of X and Y we have

$$\forall x, y \in R \qquad F^-_{X,Y}(x, y) \leqslant F_{X,Y}(x, y) \leqslant F^+_{X,Y}(x, y)$$

where $F_{X,Y}$ is the joint distribution function of (X, Y).

Moreover, $F^+_{X,Y}$ and $F^-_{X,Y}$ should be interpreted as distributions under which the positive and respectively negative stochastic dependence between X and Y is strongest among all bivariate random variables (X, Y) with marginals distributed as X and Y. $F^+_{X,Y}$ assigns probability 1 to the set

$$B^+_{X,Y} = \{(x, y): x = F^{-1}_X(p), \; y = F^{-1}_Y(p), \; p \in (0, 1)\}$$

and $F^-_{X,Y}$ assigns probability 1 to the set

$$B^-_{X,Y} = \{(x, y): x = F^{-1}_X(p), \; y = F^{-1}_Y(1-p), \; p \in (0, 1)\}$$

where $F^{-1}_X(p) = \inf\{t: F_X(t) > p\}$ is the greatest p-quantile of X and F^{-1}_Y is defined analogously. Points (x, y) belonging to $B^+_{X,Y}$ (respectively $B^-_{X,Y}$) constitute a nondecreasing (nonincreasing) relation in the sense of Konijn [4] i.e. for any $(x_1, y_1), (x_2, y_2) \in B^+_{X,Y}$ $x_1 \leqslant x_2$ iff $y_1 \leqslant y_2$ (for any $(x_1, y_1), (x_2, y_2) \in B^-_{X,Y}$ $x_1 \leqslant x_2$ iff $y_1 \geqslant y_2$). In particular, if the graph of $B^+_{X,Y}$ ($B^-_{X,Y}$) corresponds to an a.s. nondecreasing (nonincreasing) function i.e.

(2.1.a) $F_X = F_{f^+(Y)}$

(2.1.b) $F_X = F_{-f^-(Y)}$

for some a.s. nondecreasing function f^+ (f^-) then the joint distribution function of (X, Y) is equal to $F^+_{X,Y}$ $(F^-_{X,Y})$ if and only if $X = f^+(Y)$ a.s. $(X = -f^-(Y)$ a.s.)

In the sequel BM^+ (respectively BM^-) will denote the family of all bivariate random variables (X, Y) for which (2.1.a) ((2.1.b)) is satisfied by some a.s. increasing function f^+ (f^-). Similarly BL^+ (respectively BL^-) will stand for the family of all bivariate random variables (X, Y) for which (2.1.a) ((2.1.b)) holds for some a.s. increasing linear function f^+ (f^-). Note that if the marginal distributions of (X, Y) are continuous then $(X, Y) \in BM^+ \cap BM^-$.

3. NATURAL REAL VALUED PARAMETERS OF STOCHASTIC DEPENDENCE IN THE MODEL L AND ITS SUBMODELS

A rough characterization of the three types of stochastic dependence mentioned in Section 1 may be obtained by taking into the account the distribution of the conditional expectation $E(X|Y)$ or the form of the regression function $E(X|Y = y)$. Thus let us denote by E the family of all bivariate random variables (X, Y) for which the distribution of X is nondegenerate and EX is finite. Further, let $MR = MR^+ \cup MR^- \cup MR^0$ be the family of all bivariate random variables $(X, Y) \in E$ for which the regression $E(X|Y = y)$ is either a.s. increasing (MR^+) or a.s. decreasing (MR^-) or a.s. constant (MR^0). Similarly, let $LR = LR^+ \cup LR^- \cup LR^0$ be the family of all $(X, Y) \in E$ for which $E(X|Y = y)$ is either a.s. increasing linear (LR^+) or a.s. decreasing linear (LR^-) or a.s. constant $(LR^0 = MR^0)$.

Let us consider the family L of all $(X, Y) \in E$ such that for some $a(X, Y) \geqslant 0$ and $b(X, Y) \in R$ the distribution of $E(X|Y)$ is equal to that of $a(X, Y)X + b(X, Y)$. If $(X, Y) \in L$ then

(i) $a(X, Y) \in [0, 1]$, $b(X, Y) = [1 - a(X, Y)]EX$;

(ii) $a(X, Y) = 0$ if and only if $E(X|Y) = EX$ a.s.;

(iii) $a(X, Y) = 1$ if and only if $\mathsf{E}(X \mid Y) = X$ a.s.;

(iv) if $\mathsf{E} X^2 < \infty$ then $a(X, Y) = \eta(X, Y)$ where η is the correlation ratio;

(v) if l is a nonconstant linear function and g is a one-to-one measurable function then $(l(X), g(Y)) \in L$ and $a(l(X), g(Y)) = a(X, Y)$.

Further

(vi) if $(X_1, Y_1), \ldots, (X_n, Y_n) \in L$ are independent, $X_1, \ldots, X_n$ are identically distributed and if (X_0, Y_0) is a mixture of $(X_1, Y_1), \ldots, (X_n, Y_n)$ with weighting constants $\lambda_1, \ldots, \lambda_n$ then $(X_0, Y_0) \in L$ and $a(X_0, Y_0) = \sum_{i=1}^{n} \lambda_i a(X_i, Y_i)$.

The importance of the model L may be explained in the following way. Intuitively, the more is the distribution of $\mathsf{E}(X \mid Y)$ similar to that of X the stronger is the global dependence of X on Y. Two extremal cases might be easily pointed out. The first occurs when the distribution of $\mathsf{E}(X \mid Y)$ is degenerate (e.g. when X and Y are independent) while the other one takes place when $\mathsf{E}(X \mid Y)$ and X are identically distributed. The latter is equivalent to the equality $\mathsf{E}(X \mid Y) = X$ a.s. (D a b r o w s k a [2]). Apart from these two cases the distributions of $\mathsf{E}(X \mid Y)$ and X may differ considerably and usually it is not so easy to trace the similarities between these two distributions. In this context the model L is the only one contained in E where the similarity between the distribution of $\mathsf{E}(X \mid Y)$ and that of X might be fully expressed by means of a real measure. More precisely, L should be considered as a model with a natural real valued parameter of global dependence.

It is easy to see that

if $(X, Y) \in L \cap MR^+$ then $(X, Y) \in L \cap BM^+$ and

$a(X, Y) \neq 0$,

if $(X, Y) \in L \cap LR^+$ then $(X, Y) \in L \cap BL^+$ and

$a(X, Y) \neq 0$,

analogously

if $(X, Y) \in L \cap MR^-$ then $(X, Y) \in L \cap BM^-$ and

$a(X, Y) \neq 0$,

if $(X, Y) \in L \cap LR^-$ then $(X, Y) \in L \cap BL^-$ and

$a(X, Y) \neq 0$.

Monotonicity of the regression function $E(X \mid Y = y)$ belongs to the possible formalizations of signed dependence of a random variable X on a random variable Y. If $(X, Y) \in L \cap MR^+$ (respectively $(X, Y) \in L \cap MR^-$) then information about the strength of positive (negative) dependence might be obtained by comparing the graphs of $E(X \mid Y = y)$ and $f^+(y)$ $(-f^-(y))$ where f^+ (f^-) is an a.s. increasing function such that (2.1.a) ((2.1.b)) is satisfied. Namely, if $(X, Y) \in L \cap MR^+$ $((X, Y) \in L \cap MR^-)$ then the distributions of $E(X \mid Y)$ and $f^+(Y)$ $(-f^-(Y))$ differ only in location and scale, so the graph of $E(X \mid Y = y)$ coincides with that of $a(X, Y)f^+(y) + b(X, Y)$ $(-a(X, Y)f^-(y) + b(X, Y))$. As a result $L \cap MR$ is a model with a natural real valued parameter of signed dependence equal to $a(X, Y)$ when $(X, Y) \in L \cap MR^+$, $-a(X, Y)$ when $(X, Y) \in L \cap MR^-$ and 0 when $(X, Y) \in MR^0$. Similar situation occurs when $(X, Y) \in L \cap LR^+$ (respectively $(X, Y) \in L \cap LR^-$) i.e. when f^+ (f^-) is an a.s. increasing linear function. In this case $L \cap LR$ might be treated as a model with a natural real valued parameter of linear dependence eqqual to $a(X, Y)$ when $(X, Y) \in L \cap LR^+$, $-a(X, Y)$ when $(X, Y) \in L \cap LR^-$ and 0 when $(X, Y) \in L \cap LR^0$. In the sequel we shall say that X is strongly linearly dependent on Y whenever $(X, Y) \in L \cap LR^+$ (positive strong linear dependence) or $(X, Y) \in L \cap LR^-$ (negative strong linear dependence).

The properties of the model $L \cap MR$ $(L \cap LR)$ may be easily formulated on the basis of (i)-(vi). In particular, if EX^2 is finite then

$$a(X, Y) = \begin{cases} \sup \rho(X, g(Y)) & \text{if} \quad (X, Y) \in L \cap MR^+, \\ -\inf \rho(X, g(Y)) & \text{if} \quad (X, Y) \in L \cap MR^-, \end{cases}$$

where ρ denotes the correlation coefficient, while supremum and infimum are taken over the set of all nondecreasing function g such that $E(g(Y)^2) < \infty$. Moreover, if $EY^2 < \infty$ then

$$a(X, Y) = \begin{cases} \rho(X, Y) & \text{if} \quad (X, Y) \in L \cap LR^+, \\ -\rho(X, Y) & \text{if} \quad (X, Y) \in L \cap LR^-. \end{cases}$$

Further, if $(X, Y) \in L \cap MR^+$ $((X, Y) \in L \cap LR^+)$ then for any non-constant linear function l and any strictly monotone function g (non-constant linear function g)

$$(l(X), g(Y)) \in L \cap MR^+ \quad ((l(X), g(Y)) \in L \cap LR^+)$$

if and only if l and g are concordant and

$$(l(X), g(Y)) \in L \cap MR^-$$

$((l(X), g(Y) \in L \cap LR^-)$ if and only if l and g

are discordant.

The model L may be examplified by some well known parametric families of distributions. Consider for instance a family of bivariate Pareto distributions (M a r d i a [7]). The joint density function of a bivariate Pareto distribution is given by

$$h(x, y) = \frac{\gamma(\gamma + 1)(\alpha\beta)^{\gamma + 1}}{(\beta x + \alpha y - \alpha\beta)^{\gamma + 2}}$$

where $x > \alpha > 0$, $y > \beta > 0$ and $\gamma > 0$. In this case X and $\frac{\alpha}{\beta} Y$ are identically distributed and, for $\gamma > 1$, $E(X \mid Y) = \alpha + \frac{\alpha}{\beta\gamma} Y$. Accordingly, if $\gamma > 1$ then $(X, Y) \in L \cap LR$ and $a(X, Y) = \frac{1}{\gamma}$. For $\gamma \in (1, 2]$ the value of $a(X, Y)$ might be considered as a generalization of the correlation coefficient. Another example is provided by the family of bivariate t-Student distributions with n degrees of freedom (D e G r o o t [3]).

The joint density function of a bivariate t-Student distribution with n degrees of freedom is given by

$$h(x, y) = c\left\{1 + \frac{1}{n(1 - \rho^2)} \times\right.$$

$$\left. \times \left[\frac{(x - m_1)^2}{\sigma_1^2} - \frac{2\rho(x - m_1)(y - m_2)}{\sigma_1 \sigma_2} + \frac{(y - m_2)^2}{\sigma_2^2}\right]\right\}^{-\frac{n}{2} - 1}$$

where $m_1, m_2 \in R$, $\sigma_1 > 0$, $\sigma_2 > 0$, $\rho \in [-1, 1]$ and

$$c = (2\pi\sigma_1\sigma_2)^{-1}(1 - \rho^2)^{-\frac{1}{2}}.$$

Here X and $m_1 + \dfrac{\sigma_1}{\sigma_2}(Y - m_2)$ are identically distributed and, for $n > 1$, $E(X \mid Y) = m_1 + \rho \dfrac{\sigma_1}{\sigma_2}(Y - m_2)$. It follows that $(X, Y) \in L \cap LR$ and $a(X, Y) = |\rho|$. For $n = 2$ the value ρ should be treated as a generalization of the correlation coefficient.

4. A NATURAL FUNCTION VALUED PARAMETER OF SIGNED DEPENDENCE

Models which have been discussed in Section 3 share the property that information about the strength of stochastic dependence might be expressed adequately by means of a natural real valued parameter. We proceed to discuss another approach leading to function valued parameters of stochastic dependence which are also natural in a specific sense.

Among numerous formalizations of signed dependence the following one, proposed by K o w a l c z y k and P l e s z c z y ń s k a [5], deserves particular attention. Namely, let EQ^+ (EQ^-) denote the family of all $(X, Y) \in E$ such that

(4.1.a) $\forall y \in R \quad E(X \mid Y > y) \geqslant EX,$

(4.1.b) $\forall y \in R \quad E(X \mid Y > y) \leqslant EX.$

For any given y_0 the inequality (4.1.a) ((4.1.b)) characterizes the tendency of values of Y greater than y_0 to associate with possibly large (small) values of X in the sense that the expectation of the marginal distribution of the first component of (X, Y) is not greater (not less) than that of (X, Y) truncated to the set $\{(x, y)\colon y > y_0\}$. Note that the model EQ^+ (EQ^-) is an extension of $MR^+ \cup MR^0$ $(MR^- \cup MR^0)$. Moreover, EQ^+ (EQ^-) contains the family of all $(X, Y) \in E$ for which X and Y are positively (negatively) quadrant dependent.

The distributions belonging to EQ^+ (EQ^-) might be easily ordered with respect to the strength of positive (negative) dependence. Moreover, a suitable normalization of $E(X \mid Y > y_p) - EX$, where y_p is a p-quantile of Y and p ranges over $(0, 1)$, leads to a natural function valued parameter of signed dependence for the model $EQ^+ \cup EQ^-$. If the marginal distributions of (X, Y) are continuous then the parameter in question is given by

$$\mu(X, Y; p) = \begin{cases} \dfrac{E(X \mid Y > y_p) - EX}{E(X \mid X > x_p) - EX} & \text{if } E(X \mid Y > y_p) \geqslant EX, \\[2ex] \dfrac{E(X \mid Y > y_p) - EX}{EX - E(X \mid X < x_{1-p})} & \text{otherwise.} \end{cases}$$

The function $\mu(X, Y)$ might be also treated as a measure of signed dependence for the model E. In this case the lack of monotonicity is reflected by the change of sign of the function $\mu(X, Y)$ at some $p_0 \in (0, 1)$.

The general definition of the function $\mu(X, Y)$, known as the monotone dependence function, as well as its properties are given in K o w a l c z y k [6]. Let us only recall that if $(X, Y) \in BM^+$ $((X, Y) \in BM^-)$ i.e. if there exists an a.s. increasing function f such that $F_X = F_{f(Y)}$ $(F_X = F_{-f(Y)})$ then

$$\mu(X, Y) = \xi \text{ if and only if } E(X \mid Y) = \xi f(Y) + (1 - |\xi|) EX \text{ a.s.}$$

Accordingly, if $(X, Y) \in L \cap MR$ or $(X, Y) \in L \cap LR$ then $\mu(X, Y)$ reduces to the previously mentioned natural parameter of monotone or linear stochastic dependence.

If $(X, Y) \in LR$ then $\mu(X, Y)$ may be used to characterize the discrepancy of (X, Y) from the model of strong linear dependence of X

on Y. Namely, if the discrepancy of $\mu(X, Y)$ from a constant function (equal to $\rho(X, Y)$ in case the correlation coefficient exists) is rather small then the distribution of (X, Y) might be considered as belonging to a neighbourhood of the model of strong linear dependence of X on Y. To illustrate let us consider for instance the family of bivariate χ^2-distributions (M a r d i a [7]) defined as follows.

Assume that T_1, T_2 and T_3 are independent and have χ^2-distributions with n_1, n_2 and n_3 degrees of freedom. Set $X = T_1 + T_3$ and $Y = T_2 + T_3$. Then X and Y have χ^2-distributions with $n_1 + n_3$ and $n_2 + n_3$ degrees of freedom respectively and the joint density function of (X, Y) is given by

$$h \quad h(x, y) = \frac{\exp\left(-\dfrac{x + y}{2}\right)}{\prod\limits_{i=1}^{3} 2^{\frac{n_i}{2}} \Gamma\left(\dfrac{n_i}{2}\right)} I(x, y)$$

where $x, y > 0$ and

$$I(x, y) = \int\limits_{0}^{\min(x, y)} t^{\frac{n_3}{2}} (x - t)^{\frac{n_1}{2} - 1} (y - t)^{\frac{n_2}{2} - 1} e^{\frac{t}{2}} dt.$$

It can be shown that $\rho(X, Y) = n_3[(n_1 + n_3)(n_2 + n_3)]^{-\frac{1}{2}}$, $\mathsf{E}(X \mid Y) =$
$= n_1 + \left(\dfrac{n_3}{n_2 + n_3}\right) Y$ and $\mathsf{E}(Y \mid X) = n_2 + \left(\dfrac{n_3}{n_1 + n_3}\right) X$. If $n_1 = n_2$ then $(X, Y) \in L \cap LR$ and $a(X, Y) = \rho(X, Y)$. If $n_1 \neq n_2$ then both regressions are still linear but $(X, Y) \notin BL^+$. The corresponding monotone dependence function $\mu(X, Y)$ is given by

$$\mu(X, Y; p) = \frac{n_3}{n_2 + n_3} \frac{\Gamma\left(\dfrac{n_1 + n_3}{2}\right)}{\Gamma\left(\dfrac{n_2 + n_3}{2}\right)} \exp \varphi(p)$$

where

$$\varphi(p) = \frac{x_p}{2} - \frac{y_p}{2} + \frac{n_2 + n_3}{2} \ln\left(\frac{y_p}{2}\right) - \frac{n_1 + n_3}{2} \ln\left(\frac{x_p}{2}\right)$$

and $\mu(Y, X)$ is given by

$$\mu(Y, X; p) = \frac{n_3}{n_1 + n_3} \frac{\Gamma\left(\dfrac{n_2 + n_3}{2}\right)}{\Gamma\left(\dfrac{n_1 + n_3}{2}\right)} \exp\left(-\varphi(p)\right)$$

If $n_2 > n_1$ then $\mu(X, Y)$ is positively decreasing which means that for $p \in \left(0, \dfrac{1}{2}\right)$ values of Y smaller than y_p tend to associate with small values of X and this tendency is stronger than that of values of Y greater than y_{1-p} to associate with large values of X. On the other hand, $\mu(Y, X)$ is positively increasing thus for $p \in \left(0, \dfrac{1}{2}\right)$ the values of X smaller than x_p tend to coappear with small values of Y and this tendency is weaker than that of values of X greater than x_{1-p} to coappear with large values of Y. For any $n_1 \neq n_2$ the discrepancies of $\mu(X, Y)$ and $\mu(Y, X)$ from the constant function $\tilde{\mu} \equiv \rho(X, Y)$ are rather small in spite of the fact that the upper Frechet bound $F_{X,Y}^+$ is concentrated on an increasing function which differs considerably from a linear function.

It is clearly possible to indicate cases of $(X, Y) \in LR$ for which $\mu(X, Y)$ and $\mu(Y, X)$ are monotone in the opposite directions and both differ considerably from the constant function $\tilde{\mu} \equiv \rho(X, Y)$.

Moreover, any other pattern of discrepancy (large or small) is also possible.

Acknowledgement. I wish to thank Dr. E. Pleszczyńska for her help in preparation of this paper.

REFERENCE

[1] P.J. Bickel – E.L. Lehmann, Descriptive statistics for non-parametric models, I. Introduction, *Ann. Statist.*, 3 (1975), 1038-1041.

[2] D. Dabrowska, Regression-based orderings and measures of stochastic dependence, *Math. Operationsforsch. Statist.*, 12 (1981).

[3] M.H. De Groot, *Optimal Statistical Decisions*, McGraw Hill, New York, 1970.

[4] S. Konijn, Positive and negative dependence of two random variables, *Sankhyā*, 21 (1959), 269-280.

[5] T. Kowalczyk — E. Pleszczyńska, Monotonic dependence functions of bivariate distributions, *Ann. Statist.*, 5 (1977), 1221-1227.

[6] T. Kowalczyk, General definition and sample counterparts of monotonic dependence functions of bivariate distributions, *Math. Operationsforsch. Statist.*, 8 (1977), 351-369.

[7] K.V. Mardia, *Families of bivariate distributions*, Hafner, Darien, 1970.

[8] T. Yanagimoto, Families of positively dependent random variables, *Ann. Inst. Stat. Math.*, 24 (1972), 559-573.

D. Dabrowska

Institute of Computer Science, Polish Academy of Sciences, Warsaw, Poland.

COLLOQUIA MATHEMATICA SOCIETATIS JÁNOS BOLYAI

32. NONPARAMETRIC STATISTICAL INFERENCE,

BUDAPEST (HUNGARY), 1980.

SOME APPLICATIONS OF THE DEPENDENCE FUNCTIONS TO STATISTICAL INFERENCE: NONPARAMETRIC ESTIMATES OF EXTREME VALUES DISTRIBUTIONS, AND A KIEFER TYPE UNIVERSAL BOUND FOR THE UNIFORM TEST OF INDEPENDENCE

P. DEHEUVELS

1. INTRODUCTION

Let $X_n = \{X_n(1), \ldots, X_n(p)\}$, $n = 1, 2, \ldots$ be a sequence of i.i.d. p-dimensional random vectors, with common cumulative distribution function F, and marginal cumulative distribution functions $F^{(i)}$ $(1 \leqslant i \leqslant p)$. Let likewise F_n and $F_n^{(i)}$ $(1 \leqslant i \leqslant p)$, be the corresponding elements for the empirical measure associated to $X_1, \ldots, X_n$.

We define the *dependence function* D of F and the *empirical dependence function* D_n of F_n the identities:

$$D(F^{(1)}(x_1), \ldots, F^{(p)}(x_p)) = F(x_1, \ldots, x_p),$$

$$D_n(F_n^{(1)}(x_1), \ldots, F_n^{(p)}(x_p)) = F_n(x_1, \ldots, x_p),$$

satisfied for $x_1, \ldots, x_p$ continuity points of $F^{(1)}, \ldots, F^{(p)}$, and

$F_n^{(1)}, \ldots, F_n^{(p)}$ respectively.

D is not necessarily uniquely defined, except when $F^{(1)}, \ldots, F^{(p)}$ are continuous; D_n is uniquely defined on the lattice

$$I_n = \left\{ \left(\frac{i_1}{n}, \ldots, \frac{i_p}{n} \right), \; 1 \leqslant i_1, \ldots, i_p \leqslant n \right\}$$

and can be extended (see [3]) to the distribution function of a probability distribution on $[0, 1]^p$ with uniform marginals on $[0, 1]$ (which we will call a *dependence function*).

The empirical dependence function D_n was introduced in [6]. Its use to derive distribution-free tests of independence was made in [7] and [8].

The ideas of the dependence function goes back to F r e c h e t [10]. Its theory has been extended (see e.g. [3], [4], [5]) with emphasis on extreme values theory. In general any problem, involving a multivariate distribution in a space with a given coordinate system, can be factorized in a problem concerning dependence functions and a problem involving the marginal distributions.

The aim of this paper is to give some results on the applications of dependence functions to statistical inference.

We will first study the problem of the estimation *of extreme values distributions*, showing how to construct admissible extreme values dependence functions from the empirical dependence function D_n. We obtain two estimates $D_{n,r}$ and $D_{n,r}^{(R)}$, for which we prove the strong consistency under weak conditions. We derive from these estimates a new test procedure to verify the hypothesis that the distribution is extreme (Theorems 1, 2, 3).

In the second part, we discuss the use of the dependence functions to test the hypothesis of independence of the marginals of a distribution.

We obtain, in the bivariate case, the main result that, for each $\epsilon > 0$, there exist an absolute constant B_ϵ, such that $\forall t = o(\sqrt{n})$, $\forall n \geqslant 2$,

$$P[\sqrt{n} \max_{x,y} |F_n(x, y) - F_n^{(1)}(x)F_n^{(2)}(y)| > t] \leq$$

$$\leq B_\epsilon [\exp - (8 - \epsilon)t^2]$$

which can be written in terms of dependence functions by

$$\max_{x,y} |F_n(x, y) - F_n^{(1)}(x)F_n^{(1)}(y)| = \max_{i,j} \left| D_n\left(\frac{i}{n}, \frac{j}{n}\right) - \frac{ij}{n^2} \right|,$$

the best possible bound for $t > 0$ beeing $B_\epsilon \exp (- (4 \log 4 - \epsilon)t^2)$ Theorems 4, 5); the preceding results are sharp and give a result analogous to K i e f e r 's [13] for empirical distribution functions.

2. STATISTICAL INFERENCE FOR MULTIVARIATE EXTREME VALUE DISTRIBUTIONS

We have (as far as we now) introduced in [5] the *order* of a cumulative distribution function F, as the smallest $r > 0$, such that $F^{\frac{1}{r}}$ is still the distribution function of a probability neasure. (We had, at that time, been unaware of the paper [1], of B a l k e m a and R e s n i c k, who had defined the notion of *max-infinite divisibility*, identical to what we called *infinite order distribution functions*, and we gave unfortunately no reference of their work.)

Distribution functions of infinite oder have always dependence functions of infinite order ([5], Pr. (1.2)). We will therefore be concerned with the estimation of such a dependence function.

The distribution functions of infinite order give, as particular cases, the extreme values distributions; it can be shown (G a l a m b o s [11], p. 251-259) that the dependence function D of an extreme value distribution is characterized by

(1) $\qquad \forall r > 0, \quad D^r(u_1^{\frac{1}{r}}, \ldots, u_p^{\frac{1}{r}}) = D(u_1, \ldots, u_p),$

so that D is of infinite order.

In [3] and [5] we gave a constructive characterization of extreme value distributions (we must here mention that our results extend those of de Haan and Resnick [2], of which we were uniformed when we submitted our own paper, and therefore did not refer to it). Let us point out that the fact that a distribution has an extreme dependence function (satisfying (1)) does not imply that the distribution is extreme unless the marginal distributions are of Frechet, Gumbel or Weibull types (see [11]).

Let us now assume that F has continuous marginals and is of infinite order; we would like to estimate its dependence function D, using the fact that $\forall r > 0$, $D^r(u_1^{\frac{1}{r}}, \ldots, u_p^{\frac{1}{r}})$ is a dependence function (see [5]).

It is natural to think of using the empirical dependence function D_n to do this; but it is easily seen that, though $D_n \to D$ uniformly with probability 1 (see [6]), even when D is an extreme value dependence function, D_n *is generally not an infinite order distribution function.* This can be showed by the Frechet bounds ([3], [5], Theorem 2.2) which give (assuming that $(u_1, \ldots, u_p) \in I_n$):

$$\max \left(0, 1 - p + \sum_{i=1}^{p} u_i \right) \leqslant D_n(u_1, \ldots, u_p) \leqslant \min_{1 \leqslant i \leqslant p} u_i,$$

which is a sharp bound, as for an infinite order dependence function,

(2) $\qquad u_1 \ldots u_p \leqslant D_n(u_1, \ldots, u_p) \leqslant \min_{1 \leqslant i \leqslant p} u_i.$

To get over this non admissibility of D_n, we will use the following estimate of D:

$$D_{n,r}(u_1, \ldots, u_p) =$$

(3)

$$= \exp \left[r \left\{ D_n \left(\frac{1}{r} \log u_1 + 1, \ldots, \frac{1}{r} \log u_p + 1 \right) - 1 \right\} \right],$$

if $\forall 1 \leqslant i \leqslant p$, $e^{-r} \leqslant u_i \leqslant 1$.

This estimate will be extended for the remaining part of $[0, 1]^p$ (see [5], Theorem 1.1) by $\min(u_1, \ldots, u_p)$. (We will choose r such that $r \to \infty$, and that will be enough to ensure, using the continuity properties of the dependence functions ([3], Theorem 2.2), that the consistency of

$D_{n,r}$ for D on the set $S_r = \{(u_1, \ldots, u_p); \ e^{-r} \leqslant u_i \leqslant 1\}$ will imply the consistency of *any* extension of $D_{n,r}$, as a dependence function on $[0, 1]^p - S_r$, for D .) This will be sufficient for $D_{n,r}$ to be a dependence function on $[0, 1]^p$.

Theorem 1.

(1°) *For each* $r > 0$, $D_{n,r}$ *is an infinite order dependence function.*

(2°) $D_{n,r}$ *is uniquely defined on the grid*

$$G_{n,r} =$$

$$= \left\{ \left(\exp\left(-\frac{ri_1}{n} \right), \ldots, \exp\left(-\frac{ri_p}{n} \right) \right); \ 0 \leqslant i_1, \ldots, i_p \leqslant n \right\} =$$

$$= \{ (\alpha_{r,n}^{i_1}, \ldots, \alpha_{r,n}^{i_p}); \ 0 \leqslant i_1, \ldots, i_p \leqslant n \}$$

where $\alpha_{r,n} = e^{-\frac{r}{n}}$.

(3°) *If* $r \to \infty$, $\dfrac{r\sqrt{\log \log n}}{\sqrt{n}} \to 0$, *then if* D *is an extreme value dependence function,*

$$\lim_{n \to \infty} \left(\sup_{0 \leqslant u_1, \ldots, u_p \leqslant 1} |D_{n,r}(u_1, \ldots, u_p) - D(u_1, \ldots, u_p)| \right) = 0$$

with probability 1.

Proof. We will use Theorem (1.1) of [5], which ensures that $D_{n,r}$ is an infinite order distribution function, and also that

$$\lim_{r \to \infty} \exp[r\{D^{\frac{1}{r}}(u_1, \ldots, u_p) - 1\}] = D(u_1, \ldots, u_p),$$

and therefore, D being a dependence function, that

$$\lim_{r \to \infty} \exp\left[r\{ D^{\frac{1}{r}} \left(\left[\frac{1}{r} \log u_1 + 1 \right]^r, \ldots, \left[\frac{1}{r} \log u_p + 1 \right]^r \right) - 1 \} \right] =$$

$$= D[u_1, \ldots, u_p].$$

Therefore, by (1), if D is an extreme dependence function,

$$\lim_{r \to \infty} \exp\left[r\left\{D\left(\frac{1}{r}\log u_1 + 1, \ldots, \frac{1}{r}\log u_p + 1\right) - 1\right\}\right] =$$

$$= D(u_1, \ldots, u_p).$$

Using the fact ([3], Theorem (2.3)) that pointwise consistency of dependence functions is equivalent to uniform consistency, it is sufficient to prove that if $0 < u_1, \ldots, u_p < 1$, with probability 1,

$$\lim_{n \to \infty} \left(\exp\left[r\left\{D_n\left(\frac{1}{r}\log u_1 + 1, \ldots, \frac{1}{r}\log u_p + 1\right) - 1\right\}\right] - \right.$$

$$\left. - \exp\left[r\left\{D\left(\frac{1}{r}\log u_1 + 1, \ldots, \frac{1}{r}\log u_p + 1\right) - 1\right\}\right]\right) = 0.$$

But since by (2), $D(u_1, \ldots, u_p) \geq u_1 \ldots u_p > 0$, letting $\alpha_i = -\log u_i > 0$ $(i = 1, 2, \ldots, p)$, it is sufficient to prove that

$$\lim_{n \to \infty} r\left\{D_n\left(1 - \frac{\alpha_1}{r}, \ldots, 1 - \frac{\alpha_p}{r}\right) - \right.$$

$$\left. - D\left(1 - \frac{\alpha_1}{r}, \ldots, 1 - \frac{\alpha_p}{r}\right)\right\} = 0.$$

But now it has been proved ([6], Theorem (3.1)) that, with probability 1,

$$\max_{0 \leq u_1, \ldots, u_p \leq 1} |D_n(u_1, \ldots, u_p) - D(u_1, \ldots, u_p)| =$$

$$= O\left[\frac{\sqrt{\log \log n}}{\sqrt{n}}\right]$$

so the result follows readily.

The preceding proof yields deeper results in the case when D is not an extreme value dependence function. Since, even in that case, we obtain that

$$\lim_{n \to \infty} \left(\exp\left[r\left\{D_n\left(\frac{1}{r}\log u_1 + 1, \ldots, \frac{1}{r}\log u_p + 1\right) - 1\right\}\right] - \right.$$

$$\left. - \exp\left[r\left\{D\left(\frac{1}{r}\log u_1 + 1, \ldots, \frac{1}{r}\log u_p + 1\right) - 1\right\}\right]\right) = 0$$

with probability 1

if there exists a dependence function $\tilde{D}$ such that with probability 1 $\lim_{n \to \infty} D_{n,r} = \tilde{D}$, then,

$$\tilde{D}(u_1, \ldots, u_p) =$$

$$= \lim_{r \to \infty} \left(\exp \left[r \left\{ D \left(\frac{1}{r} \log u_1 + 1, \ldots, \frac{1}{r} \log u_p + 1 \right) - 1 \right\} \right] \right),$$

and, by [3], Theorem (3.4), it implies that

$$\tilde{D}(u_1, \ldots, u_p) = \lim_{r \to \infty} D^r(u_1^{\frac{1}{r}}, \ldots, u_p^{\frac{1}{r}}).$$

This is summed up by:

Theorem 2. *If there exists a dependence function* $\tilde{D}$, *such that*

$$\lim_{n \to \infty} D_{n,r}(u_1, \ldots, u_n) = \tilde{D}(u_1, \ldots, u_n) \quad \text{with probability } 1,$$

with $r = r(n)$ *satisfying*

$$r \to \infty, \quad \frac{r\sqrt{\log \log n}}{\sqrt{n}} \to 0, \quad \text{then, with probability } 1,$$

D belongs to the domain of attraction of $\tilde{D}$ which is an extreme value distribution function, that is

$$\lim_{n \to \infty} D^n(u_1^{\frac{1}{n}}, \ldots, u_p^{\frac{1}{n}}) = \tilde{D}(u_1, \ldots, u_p).$$

Let us now study in greater detail $D_{n,r}$; by Theorem 1, $D_{n,r}$ is uniquely defined on

$$G_{n,r} = \{(\alpha_{r,n}^{i_1}, \ldots, \alpha_{r,n}^{i_p}), \; 0 \leqslant i_1, \ldots, i_p \leqslant n\}.$$

Furthermore, $D_{n,r}$ is generally not an extreme value dependence function, though it is of infinite order.

In general the dependence functions of infinite order can be shown ([5], Theorem (2.2)) to follow the representation formula:

$$D(u_1, \ldots, u_p) =$$

$$(4) \qquad = u_1 \ldots u_p \exp \left\{ \sum_{k=2}^{p} (-1)^k \sum_{1 \leqslant i_1 < \ldots < i_k \leqslant p} \int_0^{-\log u_{i_1}} \ldots \right.$$

$$\left. \ldots \int_0^{-\log u_{i_k}} d\mu_{k;i_1,\ldots,i_k} \right\}$$

where $\{\mu_{k;i_1,\ldots,i_k}\}$ is a set of positive Radon measures satisfying the necessary and sufficient global conditions given in [5], §. 2.B.

A necessary and sufficient condition in order that D be an extreme value dependence function is that the measures $\{\mu_{k;i_1 \ldots i_k}\}$ be homogenous, that is, if

$$\forall \lambda > 0, \quad \int_0^{\lambda z_1} \ldots \int_0^{\lambda z_p} d\mu_{k;i_1 \ldots i_p} = \lambda \int_0^{z_1} \ldots \int_0^{z_p} d\mu_{k;i_1 \ldots i_p}$$

(see [5], Remark after Theorem (2.2)). It can therefore be seen that if we project the measure on the simplex

$$S_{k;i_1 \ldots i_k}^{R} = \{(z_1, \ldots, z_p); \ z_{i_1} + \ldots + z_{i_k} = R, \ z_i \geqslant 0\}$$

by setting

$$\int_0^{z_1} \ldots \int_0^{z_p} d\mu_{k;i_1 \ldots i_k}^{(R)} =$$

$$(5)$$

$$= \frac{z_{i_1} + \ldots + z_{i_k}}{R} \int_0^{\frac{R z_{i_1}}{z_{i_1} + \ldots + z_{i_k}}} \ldots \int_0^{\frac{R z_{i_k}}{z_{i_1} + \ldots + z_{i_k}}} d\mu_{k;i_1 \ldots i_p},$$

we would obtain $\mu_{k;i_1 \ldots i_p} = \mu_{k;i_1 \ldots i_p}^{(R)}$ in the case of an extreme value dependence function. We deduce from this the estimate of D given by

$$(6) \qquad D_{n,r}^{(R)}(u_1, \ldots, u_p) =$$

$$= u_1 \ldots u_p \exp\left\{ \sum_{k=2}^{p} (-1)^k \sum_{1 \leqslant i_1 < \ldots < i_k \leqslant p} \int_0^{-\log u_{i_1}} \ldots \right.$$

$$\left. \ldots \int_0^{-\log u_{i_k}} d\mu_{k;i_1\ldots i_k}^{(R)} \right\}$$

where D stands for $D_{n,r}$ in (4) and $\mu_{k;i_1\ldots i_k}^{(R)}$ is defined by (5).

In a less abstract way, it can be seen that, if $(u_1, \ldots, u_p) \in G_{n,r}$, the condition

$$\log u_{i_1} + \ldots + \log u_{i_p} = -R$$

becomes

$$\frac{k_{i_1}}{n} + \ldots + \frac{k_{i_j}}{n} = \frac{R}{r}$$

by setting $u_i = e^{-\frac{k_i r}{n}}$. According to this, we can deduce, assuming that $\frac{nR}{r}$ is integer, that $D_{n,r}^{(R)}$ is uniquely defined on the set

$$H_{n,r}^{(R)} = \left\{ \left(\left(\frac{k_1}{n}\right)^s, \ldots, \left(\frac{k_p}{n}\right)^s \right); \right.$$

$$\left. s > 0, \ k_i \geqslant 0, \ \frac{k_1}{n} + \ldots + \frac{k_p}{n} = \frac{R}{r} \right\} \cap [0, 1]^p$$

by the values of $D_{n,r}$ on the set

$$\left\{ \left(\frac{k_1}{n}, \ldots, \frac{k_p}{n} \right); \ k_i \geqslant 0, \ \frac{k_1}{n} + \ldots + \frac{k_p}{n} = \frac{R}{r} \right\}.$$

We now give a consistency result for $D_{n,r}^{(R)}$.

Theorem 3. *Let* $D_{n,r}^{(R)}$ *be an extreme value dependence function such that if* $n \to \infty$,

$$r \to \infty, \quad \frac{r\sqrt{\log \log n}}{\sqrt{n}} \to 0, \quad R \to \infty, \quad \frac{R}{r} \to 0.$$

Then, if D *belongs to the domain of attraction of an extreme dependence function* $\tilde{D}$,

$$\lim_{\substack{n \to \infty \\ 0 \leqslant u_1, \ldots, u_p \leqslant 1}} \max \left| D_{n,r}^{(R)}(u_1 \ldots u_p) - \tilde{D}(u_1, \ldots, u_p) \right| = 0.$$

Proof. As in Theorem 2, using [3], Theorem (3.4)

We have, in the preceding developments, given two transformations:

$$D_n \xrightarrow{(r)} D_{n,r} \xrightarrow{(R)} D_{n,r}^{(R)},$$

which yields successively an extreme order dependence function and an extreme value dependence function. The last estimate is admissible for extreme underlying dependence functions.

This gives the idea of a test to verify the hypothesis that a dependence function D is extreme, by computing some measure of discrepancy between D_n, which is consistent for D, and $D_{n,r}$ or $D_{n,r}^{(R)}$ which are consistent (under mild hypothesis on r and R) for $\tilde{D}$.

Any test based on a statistic such as $\max_u |D_{n,r}(u) - D_n(u)|$ can be shown to be consistent. The computation of the distributions of these tests seems, however, to be a considerable task.

The most interesting application could be when

$$D(u_1 \ldots u_p) = u_1 \ldots u_p,$$

that is in the case of independence of the marginals, since then, by (2),

$$D(u_1, \ldots, u_p) \leqslant D_{n,r}(u_1 \ldots u_p),$$

and the test would reduce to

$$\max_u \{D_{n,r}(u) - D(u)\}.$$

These questions are, therefore, open problems.

3. THE USE OF DEPENDENCE FUNCTIONS IN TESTS OF INDEPENDENCE

The hypothesis of independence (of the marginals) can be characterized by $D(u_1, \ldots, u_p) = u_1 \ldots u_p$. This can be, as a matter of fact, considered as a particular (often occurring) case of extreme value dependence function. Still, the methods used to test this hypothesis are specific. We consider test based on discrepancy measures based on the difference between $D_n(u_1 \ldots u_p)$ and $u_1 \ldots u_p$.

In [6] and [7], we introduced the following procedure. Let

$$\Delta_{i,h} F(x_1, \ldots, x_p) =$$

$$= F(x_1, \ldots, x_i + h, \ldots, x_p) - F(x_1, \ldots, x_p),$$

and

$$\Delta_h F(x_1, \ldots, x_p) = \Delta_{1,h} \circ \ldots \circ \Delta_{p,h} F(x_1, \ldots, x_p).$$

If we set h to be $\dfrac{1}{R}$ where R is a positive integer, we use the statistic

$$T_{n,R} = \sum_{0 \leqslant j_1, \ldots, j_p \leqslant R-1} \left(\Delta_{\frac{1}{R}} \left\{ D_n \left(\frac{j_1}{R}, \ldots, \frac{j_p}{r} \right) - \frac{j_1 \ldots j_p}{R^p} \right\} \right)^2,$$

(or some slight modification of this statistic obtained by calculating it at $\left(\frac{1}{n} \left[\frac{nj_1}{R} \right], \ldots, \frac{1}{n} \left[\frac{nj_p}{R} \right] \right)$ instead of $\left(\frac{j_1}{R}, \ldots, \frac{j_p}{R} \right)$ for sake of unicity in terms of the sample, D_n beeing uniquely defined on I_n).

It can be proved (see [7]) that if R is fixed, or slowly increasing, the approximation of the distribution of $T_{n,r}$ by a χ^2 is uniform when $n \to \infty$, under the null hypothesis of independence of the marginals.

Quite independently of this idea, the empirical dependence function can be used to derive Kolmogorov – Smirnov type tests such as:

$$T_n(1) = \max_{0 \leqslant i_1, \ldots, i_p \leqslant n} \left| D_n \left(\frac{i_1}{n}, \ldots, \frac{i_p}{n} \right) - \frac{i_1 \ldots i_p}{n} \right|.$$

This test was considered in [7], [8]. We will now give further details on this topic, limiting ourselves mainly to the bivariate case.

We set, for sake of simplicity, $F_n^{(1)} = G_n$, $F_n^{(2)} = H_n$, $F^{(1)} = G$, $F^{(2)} = H$, G and H beeing continuous, $F(x, y) = G(x)H(y)$, and

$D(u, v) = uv$ in the independence case. Consider

$$D_n = \max_{x,y} |F_n(x, y) - G(x)H(y)|,$$

$$D_n^{(1)} = \max_x |G_n(x) - G(x)|,$$

$$D_n^{(2)} = \max_y |H_n(y) - H(y)|,$$

$$T_n = \max_{i, j} \left| D_n\left(\frac{i}{n}, \frac{j}{n}\right) - \frac{ij}{n^2} \right|.$$

The following bounds are straightforward:

$$D_n \leqslant T_n + D_n^{(1)} + D_n^{(2)},$$

$$T_n \leqslant D_n + D_n^{(1)} + D_n^{(2)} \leqslant 3D_n.$$

We can therefore use results of Kiefer [13], and Dvoretzky – Kiefer – Wolfowitz [9] (see the review of Gaenssler and Stute [12]), showing that

$$\forall \epsilon > 0, \ \exists c_\epsilon > 0; \ \forall n, \ P[\sqrt{n} D_n > t] \leqslant c_\epsilon \exp\left[-(2 - \epsilon)t^2\right]$$

$$\exists d > 0; \ \forall n, \ P[\sqrt{n} D_n^{(i)} > t] \leqslant d \exp\left[-2t^2\right] \qquad (i = 1, 2).$$

This gives

$$P[\sqrt{n} T_n > t] \leqslant P[3\sqrt{n} D_n > t] \leqslant c_\epsilon \exp\left[-\frac{2 - \epsilon}{9} t^2\right].$$

This bound is very rough and can be considerably sharpened as we will now see.

First, we can deduce from it the lemma:

Lemma 1. *There exist an absolute constant A, such that, if $\epsilon > 0$, there exist B_ϵ, such that*

(7) $\forall n, \ \forall t, \ P[\sqrt{n} T_n > t] \leqslant B_\epsilon \exp\left[-(A - \epsilon)t^2\right].$

Lemma 2. $\frac{2}{9} \leqslant A \leqslant 8.$

Proof. In [8], we have derived the test statistic

$$\max_{0 \leqslant i \leqslant n} \quad D_n\left(\frac{i}{n}, \frac{1}{n}\left[\frac{n}{2}\right]\right) - \frac{1}{n^2}\left(i\left[\frac{n}{2}\right]\right)$$

which has been shown to follow the distribution of a two sample Kolmogorov – Smirnov test. It has been obtained also that

(8)
$$\lim_{n \to \infty} \ \mathbf{P}\Big[\frac{\sqrt{n}}{\sqrt{u(1-u)}} \sup_{0 \leqslant v \leqslant 1} |D_n(u, v) - uv| > t\Big] =$$

$$= 2 \sum_{k=1}^{\infty} (-1)^{k+1} e^{-2k^2 t^2} \sim 2e^{-2t^2}, \qquad t \to \infty.$$

So that

$$B_\epsilon \exp\left[-(A - \epsilon)t^2\right] \geqslant 2e^{-8t^2}$$

(taking in (8), $u = \frac{1}{2}$), giving the result.

We may now proceed with

Theorem 4. *For each $\epsilon > 0$ there exists an absolute constant B_ϵ such that, for each n, in the independence case, if $t = o(\sqrt{n})$, then*

$$\mathbf{P}[\sqrt{n} \max_{x, y} |F_n(x, y) - G_n(x)H_n(y)| > t] \leqslant B_\epsilon \exp\left[-(8 - \epsilon)t^2\right]$$

uniformly in α and t. The bound is the best possible. (Note that

$$\max_{x, y} |F_n(x, y) - G_n(x)H_n(y)| = \max_{i, j}\Big|D_n\left(\frac{i}{n}, \frac{j}{n}\right) - \frac{ij}{n^2}\Big|\text{).}$$

Proof. Let us consider the probability

$$P = \mathbf{P}\Big[D_n\left(\frac{i+k}{n}, \frac{j}{n}\right) - \frac{(i+k)j}{n^2} \geqslant \frac{c-\epsilon}{n^2}\,\Big|\,D_n\left(\frac{i}{n}, \frac{j}{n}\right) = \frac{ij+c}{n^2}\Big].$$

If we let

$$C = \left[\frac{ij+c}{n^2}\right], \quad N = n\left[D_n\left(\frac{i+k}{n}, \frac{j}{n}\right) - D_n\left(\frac{i}{n}, \frac{j}{n}\right)\right],$$

$$M = nD_n\left(\frac{i}{n}, \frac{j}{n}\right), \quad P = \mathbf{P}\Big[N \geqslant \frac{kj-\epsilon}{n}\,\Big|\,M = C\Big].$$

Now, using [6], Theorem (5.3), we get that

$$P[N = m \mid M = C] =$$

$$= \frac{k!\,(n-i-k)!\,(j-C)!\,(n-j-i+C)!}{(n-i)!\,m!\,(k-m)!\,(j-m-C)!\,(n-i-j+C-k+m)!},$$

which gives

$$\mathsf{E}[N \mid M = C] = \frac{k(j-C)}{n-i},$$

$$\mathsf{D}^2[N \mid M = C] = \frac{k(j-C)(n-i-k)(n-i-j+C)}{(n-i)^2(n-i-1)}.$$

By Tchebycheff's inequality, we obtain that

$$P\left[N \geqslant \frac{k(j-C)}{n-i} - \lambda\left(\frac{k(j-C)(n-i-k)(n-i-j+C)}{(n-i)^2(n-i-1)}\right)^{\frac{1}{2}}\right|$$

$$M = C\Big] \geqslant 1 - \frac{1}{\lambda^2}, \quad \text{for} \ \ \lambda > 0.$$

As now, $\dfrac{k(j-C)(n-i-k)(n-i-j+C)}{(n-i)^2(n-i-1)} \leqslant \dfrac{k}{4}$, for all possible

values of i, j, n, C, we get, noting that $\dfrac{k(j-C)}{n-i} = \dfrac{kj}{n} - \dfrac{kc}{n(n-i)}$.

$$P\left[N \geqslant \frac{kj}{n} - \frac{kc}{n(n-i)} - \frac{1}{2}\,\lambda\sqrt{k} \mid M = C\right] \geqslant 1 - \frac{1}{\lambda^2}.$$

We obtain accordingly that if $\dfrac{\epsilon}{n} - \dfrac{kc}{n(n-i)} - \dfrac{1}{2}\,\lambda\sqrt{k} \geqslant 0$, then

$P \geqslant 1 - \dfrac{1}{\lambda^2}$.

Let us now consider $M_n\left(\dfrac{i}{n}\right) = \max_{0 \leqslant j \leqslant n}\left\{D_n\left(\dfrac{i}{n}, \dfrac{j}{n}\right) - \dfrac{ij}{n^2}\right\}$.

If the event

$$\sup_{m \leqslant i \leqslant p} M_n\left(\frac{i}{n}\right) \geqslant \frac{\alpha}{\sqrt{n}}$$

has occurred, there is necessarily one point $\dfrac{i_0}{n}$ such that $M_n\left(\dfrac{i}{n}\right)$

$(i = m, m+1, \ldots)$, exceeds $\dfrac{a}{\sqrt{n}}$ for the first time, and likewise for

$D_n\left(\dfrac{i_0}{n}, \dfrac{j_0}{n}\right) - \dfrac{i_0 j_0}{n^2}$ which can be taken to be $\dfrac{\alpha}{\sqrt{n}}$ (assuming that $\alpha n^{\frac{3}{2}}$ is integer). Considering the disjoints events

$$\left\{ D_n\left(\frac{i_0}{n}, \frac{j_0}{n}\right) - \frac{i_0 j_0}{n^2} = \frac{\alpha}{\sqrt{n}} \quad \text{for the first time} \right\},$$

we can deduce from the preceding reasoning that: if $\beta\alpha - \dfrac{p-m}{n-p}\alpha \geqslant$

$\geqslant \dfrac{1}{2}\lambda\sqrt{\dfrac{p-m}{n}}$, then

$$P\left[M_n\left(\frac{p}{n}\right) \geqslant \frac{\alpha(1-\beta)}{\sqrt{n}}\right] \geqslant \left(1 - \frac{1}{\lambda^2}\right) P\left[\sup_{m \leqslant i \leqslant p} M_n\left(\frac{i}{n}\right) \geqslant \frac{\alpha}{\sqrt{n}}\right].$$

Let now β be fixed $(0 < \beta < 1)$, and consider a subdivision of $\left[0, \left[\frac{n}{2}\right]\right]$, by $0 < i_1 < \ldots < i_R = \left[\frac{n}{2}\right]$, with $i_j = \left[\frac{nj}{2R}\right]$. Setting each time $m = i_{r-1}$, $p = i_r$, we see that

(9)
$$P\left[\sup_{0 \leqslant i \leqslant \left[\frac{n}{2}\right]} M_n\left(\frac{i}{n}\right) \geqslant \frac{\alpha}{\sqrt{n}}\right] \leqslant$$

$$\leqslant \frac{\lambda^2}{\lambda^2 - 1} \sum_{j=1}^{R} P\left[M_n\left(\frac{i_j}{n}\right) \geqslant \frac{\alpha(1-\beta)}{\sqrt{n}}\right],$$

if $\alpha\left[\beta - \left(\frac{1}{R} + \frac{4}{n}\right)\right] \geqslant \lambda 2^{-\frac{3}{2}}$.

If we choose R and n big enough, the formula will be true for $\alpha \geqslant \dfrac{\lambda}{\beta\sqrt{2}}$.

We will now use the fact ([8], Theorem 2) that

$$\lim_{n \to \infty} P\left[M_n\left(\frac{i_j}{n}\right) \geqslant \frac{u}{\sqrt{n}}\right] = \exp\left[-2u^2\left(\frac{j}{2R}\left(1 - \frac{j}{2R}\right)\right)^{-1}\right].$$

A comparison of the exact formulas obtained in [8] to the distribution of the two sample $D_{i_j, n-i_j}$ test shows the identity of the distribution of $M_n\left(\frac{i_j}{n}\right)$ and of the distribution of $D_{i_j, n-i_j}$. We can, accordingly, use the fact that

(10) $P\left[M_n\left(\frac{i_j}{n}\right) \geqslant u\right] \leqslant P\left[M_n\left(\frac{i_R}{n}\right) \geqslant u\right]$,   uniformly in u.

It is thus sufficient to show that

$$P\left[D_{N,N}^+ \geqslant \frac{\alpha}{\sqrt{N}}\right] = \frac{1}{\binom{2N}{N}}\left(N + \frac{2N}{[\alpha\sqrt{N}]}\right) \leqslant C_\epsilon \exp\left[-(2-\epsilon)\alpha^2\right]$$

for some universal bound C_ϵ, if $\epsilon > 0$ is given.

Now, by Stirling's formula,

$$P\left[D_{N,N}^+ \geqslant \frac{\alpha}{\sqrt{N}}\right] \sim \left[1 - \frac{\alpha^2}{N}\right]^{-N}\left[1 + \frac{\alpha^2}{\sqrt{N}}\right]^{-\alpha\sqrt{N}}\left[1 - \frac{\alpha^2}{N}\right]^{\alpha\sqrt{N}},$$

as long as $\alpha = o(\sqrt{n})$. For $[\alpha\sqrt{N}] = N$ since $\binom{2N}{N}^{-1} \sim \left(\frac{1}{4}\right)^N\sqrt{\pi N} \leqslant$ $\leqslant \exp\left[-N(-\epsilon + \log 4)\right]$, it is not worthwhile to seek a uniform bound of the form (7) with $A \geqslant \log 4$; for all α:

Lemma 3. $\frac{2}{9} \leqslant A \leqslant \log 4$.

Now, if $\alpha = o(\sqrt{N})$, we get that

$$P\left[D_{N,N}^+ \geqslant \frac{\alpha}{\sqrt{N}}\right] \leqslant 2 \exp\left[-2\alpha^2 + O(1)\right]$$

which will be enough to get the conclusion in Theorem 4.

If now $\alpha^2 = k\sqrt{N}$, we get similarly that

$$P[D_{N,N}^+ \geqslant kN] = O\left\{[1 - k^2]^{-N}\left[\frac{1 + k^2}{1 - k^2}\right]^{-kN}\right\}$$

and a bound of the form

$$P[D_{N,N}^+ \geqslant kN] \leqslant C \log\left\{-N\log\left[(1 + k^2)^k(1 - k^2)^{1-k}\right]\right\}.$$

We deduce from the fact that

$$(1 + k^2)^k(1 - k^2)^{1-k} \leqslant 2 \quad \text{if} \quad 0 \leqslant k \leqslant 1,$$

the next theorem:

Theorem 5. *For each* $\epsilon > 0$ *there exists an absolute constant* B_ϵ, such that, in the independency case,

$$P[\sqrt{n} \max_{x,y} |F_n(x, y) - G_n(x)H_n(y)| > t] \leqslant$$

$$\leqslant B_\epsilon \exp[-(4 \log 4 - \epsilon)t^2]$$

uniformly in n *and* t. *The bound is the best possible.*

We now end the proof of Theorem 4, for Theorem 5 having the same demonstration:

The estimate (9) can be simplified if $\alpha = o(\sqrt{n})$ in

$$P\Big[\sup_{0 \leqslant i \leqslant \frac{n}{2}} M_n\Big(\frac{i}{n}\Big) \geqslant \frac{\alpha}{\sqrt{n}}\Big] \leqslant 3 \frac{\lambda^2 R}{\lambda^2 - 1} \exp[-8\alpha^2(1 - \beta)^2],$$

which is precisely the result we seek, if we use the symmetry of D_n.

Remarks. According to (8), and Lemmas 1, 2, 3, Theorem 4, 5 give sharp results.

Furthermore, the lower bound (8) can be made precise. This will be the subject of a forthcoming paper.

The intimate relationship between the empirical dependence function and the empirical distribution function suggest the use of the latter to get further information on the (still) unknown distribution of

$$\max_{x,y} |F_n(x, y) - F(x)G(y)|.$$

The preceding results can be extended in R^p. This will be done in a next paper. Let us remark though that the proofs become very intricate when p gets bigger!

4. CONCLUSION

We conclude that the exact and empirical dependence functions are a very useful concept in multivariate statistics, opening a promising field for investigation.

REFERENCES

[1]   A.A. Balkema – S.I. Resnick, Max-infinite divisibility, *J. App. Prob.*, 14 (1977), 309-319.

[2]   L. de Haan – S.I. Resnick, Limit theory for multi-variable sample extremes, *Z. Wahrscheinlichkeitstheorie verw. Geb.*, 40 (1977), 317-337.

[3]   P. Deheuvels, Caractérisation complète des lois extrêmes multivariées et de la convergence des types extrêmes, Publ. Inst. Statist. Univ. Paris, 3.4 (1978), 1-36.

[4]   P. Deheuvels, Détermination des lois jointes de l'ensemble des points extrêmes d'un échantillon multivarié, *C.R. Acad. Sci. Paris*, sér. A, 288 (1979), 631-634.

[5]   P. Deheuvels, The decomposition of infinite order and extreme multivariate distributions, *Asymptotic theory of statistical tests and estimation,* I.M. Chakravarti edit., Academic Press, 1980, 259-286.

[6]   P. Deheuvels, La fonction de dépendance empirique et ses propriétés, *Bulletin de l'Académie Royale de Belgique,* Classe des Sciences. 5e Ser. LXV, 6 (1979), 274-292.

[7]   P. Deheuvels, A non parametric test for independence, *Publ. ISUP,* 1981 (in print).

[8]   P. Deheuvels, A Kolmogorov – Smirnov test for independence, *Revue Roumaine de Mathématiques pures et appliquées,* XXVI, n° 2 (1981), 213-226.

[9]   A. Dvoretzky – J. Kiefer – J. Wolfowitz, Asymptotic minimax character of the sample distribution function and of the classical multinomial operator, *Ann. Math. Statist.,* 27 (1956), 642-669.

[10] M. Frechet, Sur les tableaux de corrélation dont les marges sont données, *Ann. Univ. Lyon, Sec. A,* Ser. 3, 14 (1951), 53-77.

[11]   J . G a l a m b o s,   *The asymptotic theory of extreme order sta-tistics,*   Wiley, New York, 1978.

[12]   P . G a e n s s l e r − W . S t u t e,   Empirical processes: a survey of results for independent and identically distributed random varia-bles, *Ann. Prob.,* 7 (2) (1979), 193-243.

[13]   J . K i e f e r,   On large deviation of the empiric D.F. of vector chance variables and a law of iterated logarithm, *Pacific J. Math.,* 11 (1961), 649-660.

P . D e h e u v e l s
Inst. de Stat. des Univ. de Paris, Tour 45.55-3e Étage, 4 Place Jussieu, F-75230 Paris, France.

COLLOQUIA MATHEMATICA SOCIETATIS JÁNOS BOLYAI

32. NONPARAMETRIC STATISTICAL INFERENCE,

BUDAPEST (HUNGARY), 1980.

EDGEWORTH EXPANSIONS FOR FUNCTIONS OF UNIFORM SPACINGS*

R.J.M.M. DOES — R. HELMERS

1. INTRODUCTION

Let $U_1, U_2, \ldots$ be a sequence of independent uniform $(0, 1)$ random variables (r.v.'s). For $n = 1, 2, \ldots,$ $U_{1:n} \leqslant U_{2:n} \leqslant \ldots \leqslant U_{n:n}$ denote the ordered $U_1, U_2, \ldots, U_n$. Let $U_{0:n} = 0$ and $U_{n+1:n} = 1$. Uniform spacings are defined by

$$(1.1) \qquad D_{in} = U_{i:n} - U_{i-1:n}$$

for $i = 1, 2, \ldots, n + 1$. Let g be a fixed real-valued nonlinear measurable function and define statistics T_n $(n = 1, 2, \ldots)$ by

$$(1.2) \qquad T_n = \sum_{i=1}^{n+1} g((n + 1)D_{in}).$$

It is well-known that suitably normalized statistics of the form (1.2) are asymptotically $\mathcal{N}(0, 1)$ distributed under quite general conditions. We refer to the papers of Darling [3], Le Cam [6], Shorack [10],

---

*Report SW 55/80 Mathematisch Centrum, Amsterdam.

Pyke [9] and Koziol [5] (see also the theorem contained in the discussion between Kingman and Pyke in Pyke [8]). A survey of the general area of limit theorems for spacings may be found in Pyke [9].

The purpose of this paper is to establish Edgeworth expansions for statistics of the form (1.2) with remainder $o\left(\frac{1}{n}\right)$. Using a well-known characterization of the joint distribution of uniform spacings the problem is reduced to that of deriving an asymptotic expansion for the distribution function (d.f.) of a normalized sum of independent r.v.'s conditioned on another sum of independent r.v.'s. This latter problem is then dealt with in a standard manner by applying classical results for sums of independent random vectors.

After the result of the present paper was obtained, a related paper of Michel [7] appeared. In his paper he proves, using a similar method of proof, a general theorem on asymptotic expansions for conditional distributions. However, Michel's result is, though more general in scope, less explicit than ours when applied to the conditional distributions we have to consider. Another related paper is that of Van Zwet [11]. He derives the Edgeworth expansion for linear combinations of uniform order statistics under minimal conditions.

2. THE THEOREM

For any r.v. X let us denote by X^* the r.v. $X^* = \dfrac{X - \mathsf{E}(X)}{\sigma(X)}$ whenever $0 < \sigma(X) < \infty$. Let Y be an exponential r.v. with expectation 1 and let g be a fixed real-valued nonlinear measurable function defined on R^+. Introduce, whenever well-defined, for integers i and j, quantities m_{ij} by

(2.1) $m_{ij} = \mathsf{E}((g^*(Y))^i (Y^*)^j).$

We shall establish an asymptotic expansion with remainder $o\left(\frac{1}{n}\right)$ for the d.f.

(2.2) $G_n(x) = \mathsf{P}\left(\dfrac{1}{\sigma\tau\sqrt{n}} (T_n - (n + 1)\mu) \leq x\right)$ $(-\infty < x < \infty),$

where

$$(2.3) \qquad \mu = \int_0^\infty g(y)e^{-y}\, dy, \qquad \sigma^2 = \int_0^\infty (g(y) - \mu)^2 e^{-y}\, dy$$

and

$$(2.4) \qquad \tau = \sqrt{1 - m_{11}^2}\,.$$

Note that $\sigma\tau > 0$ if and only if g is not linear.

Let Φ and φ denote the d.f. and the density of the standard normal distribution and let ρ denote the characteristic function (ch. f.) of $(g^*(Y), Y^*)$; i.e.

$$(2.5) \qquad \rho(s, t) = Ee^{isg^*(Y) + itY^*} \quad \text{for} \quad (s, t) \in R^2.$$

Theorem 1. *Suppose that the assumptions*

(i) $\displaystyle\int_0^\infty g^4(y)e^{-y}\,dy < \infty$;

(ii) $\displaystyle\int_{-\infty}^\infty \int_{-\infty}^\infty |\rho(s, t)|^p\, ds\,dt < \infty$ *for some* $p \geqslant 1$

are satisfied. Then we have that

$$(2.6) \qquad \sup_x |G_n(x) - \tilde{G}_n(x)| = o\left(\frac{1}{n}\right) \quad \text{as} \quad n \to \infty,$$

where

$$\tilde{G}_n(x) = \Phi(x) - \varphi(x)\left\{\frac{1}{\sqrt{n}}\left(\frac{1}{6}\kappa_3(x^2 - 1) + a\right) + \right.$$

$$+ \frac{1}{n}\left(\frac{1}{24}\kappa_4(x^3 - 3x) + \frac{1}{72}\kappa_3^2(x^5 - 10x^3 + 15x) + \right.$$

$$\left. + \frac{1}{8}(-4a\kappa_3 + b)x + \frac{1}{6}a\kappa_3 x^3\right)\right\}$$

with

$$\kappa_3 = \frac{1}{\tau^3}(m_{30} - 3m_{21}m_{11} + 3m_{12}m_{11}^2 - 2m_{11}^3),$$

$$\kappa_4 = \frac{1}{\tau^4}\,(m_{40} - 3 - 3m_{21}^2 + 12m_{12}m_{21}m_{11} - 4m_{31}m_{11} +$$

$$+ 6m_{11}^2 + 6m_{22}m_{11}^2 - 12m_{12}^2 m_{11}^2 - 12m_{21}m_{11}^2 +$$

$$+ 24m_{12}m_{11}^3 - 4m_{13}m_{11}^3 - 6m_{11}^4),$$

$$a = \frac{1}{2\tau}\,(-m_{12} + 2m_{11})$$

and

$$b = \frac{1}{\tau^2}\,(6 + 4m_{21} + 3m_{12}^2 - 2m_{22} - 20m_{12}m_{11} +$$

$$+ 4m_{13}m_{11} - 4m_{11}^2).$$

In Theorem 1 we establish an asymptotic expansion for the d.f. of suitably normalized sums of a function of uniform spacings with remainder $o\left(\frac{1}{n}\right)$. In fact it can be proved that under the assumptions of Theorem 1 there exists an expansion for $P\left(\frac{1}{\sigma\tau\sqrt{n}}\,(T_n - (n+1)\mu) \in A\right)$ for every Borel set A, the remainder being $o\left(\frac{1}{n}\right)$ uniform over all Borel sets. However, for statistical applications, the theorem in its present form will be of sufficient generality. The theorem requires a natural moment condition (i). Moreover, the ch.f. ρ (cf. (2.5)) must satisfy an integrability assumption (ii) which is commonly encountered in problems of establishing expansions for conditional distributions (see, e.g., A l b e r s [1], Ch. 2, and M i c h e l [7]). An assumption, equivalent to (ii) and obviously implying that g is nonlinear, is that the n-th convolution of the distribution of $(g^*(Y), Y^*)$ possesses almost everywhere on R^2 a bounded density for all sufficiently large n (see B h a t t a c h a r y a and R a o [2], Th. 19.1). The verification of this assumption is a problem in itself which we have not attempted to solve.

We conclude this section with some remarks. In the first place we note that it is not difficult to check that under the assumptions of Theorem 1 not only (2.6) holds but that also

$$\sup_x |F_n^*(x) - \tilde{F}_n(x)| = o\left(\frac{1}{n}\right) \quad \text{as} \quad n \to \infty,$$

is true, where $F_n^*(x) = P(T_n^* \leqslant x)$ and

$$\tilde{F}_n(x) = \Phi(x) - \varphi(x)\left\{\frac{1}{6\sqrt{n}}\kappa_3(x^2 - 1) + \frac{1}{24n}\kappa_4(x^3 - 3x) + \right.$$

$$\left. + \frac{1}{72n}\kappa_3^2(x^5 - 10x^3 + 15x)\right\}.$$

Note that $\tilde{F}_n$ does not depend on the quantities a and b appearing in the expansion $\tilde{G}_n$. This is due to the exact standardization we have employed here. Secondly we remark that Theorem 1 provides a partial answer to a question posed by P y k e [9] (his problem 5) concerning the rate of convergence to normality for functions of uniform spacings of the form T_n. Typically the error committed when the normal approximation is applied is of order $\frac{1}{\sqrt{n}}$. Finally, we note that, as the error of the expansion given in (2.6) is $o\left(\frac{1}{n}\right)$, we may expect, that, at least for not too small sample sizes, the expansion will yield a better numerical approximation for the distribution function of T_n than can be provided by the normal approximation.

3. PROOF OF THE THEOREM

As indicated in the introduction we begin the proof by reducing our problem to that of deriving an asymptotic expansion for the d.f. of a sum of independent r.v.'s, conditioned on another sum of independent r.v.'s. Let $Y_1, Y_2, \ldots$ be independent exponentially distributed r.v.'s with expectation 1 and define r.v.'s X_n and S_n ($n = 1, 2, \ldots$), by

$$(3.1) \qquad X_n = \frac{1}{\sqrt{n}}\sum_{i=1}^{n+1} g^*(Y_i), \quad S_n = \frac{1}{\sqrt{n}}\sum_{i=1}^{n+1} Y_i^*.$$

A well-known characterization of the joint distribution of uniform spacings (see, e.g., P y k e [8]) implies that the distribution of

$$\frac{1}{\sigma\sqrt{n}}(T_n - (n+1)\mu) \qquad (n = 1, 2, \ldots),$$

is the same as the conditional distribution of X_n given $S_n = 0$ $(n = 1, 2, \ldots)$; i.e.

$$(3.2) \qquad G_n(x) = \mathsf{P}\left(\frac{1}{\sigma\tau\sqrt{n}}(T_n - (n+1)\mu) \leqslant x\right) = \mathsf{P}(X_n \leqslant x\tau \mid S_n = 0).$$

The problem to establish an asymptotic expansion for the conditional probabilities in (3.2) will now be solved in a number of steps. In the first place we establish an expansion for the ch.f. ψ_n of (X_n, S_n); i.e. of

$$(3.3) \qquad \psi_n(s, t) = \mathsf{E}e^{isX_n + itS_n} = \rho^{n+1}\left(\frac{s}{\sqrt{n}}, \frac{t}{\sqrt{n}}\right),$$

where ρ is defined in (2.5). We shall prove that for some $\delta > 0$ and uniformly for all $(s, t) \in R^2$ satisfying $s^2 + t^2 \leqslant \delta n$ we have that

$$(3.4) \qquad \psi_n(s, t) = \tilde{\psi}_n(s, t) + o\left(\frac{P(s, t)e^{-\frac{1}{4}Q(s,t)}}{n}\right),$$

where P is a polynomial of the sixth degree in s and t, without constant term and $Q(s, t) = s^2 + 2m_{11}st + t^2$. The function $\tilde{\psi}_n$ is given by

$$\tilde{\psi}_n(s, t) = e^{-\frac{1}{2}(s^2 + 2m_{11}st + t^2)} \times$$

$$\times \left\{1 - \frac{i}{6\sqrt{n}}(m_{30}s^3 + 3m_{21}s^2 t + 3m_{12}st^2 + 2t^3) + \right.$$

$$+ \frac{1}{24n}((m_{40} - 3)s^4 + 4(m_{31} - 3m_{11})s^3 t +$$

$$(3.5) \qquad + 6(m_{22} - 2m_{11}^2 - 1)s^2 t^2 + 4(m_{13} - 3m_{11})st^3 + 6t^4) -$$

$$- \frac{1}{72n}(m_{30}s^3 + 3m_{21}s^2 t + 3m_{12}st^2 + 2t^3)^2 -$$

$$\left. - \frac{1}{2n}(s^2 + 2m_{11}st + t^2)\right\}.$$

To check this we first note that because of assumption (i) we can expand the ch.f. of $(g^*(Y_1), Y_1^*)$ in a neighbourhood of the origin. A simple computation yields that uniformly for all $(s, t) \in R^2$ satisfying $s^2 + t^2 = o(n)$

$$\rho\left(\frac{s}{\sqrt{n}}, \frac{t}{\sqrt{n}}\right) = 1 - \frac{1}{2n}(s^2 + 2m_{11}st + t^2) -$$

$$- \frac{i}{6n^{\frac{3}{2}}}(m_{30}s^3 + 3m_{21}s^2t + 3m_{12}st^2 + 2t^3) +$$

(3.6)

$$+ \frac{1}{24n^2}(m_{40}s^4 + 4m_{31}s^3t + 6m_{22}s^2t^2 + 4m_{13}st^3 + 9t^4) +$$

$$+ o\left(\frac{R(s, t)}{n^2}\right),$$

where R is a polynomial of the fourth degree in s and t without constant term. Taking now the $(n + 1)$-th power of the expansion in (3.6) and expanding further we arrive at (3.4) (see Ch. 2, Sec. 9 in Bhattacharya and Rao [2] for more details).

Next we convert the expansion (3.4) into an expansion for the joint density f_n of (X_n, S_n) in the point $(x, 0)$. After long but easy computations we find that for all sufficiently large n

(3.7) $f_n(x, 0) = \tilde{f}_n(x, 0) + r_n(x)$,

where

$$\tilde{f}_n(x, 0) = \frac{1}{(2\pi)^2} \int\limits_{-\infty}^{\infty} \int\limits_{-\infty}^{\infty} e^{-isx} \tilde{\psi}_n(s, t)\, ds dt =$$

$$= \frac{1}{2\pi\tau} e^{-\frac{1}{2}x^2\tau^{-2}}\left\{1 - \frac{\kappa_3}{6\sqrt{n}}\left(\frac{3x}{\tau} - \frac{x^3}{\tau^3}\right) + \right.$$

(3.8)

$$+ \frac{\kappa_4}{24n}\left(3 - \frac{6x^2}{\tau^2} + \frac{x^4}{\tau^4}\right) -$$

$$- \frac{\kappa_3^2}{72n}\left(15 - \frac{45x^2}{\tau^2} + \frac{15x^4}{\tau^4} - \frac{x^6}{\tau^6}\right) +$$

$$\left. + \frac{ax}{\tau\sqrt{n}} + \frac{a\kappa_3}{6n}\left(3 - \frac{6x^2}{\tau^2} + \frac{x^4}{\tau^4}\right) - \frac{b}{8n}\left(1 - \frac{x^2}{\tau^2}\right) - \frac{7}{12n}\right\}$$

and

(3.9) $(1 + x^4)r_n(x) = o\left(\frac{1}{n}\right)$

– 209 –

uniformly in x. Here we have used assumption (ii) to guarantee that f_n exists for all sufficiently large n (see Th. 19.1 of Bhattacharya and Rao [2]) and to validate the application of the Fourier inversion theorem in R^2 (see Feller [4], p. 524). The nonuniform estimate of the remainder $r_n(x)$ in (3.9) follows directly from Theorem 19.2 of Bhattacharya and Rao [2]. The validity of this application of Theorem 19.2 of Bhattacharya and Rao [2] can be inferred from the assumptions of Theorem 1.

We also need an expansion for the marginal density f_{n2} of S_n in the point 0. Since $S_n = \frac{1}{\sqrt{n}} \sum_{i=1}^{n+1} (Y_i - 1)$ (cf. (3.1)) the theory of asymptotic expansions for the densities of sums of i.i.d. r.v.'s can be applied (see e.g. Feller [4], Th. XVI. 2.2). It follows that

$$(3.10) \qquad f_{n2}(0) = \frac{1}{\sqrt{2\pi}} \left(1 - \frac{7}{12n}\right) + O\left(\frac{1}{n^{\frac{3}{2}}}\right).$$

We are now in a position to prove (2.6). Remark first that using (3.2) and the fact that the conditional density of X_n, given $S_n = 0$, in the point x is obviously equal to $\dfrac{f_n(x, 0)}{f_{n2}(0)}$, yields that for all sufficiently large n

$$(3.11) \qquad G_n(x) = \int_{-\infty}^{x\tau} \frac{f_n(y, 0)}{f_{n2}(0)}\, dy.$$

The relations (3.7)-(3.10) provide expansions for the numerator and denominator of the integrand in (3.11). To find an expansion for $G_n(x)$ from these results we note that

$$\frac{f_n(y, 0)}{f_{n2}(0)} = \sqrt{2\pi} \left(1 + \frac{7}{12n} + O\left(\frac{1}{n^{\frac{3}{2}}}\right)\right) (\tilde{f}_n(y, 0) + r_n(y)) =$$

$$(3.12) \qquad = \sqrt{2\pi}\tilde{f}_n(y, 0) + \frac{7\varphi\left(\frac{y}{\tau}\right)}{12n\tau} + \frac{o\left(\frac{1}{n}\right)}{1 + y^2} + \frac{o\left(\frac{1}{n}\right)}{1 + y^4} +$$

$$+ O\left(\frac{1}{n^{\frac{3}{2}}}\right)\tilde{f}_n(y, 0)$$

uniformly in y. Note that to obtain the third term in the second line we have used the remark following Theorem 19.2 of B h a t t a c h a r y a and

R a o [2] to check that $\sqrt{2\pi} f_n(y, 0) = \dfrac{\varphi\left(\frac{y}{\tau}\right)}{\tau} + \dfrac{o(1)}{1 + y^2}$ uniformly in y.

Combining now (3.8) and (3.12) with (3.11) we find that

$$(3.13) \qquad G_n(x) = \int_{-\infty}^{x\tau} \left\{ \sqrt{2\pi} \tilde{f}_n(y, 0) + \frac{7\varphi\left(\frac{y}{\tau}\right)}{12 n\tau} \right\} dy + o\left(\frac{1}{n}\right)$$

uniformly in x. The theorem now follows from (3.8) and (3.13) after a number of elementary integrations. ∎

Acknowledgements. The authors are grateful to F.H. R u y m g a a r t for suggesting the problem and to W.R. V a n Z w e t for his valuable comments.

REFERENCES

[1] W. A l b e r s, *Asymptotic expansions and the deficiency concept in statistics,* Mathematical Centre Tract 58, Amsterdam, 1974.

[2] R.N. B h a t t a c h a r y a — R.R. R a o, *Normal approximation and asymptotic expansions,* Wiley, New York, 1976.

[3] D.A. D a r l i n g, On a class of problems related to the random division of an interval, *Ann. Math. Statist.,* 24 (1953), 239-253.

[4] W. F e l l e r, *An introduction to probability theory and its applications* 2, 2nd ed., Wiley, New York, 1971.

[5] J.A. K o z i o l, A note on limiting distributions for spacings statistics, *Z. Wahrscheinlichkeitstheorie verw. Geb.,* 51 (1980), 55-62.

[6] L. L e C a m, Un théorème sur la division d'un intervalle par des points pris au hasard, *Publ. Inst. Statist. Univ. Paris,* 7 (1958), 7-16.

[7] R. M i c h e l, Asymptotic expansions for conditional distributions, *J. Multivariate Anal.,* 9 (1979), 393-400.

[8]   R. P y k e, Spacings, *J. Roy. Statist. Soc., Ser. B,* 27 (1965), 359-449.

[9]   R. P y k e, Spacings revisited, *Proc. Sixth Berkeley Symp. Math. Statist. Prob.,* 1 (1972), 417-427.

[10] G.R. S h o r a c k, Convergence of quantile and spacings processes with applications, *Ann. Math. Statist.,* 43 (1972), 1400-1411.

[11] W.R. V a n Z w e t, The Edgeworth expansion for linear combinations of uniform order statistics, *Proc. Second Prague Symp. Asymp. Statist.,* (P. Mandl and M. Hušková (eds.)), 93-101, North Holland, Amsterdam, 1979.

R.J.M.M. D o e s

University of Limburg, Data Processing Department, P.O. Box 616, 6200 MD Maastricht, The Netherlands.

R. H e l m e r s

Mathematical Centre, Dept. of Statistics, Kruislaan 413, 1098 SJ Amsterdam, The Netherlands.

COLLOQUIA MATHEMATICA SOCIETATIS JÁNOS BOLYAI

32. NONPARAMETRIC STATISTICAL INFERENCE,

BUDAPEST (HUNGARY), 1980.

NOTES ON THE THEIL TEST FOR THE HYPOTHESIS OF LINEARITY FOR THE MODEL WITH TWO EXPLANATORY VARIABLES

C. DOMAŃSKI

1. INTRODUCTION

H. Theil [6] (cf. also [7]) presented a nonparametric test by means of which the hypothesis $H_0: \beta = \beta_0$ can be verified in the model

(1) $\qquad y_i = \alpha + \beta x_i + \epsilon_i,$

where β_0 is specified. Especially this test can be used to verify the hypothesis $H_0: E(Y \mid X) = \alpha + \beta X$, i.e. the hypothesis that the relationship between the variables Y and X is linear.

The test statistic of the Theil test is the following:

(2) $\qquad T_n = \sum_{i=1}^{n} \sum_{j=i+1}^{n} \text{sign} \, (R_{u(j)} - R_{u(i)})$

where

$$R_{u(i)} = y_{u(i)} - \beta_0 x_{u(i)}$$

and u is a permutation of the sequence $\{x_i\}$, namely

$$x_{u(1)} \leqslant x_{u(2)} \leqslant \ldots \leqslant x_{u(n)}.$$

We reject the hypothesis H_0 on the significance level α, and apply the alternative hypothesis

$$H_1: \beta > \beta_0 \quad \text{if} \quad T_n \geqslant k_\alpha,$$
$$H_1': \beta < \beta_0 \quad \text{if} \quad T_n \leqslant -k_\alpha,$$
$$H_1'': \beta \neq \beta_0 \quad \text{if} \quad T_n \geqslant k_{\frac{\alpha}{2}} \quad \text{or} \quad T_n \leqslant -k_{\frac{\alpha}{2}}.$$

where k_α and $k_{\frac{\alpha}{2}}$ are critical values which were given by L. K a a r s e m a k e r and A. v a n W i j n g a a r d e n [5] (cf. also [2] and [4]).

2. EXTENSION OF THE THEIL TEST

Let the model

$$(3) \qquad Y = \alpha_0 + \alpha_1 X_1 + \alpha_2 X_2 + \epsilon$$

be given at usually postulated assumptions concerning ϵ, that the vector ϵ has n-dimensional normal distribution $N(0, \sigma^2 I)$. Based on the sample

$$(x_{11}, x_{12}, y_1), (x_{21}, x_{22}, y_2), \ldots, (x_{n1}, x_{n2}, y_n)$$

consisting of n independent observations we should test the hypothesis $H_0: E(Y \mid X_1, X_2) = \alpha_0 + \alpha_1 X_1 + \alpha_2 X_2$.

In order to test the hypothesis about linearity of the model with two explanatory variables by means of the Theil test we find different forms of

$$(4) \qquad R_{s(i)} = y_{s(i)} - a_1 x_{s(i),1} - a_2 x_{s(i),2}$$

where a_1 and a_2 are the estimators of parameters α_1 and α_2, respectively, in the model (3) obtained using LSM, and s is a permutation of the values y_i in the following way:

$$y_{s(1)} \leqslant y_{s(3)} \leqslant y_{s(5)} \leqslant \ldots \leqslant y_{s(4)} \leqslant y_{s(2)}.$$

The test statistic of the extended Theil test is given by the formula

$$(5) \qquad T_n = \sum_{i=1}^{n} \sum_{j=i+1}^{n} \text{sign} \, (R_{s(j)} - R_{s(i)}).$$

3. PROBLEM FORMULATION AND THE METHOD OF INVESTIGATION

The paper will present some conclusions concerning the comparison of powers of Theil test and F test of Fisher – Snedecor [3] which verify the hypothesis about the linearity of a model with two explanatory variables. Besides, some practical indications will be given to enable the investigated tests to be applied in the case of verification of the hypothesis about the linearity of a model with two explanatory variables.

Note that the Theil test is based on a statistic with discrete distribution, while the F test statistic has a continuous distribution. In this case the comparison of test powers is not precise because error probability of the first type is not the same for particular tests. To overcome this difficulty a randomized Theil test was investigated. To evaluate the powers of the tests considered, we used a Monte Carlo experiment, similarly as in [1]. The procedure of the experiment is as follows.

For given sample sizes $n = 10, 20, 40, 60, 100$ we generate two variants (w_1, w_2) of sequence $\{x_{i1}^*\}$ $(i = 1, 2, \ldots, n)$. The first variant w_1 from uniform distribution and the second one w_2 from normal distribution. The needed sequence $\{x_{i1}^*\}$ is obtained by standardization of $\{x_{i1}\}$. It is analysed further on. The sequence $\{x_{i2}\}$ in both variants is generated from the distribution $N(\rho x_{i1}, 1)$, where ρ is a predetermined correlation coefficient between $\{x_{i1}\}$ and $\{x_{i2}\}$. In the paper $\rho = 0.5, 0.9$ were considered.

For each fixed sample size n, both variants of the sequence $\{(x_{i1}, x_{i2})\}$ and nonlinear function g defining the alternative hypothesis, we generate another sequence $\{y_i\}$, where

$$(6) \qquad y_i = g(x_{i1}, x_{i2}) + \xi_i \qquad (i = 1, 2, \ldots, n),$$

ξ_i independent normal variables generated from $N(0, \sigma_\xi)$.

(Note that σ_ξ determines the scatter of empirical points around the surface defined by the function g.)

Similarly to [1], the function g is defined

(7) $\qquad g(x_1, x_2) = c_0 + c_1 x_1 + c_2 x_2 + c_3 x_1^2 + c_4 x_2^2 + c_5 x_1 x_2.$

Assuming that $c_0 = 0$ (because displacement along the y axis does not change the value and order of residuals) and $c_5 = 0$ (the $0y$ axis is then the axis of paraboloid (7), which facilitates the experiment very much and does not affect the generality of conclusions) the function g has the form

(8) $\qquad g(x_1, \dot{x}_2) = c_4 (2v^2 u_1 x_1 - 2u_2 x_2 + v^2 x_1^2 + x_2^2)$

where $v = \sqrt{\dfrac{c_3}{c_4}}$, and (u_1, u_2) are coordinates of the paraboloid extremum (8).

Thus the generated sample

(9) $\qquad (x_{11}, x_{12}, y_1), (x_{21}, x_{22}, y_2), \ldots, (x_{n1}, x_{n2}, y_n)$

is a basis on which the model

(10) $\qquad y_i = \alpha_0 + \alpha_1 x_{i1} + \alpha_2 x_{i2} + \epsilon_i \qquad (i = 1, 2, \ldots, n),$

is estimated.

The LSM estimates a_0, a_1, a_2 of the parameters $\alpha_0, \alpha_1, \alpha_2$ allow us to calculate the residuals (4), which are a basis for determining the values of statistic (5). The comparison of this number with the critical value from table [2] gives the result of the test: acceptance or rejection of H_0.

Repeating this process many times (in our experiment 200 times) we were able to find relatively accurate empirical fraction of rejecting H_0 (an empirical test power).

Note that in the experiment the sample contains points generated by paraboloid. Hence this variant allows us to reject the linearity hypothesis of the relationship between Y, X_1 and X_2.

We do not want to restrict the analysis only one form of the alternative hypothesis, determined by the function g. Thus we changed some of its parameters.

Let us consider five variants of the parameter: ($v = 0.0, 0.2, 0.5, 1.0, 3.0$), and seven variants of the position of the paraboloid extremum: $(u_1, u_2) = (0, 0), (1, 1), (1, 3), (1, 10), (3, 3), (3, 10), (10, 10)$. Due to similarity of the results we present only the variant $v = (1, 0)$ and $(u_1, u_2) = (0, 0)$. Let

$$g(x_{i1}, x_{i2}) = \beta_0 + \beta_1 x_{i1} + \beta_2 x_{i2} + \theta_i$$

where $\beta_0, \beta_1, \beta_2$ are the parameters of g approximating plane, obtained by means of the least squares method. Hence, from (6) we have

$$y_i = \beta_0 + \beta_1 x_{i1} + \beta_2 x_{i2} + \theta_i + \xi_i$$

Variants σ_ξ^2 are determined according to the value of coefficient

(11) $$\psi^2 = \frac{\sigma_\xi^2}{S_\theta^2 + \sigma_\xi^2}$$

assuming that

$$S_\theta^2 = \frac{1}{n} \sum_{i=1}^{n} \theta_i^2.$$

At such a procedure σ_ξ^2 and S_θ^2 are proportional to c_4^2. Hence, without any loss in generality we may assume $c_4 = 1$ (cf. (8)).

In the experiment $\psi^2 = 0.01, 0.05, 0.10, 0.25, 0.50, 0.75, 0.90$, although other variants were also considered. For bigger value of ψ^2, more of the paraboloids (8) become similar to a plane, and the dispersion of the empirical points is bigger around it (we can expect greater test powers for lower values of ψ^2).

In the paper we present some variants for which we took into account $n = 10, 20, 40, 60, 100$, two variants w_1 and w_2 of the distribution X_1, $\rho = 0.5, 0.9$, $(u_1, u_2) = (0, 0)$, $v = (1, 0)$, and $\psi^2 = 0.01, 0.05, 0.10, 0.20, 0.50, 0.75, 0.90$. The results are given in Table 1.

Table 1. Empirical frequencies of rejecting the hypothesis on linearity of the model $y_i = \alpha_0 + \alpha_1 x_{i1} + \alpha_2 x_{i2} + \epsilon_i$ using the Theil and F tests for $\alpha = 0.05$

n	ψ^2	$w = 1, \rho = 0.5$		$w = 2, \rho = 0.5$		$w = 1, \rho = 0.9$	
		T_n	F	T_n	F	T_n	F
10	0.01	15.78	100.00	0.00	100.00	99.13	100.00
	0.05	20.17	100.00	0.00	100.00	86.80	100.00
	0.10	21.52	100.00	0.00	100.00	77.87	100.00
10	0.25	16.17	80.50	0.00	84.50	65.74	82.50
	0.50	11.48	30.50	0.07	35.50	40.63	29.50
	0.75	6.02	13.50	0.13	10.50	18.54	12.50
	0.90	5.83	8.50	1.76	6.00	8.52	8.00
20	0.01	100.00	100.00	0.00	100.00	100.00	100.00
	0.05	100.00	100.00	0.00	100.00	100.00	100.00
	0.10	99.70	100.00	0.00	100.00	100.00	100.00
20	0.25	86.31	100.00	0.00	100.00	99.70	100.00
	0.50	54.82	91.50	0.00	91.50	89.01	92.50
	0.75	27.21	38.00	1.20	43.50	49.42	42.00
	0.90	13.90	10.50	3.50	15.50	21.31	13.50
40	0 01	100 00	100 00	0.00	100 00	100.00	100 00
	0 05	100.00	100.00	0 00	100.00	100.00	100 00
	0 10	100.00	100.00	0.00	100.00	100 00	100 00
40	0 25	89.27	100.00	0.50	100 00	100 00	100 00
	0 50	43.10	100.00	2 77	100.00	99 27	100 00
	0 75	19.27	82.50	5.50	85 50	82 77	77 00
	0 90	10.50	33.50	6.00	52 00	34 77	35 50
60	0.01	100.00	100.00	24.61	100.00	100.00	100.00
	0.05	100.00	100.00	24.02	100.00	100.00	100.00
	0.10	100.00	100.00	25.52	100.00	100.00	100.00
60	0.25	100.00	100.00	23.00	100.00	100.00	100.00
	0.50	96.00	100.00	17.50	100.00	100.00	100.00
	0.75	71.52	98.50	11.02	94.50	93.50	97.00
	0.90	27.02	54.50	7.00	57.00	54.50	51.50
100	0.01	100.00	100.00				
	0.05	100.00	100.00				
	0.10	100.00	100.00				
100	0.25	100.00	100.00				
	0.50	100.00	100.00				
	0.75	74.34	100.00				
	0.90	31.50	77.00				

4. CONCLUSIONS

On the basis of the obtained results we can draw some general conclusions on the comparison of powers of the Theil test and F test according to changing parameters.

1. The Theil test power as well as the F test power increase when the sample size increases.

2. The distribution of the variable X_1 affects significantly the Theil test power. For the variant w_2 (normal distribution of the variable X_1) the Theil test is very weak, while for the variant w_1 (uniform distribution of X_1) it is relatively strong. This results from the fact that in the first variant positive and negative residuals are better displaced.

(The problem of selecting an appropriate criterion for ordering the residuals for the Theil test is being considered.)

3. The increase of the correlation coefficient between X_1 and X_2 causes the increase of the power of the Theil test, while the F test reacts inversely to these changes.

4. As can be expected, the Theil test is weaker than the F test. However, it should be noted that the F test cannot be applied to all forms of the alternative hypothesis.

5. Great powers obtained in most cases confirm the applicability of the Theil test in practice.

REFERENCES

[1] C. Domański — K. Markowski — A. Tomaszewicz, Run test for linearity hypothesis of econometric model with two exogenous variables, *Przegląd Statystyczny*, 25 (1978), 87-93 (Polish; English and Russian summary).

[2] C. Domański — K. Markowski — A. Tomaszewicz, Table for Theil test, *Przeglad Statystyczny*, 1980, (Polish; English and Russian summery — to be published).

[3] A.S. Goldberger, *Econometric theory*, John Wiley and Sons, Inc., New York, 1964.

[4] M. Hollander – D.A. Wolfe, *Nonparametric statistical methods*, John Wiley and Sons, Inc., New York, 1973.

[5] L. Kaarsemaker – A. van Wijngaarden, Tables for use in rank correlation, *Statistica Nederlandica*, 7 (1953), 41-54.

[6] H. Theil, A rank-invariant method of linear and polynomial regression analysis I, II, III, *Proceedings of the Koninklijke Nederlandse Akademie van Wetenschappen A*, 53 (1950), 521-535, 1397-1412.

[7] H. Theil, *Principles of econometrics*, North-Holland Publ. Co., Amsterdam – London, 1971.

C. Domański

University of Łodź, Institute of Econometrics and Statistics, PL-90-214 Łodź, ul. Rewolucji 1905 r. 41, Poland.

COLLOQUIA MATHEMATICA SOCIETATIS JÁNOS BOLYAI
32. NONPARAMETRIC STATISTICAL INFERENCE,
BUDAPEST (HUNGARY), 1980.

A STATISTICAL TEST FOR EXTREME VALUE DISTRIBUTIONS*

J. GALAMBOS

1. INTRODUCTION

There is an increasing demand for more accurate techniques for dealing with stochastic problems in which the governing law is an extreme of a sequence $X_1, X_2, \ldots, X_n$ of random variables. Such diverse areas as dam building, evaluation of air quality, reliability of equipment, strength of materials and effects of food additives or other chemicals all belong to an extreme value problem. While the model building aspect of these problems has been investigated extensively in the past years, the statistical theory lags far behind. The purpose of the present paper is to point out some of the difficulties of statistical extreme value theory which are rarely encountered in other areas of statistics and to propose a new test in connection with a general model.

*This research was supported by a grant to Temple University from the Air Force Office of Scientific Research under grant No. AFOSR-78-3504.

Since Section 3.12 of the present author's recent book on extremes (G a l a m b o s [1]) describes several applied models from the mentioned areas, the discussion of some selected problems here are limited to introducing and clarifying the assumptions of the model in which a test will be proposed.

Let t be a time unit and let $X_1, X_2, \ldots$ be the highest water levels of a river R at location C during successive time intervals of unit length t. For building dikes or a dam on R at C, one would like to predict the highest water level for an extended future. Assume that initial observations $X_1 = x_1, X_2 = x_2, \ldots, X_k = x_k$ are available, which are used to "determine" the stochastic behaviour of $Z_n = \max (X_j: m < j \leqslant m + n)$, where m is an integer with $m \geqslant k$.

Notice that, in principle, the same model is applicable to determining the distribution of a single X_j as to Z_n itself. Namely, X_j itself is the maximum of successive high water levels over smaller time units. For example, if $t =$ one year, then X_j is the maximum of 12 monthly records or 365 daily floods (highest levels). This argument would in fact be completely correct if one could freely assume that successive high water levels (whatever be the time unit) are independent and identically distributed (i.i.d.). In such a case, the distribution of both X_j and Z_n would be an asymptotic distribution of a properly normalized maximum (accounting for the different time units) of i.i.d. random variables. Since water levels are nonnegative and unbounded, classical theory would imply (see Theorems 2.4.1 and 2.4.3 on p. 71 in G a l a m b o s [1]) that, apart from scale and location, the distribution of both X_j and Z_n is either

(1.1) $H_{1,\gamma}(x) = \exp (- x^{-\gamma})$ $(x > 0, \ \gamma > 0)$,

or

(1.2) $H_{3,0}(x) = \exp (- e^{-x})$ $(-\infty < x < +\infty)$.

Hence, all observed values x_j, $1 \leqslant j \leqslant k$, could be used to find the actual distribution and for estimating its parameters.

The unacceptable part of the above argument is that the assumptions may be approximately valid if the time unit t is large but they certainly fail for small time units. For example, if $t =$ one year again, then the X_j can be assumed with confidence to be identically distributed. The independence is not satisfied but further the indeces of X_j and X_t are apart, the weaker is their dependence. Therefore, Leadbetter's [4] model (which is also reproduced in the quoted book of the present author in Section 3.7) can be used as an approximation for Z_n. This implies that the asymptotic distribution of Z_n is again either (1.1) or (1.2) (with appropriate units for location and scale). However, in a model for X_j, in which we divide a year into 365 days, neither stationarity (identical distributions on successive days throughout a year) nor independence can be assumed. In other words, the X_j follow an extreme value distribution in a dependent model which may be different from the location and scale families of (1.1) and (1.2) (see the individual sections of Chapter 3 in Galambos [1] for a number of dependent extreme value models). This leads to the following general problem:

(M) Given the observations $X_1 = x_1, X_2 = x_2, \ldots, X_k = x_k$, determine the asymptotic distribution of $Z_n = \max (X_j: m < j \leqslant m + n)$, if its asymptotic distribution does exist when its location and scale are properly chosen, and if the following conditions are satisfied:

(i) the X_j are identically distributed with common distribution function $F(x)$;

(ii) $F(x) < 1$ for all x;

(iii) the distribution $H_n(x)$ of Z_n is approximately $F^n(x)$ for a sequence $x = z_n = a_n + b_n z$ such that $F^n(z_n)$ converges to a function $H(z)$.

While assumptions (i) and (ii) are clear from the preceding example, a few comments are in order in regard to assumption (iii). Since the location a_n and the scale $b_n > 0$ of Z_n are dependent on the choices of units of measurement as well as on the length of time over which Z_n is applied, we can expect to determine the asymptotic behaviour of the normalized variable $\dfrac{Z_n - a_n}{b_n}$ only. Therefore, the distribution

$H_n(x)$ of Z_n enters our investigation only at values $x = a_n + b_n z$, namely,

$$P\left(\frac{Z_n - a_n}{b_n} < z\right) = P(Z_n < a_n + b_n z) = H_n(a_n + b_n z).$$

Hence the fact that the X_j are asymptotically independent in some sense reduces to the relation

$$(1.3) \qquad H_n(a_n + b_n z) \sim F^n(a_n + b_n z)$$

for some a_n and $b_n > 0$. If we now add the assumption that the normalized form of Z_n has an asymptotic distribution, we obtain assumption (iii).

One could argue that the actual assumption of our model is equivalent to a model in which the component variables X_j are assumed to be independent. We offer the following example to show that this is not the case.

We make spot-checks and on each occasion we take d measurements $X_1, X_2, \ldots, X_d$ of a quantity X which is assumed to be normally distributed (the reader who belongs to the lognormal school of air quality data can think of X as the logarithm of a particular pollutant concentration). Since the X_j, $1 \leqslant j \leqslant d$, are successive measurements, they are dependent, and thus their distribution is d-variate normal. Now if N spot-checks are made independently, we have $n = Nd$ observations, and if we are asked to evaluate the stochastic characteristics of the maximum Z_n of these observations, we can proceed as follows. We first consider the maximum Z_{jn} of the j-th component X_j of each vector, from which

$$(1.4) \qquad Z_n = \max(Z_{1n}, Z_{2n}, \ldots, Z_{dn}).$$

We thus have that $F_n(x_1, x_2, \ldots, x_d)$ is the distribution function of the vector $(Z_{1n}, Z_{2n}, \ldots, Z_{dn})$, then

$$H_n(x) = P(Z_n < x) = F_n(x, x, \ldots, x).$$

The general theory of multivariate extremes yields (see G a l a m b o s [1], pp. 257-258) that, with suitable sequences a_n and $b_n > 0$,

$$H_n(a_n + b_n z) \sim \Phi^n(a_n + b_n z),$$

where $\Phi(x)$ is a univariate normal distribution. However, for other values of the variable x, $H_n(x)$ cannot be approximated by $\Phi^n(x)$. Namely, if $F(x_1, x_2, \ldots, x_d; R)$ is the d-dimensional normal distribution function with variance-covariance matrix R, then

$$H_n(x) = F^n(x, x, \ldots, x; R),$$

but $F(x, x, \ldots, x; R)$ can be very significantly different from $\Phi^d(x)$. Therefore, the combined observations on X are not independent even asymptotically but our assumptions are satisfied. Of course, our interest is in the case when $F(x)$ is unknown; the normality was assumed only for the purpose of illustration.

2. DISCUSSION OF STATISTICAL DIFFICULTIES

We now discuss the statistical aspects of model (M) of the Introduction in which (dependent) observations $X_1 = x_1, X_2 = x_2, \ldots, X_k = x_k$ are taken from a population with an unknown distribution $F(x)$ and the stochastic evaluation of the maximum $Z_k^* = \max(X_1, X_2, \ldots, X_k)$ is the aim. Evidently, our set of observations contains only a single observation on Z_k^* and thus the inference has to be based on as large a subset of $(x_1, x_2, \ldots, x_k)$ as possible.

One could try to "determine" $F(x)$ first (estimation by the empirical distribution function or a test for $F(x)$ and then estimation of its parameters). When a decision has been reached on $F(x)$, then (1.3) would be used to approximate the distribution of Z_k^* or a future maximum as described in model (M). This is the most dangerous course of action, however, and it has to be avoided. Namely, the empirical distribution function $F_k(x)$ based on the observations $(x_1, x_2, \ldots, x_k)$ does not give any information on $F(x)$ for any x that exceeds the observed value of Z_k^*. Therefore, we could not make a distinction between two distributions which are considerably different in their upper tails but otherwise the distributions are uniformly close to each other. Since a maximum is dependent on the upper tail only (in our model (M)), such an approach would lead to a significant misjudgement. On p. 90 of G a l a m b o s [1] an example is given when estimation or a test would accept both normality and lognormality but contradicting conclusions could arise in terms of a maximum. While that example can be considered a text-book case, it is an important signal to the need of reevaluating earlier techniques and results in applied extreme value theory. E.J. Gumbel, the pioneer of such an applied theory, repeatedly expresses his worries in his publications that the classical assumptions (the component variables are always assumed by him to be i.i.d.) can be used at best as crude approximations and thus the theory he applied can not be used to justify good fits of distributions in the final result (see particularly p. 238 of Gumbel [2]). And yet, seemingly always good fits are obtained. For example, in the case of floods discussed in the Introduction, as was pointed out there is no justification to assume that the daily highest water levels X_j have a distribution belonging to the location and scale families of either (1.1) or (1.2). All published cases, however, seem to fit one of these families. What is overlooked in this empirical justification of using the family (1.1) is the fact that the shape parameter $\gamma > 0$ makes this so large a family that any data with a single mode can be fitted to it. As a matter of fact, if we estimate A, B and γ from a set of data drawn from a normal population, we shall get a good fit with the distribution $H_{1,\gamma}(A + Bx)$. This would indicate that the asymptotic distribution of the properly normalized maximum also belongs to the family $H_{1,\gamma}(A_1 + B_1 x)$, although the theory tells us that the appropriate asymptotic distribution is $H_{3,0}(C + Dx)$ with some C and $D > 0$ (see p. 65 of Galambos [1]). This contradiction is due to the mistaken technique of trying to identify the population distribution for determining the asymptotic distribution of an extreme. The test proposed in the next section tries to correct this error. It should be added that if more than one decision is to be made from the data, only one of which is an extreme value problem (for example, the "usual" water level as well as the highest water level are to be determined), then one may proceed with standard methods to settle the nonextremal problem which may include the use of a population distribution but the additional problem related to the extreme value should be decided upon with the care described above.

3. A NEW TEST

The observations $X_1 = x_1, X_2 = x_2, \ldots, X_k = x_k$ are taken from an unknown distribution $F(x)$. The X_j can be dependent but practical considerations justify to assume that the assumptions of model (M) in the Introduction are satisfied. Then an appeal to classical theory yields that the asymptotic distribution $H(z)$ of the normalized maximum $\dfrac{Z_n - a_n}{b_n}$, where a_n and $b_n > 0$ are (unknown) sequences, belongs to a location and scale family represented by either (1.1) or (1.2). In the sequel, members of such families will be called of the same type; that is, $F(x)$ is of the type of $G(x)$ if $F(x) = G(A + Bx)$ with some constants A (location) and $B > 0$ (scale). We now want to test

H_0: $H(z)$ is of the type of $H_{3,0}(z)$.

Rejection of H_0 automatically means the acceptance of

H_1: $H(z)$ is of the type of $H_{1,\gamma}(z)$.

In the model (M), $H(z)$ is also the asymptotic distribution of $\dfrac{Z_k^* - a_k}{b_k}$, where $Z_k^* = \max(X_1, X_2, \ldots, X_k)$. Therefore a decision on H_0 is to be made on the observations $(x_1, x_2, \ldots, x_k)$. The test is as follows.

Choose a number t. Select those $X_j = x_j$, $1 \leqslant j \leqslant k$, that exceed t. For each $X_j > t$, compute the transformed value

$$Y_j = \frac{X_j - t}{E(X_j - t \mid X_j > t)}.$$

Test now the hypothesis H_0^* that the distribution of the Y_j is unit exponential: $P(Y_j < x) = 1 - e^{-x}$. If the test accepts H_0^*, accept H_0, while rejecting H_0^* means the acceptance of H_1.

The normalization $E(X_j - t \mid X_j > t)$ in Y_j is to be estimated by the arithmetical mean of those $(X_j - t)$ for which $X_j > t$. With this, we of course assume implicitly that $E(X_j)$ is finite.

Notice that the distribution of Y_j does not have any unknown parameters. Therefore the most straightforward tests are applicable (a chi square test or, in many instances, a graphic method on probability paper). For a number of possibilities of testing H_0^*, see R. Pyke [6].

The mathematical justification of the proposed test is a consequence of a result of L. de Haan [3] (which result is stated as Theorem 2.1.3 on p. 52 of Galambos [1]). Namely, if $E(X_j)$ is finite, then

$$E(X_j - t \mid X_j > t) = \frac{1}{1 - F(t)} \int_t^{+\infty} (1 - F(y)) \, dy,$$

and, for $x > 0$,

$$P(Y_j > x \mid X_j > t) = P\{X_j > t + x E(X_j - t \mid X_j > t) \mid X_j > t\} =$$

$$= \frac{P\{X_j > t + x E(X_j - t \mid X_j > t)\}}{P(X_j > t)}.$$

By the mentioned theorem, this is asymptotically e^{-x} as t increases.

As it stands so far, the number t should be preassigned. However, in order to guarantee to have several X_j which exceed t, one would like to choose t after the observations $x_1, x_2, \ldots, x_k$ have been taken. In this case, t is a function of $X_1, X_2, \ldots, X_k$ and thus it is a random variable. An easy application of a result of the present author, combined with a classical result on "large order statistics" (Theorems 4.1.1 and 2.8.1, respectively, in Galambos [1]), permits us to choose t as the value of the m-th largest order statistic where m does not increase with k. For example, if k is large and $m = 20$, then the 20 largest observations can be used in the test.

In concluding, let us make a comment on the literature. There are only two papers relevant to the present investigation. Pickands [5] was the first who recognized that a subjectivity in choosing between $H_{1,\gamma}(x)$ and $H_{3,0}(x)$ ought to be avoided by an objective statistical method. The method proposed by Pickands is not a test and thus a direct comparison is difficult. A more recent work by Weissman [7] does propose a test but it contains one basic error. Namely, the test is based on certain spacings which are i.i.d. exponential under the null hypothesis. Hence the test is not just for exponentiality as claimed but he has to test whether those few upper spacings are independent, and if yes, then whether they are exponential. The first part of this test is very crucial and it is impossible to carry out on the base of just one set of observations.

Added in proof. In the forthcoming paper "Statistical choice of univariate extreme models" (to appear in Statistical Distributions in Scientific Work, D. Reidel, Dordrecht, 1981), J. Tiago de Oliveira proposes another method to test H_0 against H_1. Although his investigations are limited to the i.i.d. case, his asymptotic test easily extends to our model (M). It would be of interest to compare tests based on our approach with those of Tiago de Oliveira.

REFERENCES

[1] J. Galambos, *The Asymptotic Theory of Extreme Order Statistics,* John Wiley & Sons, New York, 1978.

[2] E.J. Gumbel, *Statistics of Extremes,* Columbia University Press, New York, 1958.

[3] L. de Haan, A form of regular variation and its application to the weak convergence of sample extremes, *Mathematical Centre Tracts,* Vol. 32, Amsterdam, 1970.

[4] M.R. Leadbetter, On extreme values in stationary sequences, *Zeitschrift für Wahrscheinlichkeitstheorie und verw. Geb.,* 28 (1974), 289-303.

[5] J. Pickands III, Statistical inference using extreme order statistics, *Ann. Statist.,* 3 (1975), 119-131.

[6] R. Pyke, Spacings, *J. Royal Statist. Soc. Ser. B,* 27 (1965), 395-436.

[7] I. Weissman, Estimation of parameters and large quantiles based on the k largest observations, *J. Amer. Statist. Assoc.*, 73 (1978), 812-815.

J. Galambos

Department of Mathematics, Temple University, Philadelphia, Pennsylvania 19122, USA.

AN APPLICATION OF THE METHOD OF SIEVES: FUNCTIONAL ESTIMATOR FOR THE DRIFT OF A DIFFUSION

S. GEMAN*

1. INTRODUCTION

From an observation of a sample path of a diffusion process one can construct consistent estimators for the diffusion drift. If the form of the drift function is known up to a finite collection of parameters then it is possible to use maximum likelihood, and obtain consistent and asymptotically normal estimators (see Brown and Hewitt [3], Feigin [4], Lee and Kozin [8], and Liptser and Shiryayev [9]). Even when no parametric form is known, concistent (and in some cases asymptotically normal) estimators for the value of the drift at a fixed argument have been developed (Banon [1], Banon and Nguyen [2], and Nguyen and Pham [10]).

This paper is also about nonparametric estimation of the drift function. But the estimator developed here is distinguished from the nonparametric estimators cited above by:

*Supported by the Department of the Army under grant DAAG 2980-K-0006.

(1) being a *functional* estimator,

(2) being based on the principle of maximum likelihood.

By a "functional estimator" I mean that at each t the estimation procedure produces a function defined on a prescribed interval, and, as $t \to \infty$, this estimator converges (a.s.) to the drift in the sense of a function space norm. To be more precise about (2), let us look at a diffusion equation and an associated likelihood function:

$$(1.1) \qquad dx_t = g(x_t)dt + \sigma dw_t \qquad x_0 = x_o.$$

w_t is a standard (one-dimensional) Brownian motion and x_0 is a constant. g and σ are assumed to be unknown; we wish to estimate g from an observation of a sample path of x_t. It is well known that the distribution of x_s, $s \in [0, t]$ is absolutely continuous with respect to the distribution of σw_s, $s \in [0, t]$ (assuming some mild regularity condition on g). A likelihood function for the process x_s, $s \in [0, t]$ is the Radon – Nikodym derivative:

$$(1.2) \qquad \exp \left\{ \int_0^t g(x_s)\, dx_s - \frac{1}{2} \int_0^t g(x_s)^2 \, ds \right\}.$$

The "natural" estimator for g would maximize (1.2) over a suitable parameter space, most appropriately the space of uniformly Lipschitz continuous functions. But the maximum of the likelihood is not attained, either in this or in any other of the usual function spaces. Some difficulty is not unexpected: maximum likelihood typically fails in nonparametric settings. To preserve the principle of maximum likelihood in nonparametric problems, G r e n a n d e r [7] suggests a "Method of Sieves": maximize the likelihood over a subset of the parameter space, allowing the subset to grow as the observations increase. The method of sieves produces consistent estimators for a wide variety of nonparametric problems. This paper presents an application of the method to nonparametric estimation of the drift of a diffusion.

Pretty much the minimal conditions for consistent estimation of g are:

(1) conditions for the existence and uniqueness of the solution to (1.1),

(2) that the process x_t be recurrent (i.e. for every level λ there exists $t_1, t_2, \ldots$ increasing to infinity such that $x_{t_i} = \lambda$ for every i).*

It is under these conditions that the consistency of the estimator will be demonstrated.

2. STATEMENT OF MAIN RESULT

Start with the usual assumption for existence and uniqueness in (1.1):

A1. For some constant L

$$|g(x) - g(y)| \leqslant L|x - y|$$

for all $x, y \in R$.

And, an assumption which is equivalent to recurrence (see F r i e d - m a n [5], Chapter 9):

A2. If

$$\theta(x) = \int_0^x \exp\left\{-\frac{2}{\sigma^2} \int_0^z g(u)\, du\right\} dz$$

then $\theta(+\infty) = +\infty$ and $\theta(-\infty) = -\infty$.

The estimator will approximate g on a fixed, but arbitrary, interval $[\lambda_1, \lambda_2]$. For this purpose the likelihood function, (1.2), will be replaced by a function which depends only on the behavior of x_t during the time spent in the interval $[\lambda_1, \lambda_2]$. Specifically, we will seek to maximize

(2.1) $f_t(x_.; \alpha) \equiv$

$$\equiv \exp\left\{\int_0^t I_{[\lambda_1,\lambda_2]}(x_s)\alpha(x_s)\, dx_s - \frac{1}{2} \int_0^t I_{[\lambda_1,\lambda_2]}(x_s)\alpha(x_s)^2\, ds\right\} =$$

---

*The necessity of (2) derives from the fact that the measures on $C[0, t]$ induced by the solution to (1.1) and indexed by g, form an absolutely continuous family for any finite t (under some well-known regularity conditions on g). The situation for σ is quite different. σ can in principle be determined from any (arbitrarily small) interval of observation of $x_.$.

$$= \exp \left\{ \int_0^t I_{[\lambda_1, \lambda_2]}(x_s) \left(\alpha(x_s) g(x_s) - \frac{1}{2} \alpha(x_s)^2 \right) ds + \right.$$

$$\left. + \sigma \int_0^t I_{[\lambda_1, \lambda_2]}(x_s) \alpha(x_s) \, dw_s \right\}.$$

As with the full likelihood function, (1.2), nothing useful comes from an unconstrained maximization of (2.1). One remedy is to introduce a sieve, S_t, parameterized by the length of the interval of observation $[0, t]$:

$$S_t = \left\{ \sum_{j=1}^{m_t} a_j \psi_j(x) : \sum_{j=1}^{m_t} |a_j| \leqslant k_1 (\log m_t)^{k_2} \right\}$$

where

1. m_t is a nondecreasing sequence of integers governing the size of the sieve at time t, and k_1 and k_2 are arbitrary positive constants,

2. $\{\psi_j(x)\}_{j=1}^{\infty}$ is any sequence of measurable functions satisfying

a. $|\psi_j(x)| \leqslant 1$ $x \in [\lambda_1, \lambda_2]$ and all j

b. Every continuous function f on $[\lambda_1, \lambda_2]$, satisfying $f(\lambda_1) = f(\lambda_2)$, can be uniformly approximated by a linear combination of $\psi_1(x), \psi_2(x), \ldots$.

For example, $\{\psi_j(x)\}_{j=1}^{\infty}$ may be the trigonometric polynomials $\left\{ \exp \left[2\pi i l \left(\frac{x - \lambda_1}{\lambda_2 - \lambda_1} \right) \right] \right\}_{l=-\infty}^{\infty}$ (with a change of indices), or the polynomials $\left\{ \left(\frac{x}{|\lambda_1| + |\lambda_2|} \right)^l \right\}_{l=0}^{\infty}$. Note that an implication of b is that the span of $\{\psi_j(x)\}_{j=1}^{\infty}$ is dense in $L_2([\lambda_1, \lambda_2], B, dm)$, for any finite measure dm (B = Borel sets in $[\lambda_1, \lambda_2[$).

Define M_t to be the set of functions in S_t which maximize the "likelihood" (2.1):

$$M_t = \{\alpha \in S_t : f_t(x_.; \alpha) = \sup_{\beta \in S_t} f_t(x_.; \beta)\}.$$

(If we write $\alpha = \sum\limits_{j=1}^{m} a_j \psi_j$, then $f_t(x \, . \, ; \alpha)$ is continuous in $a_1, a_2, \ldots, a_m$ and it follows that M_t is not empty.) How fast should the sieve grow in order that M_t converge, in some sense, to g? (i.e. how rapidly can the sequence m_t increase to ∞?) Clearly, the growth of the sieve should be governed not directly by t, but by the amount of time that the process has spent in the interval $[\lambda_1, \lambda_2]$ up to time t. Define

$$\mathscr{I}(t) = \int_0^t I_{[\lambda_1, \lambda_2]}(x_s) \, ds.$$

The Theorem says that if m_t grows sufficiently slowly with respect to $\mathscr{I}(t)$, then

$$\sup_{\alpha \in M_t} \| \alpha - g \| \to 0 \quad \text{almost surely,}$$

where the norm is L_2 with respect to "the natural" measure for this problem:

Proposition. *Asssume* A1 *and* A2. *Then for every Borel set* A *in* $[\lambda_1, \lambda_2]$

$$(2.2) \qquad \lim_{t \to \infty} \frac{\int_0^t I_A(x_s) \, ds}{\int_0^t I_{[\lambda_1, \lambda_2]}(x_s) \, ds}$$

exists and is constant almost surely, and the set function defined by this limit is a probability measure.

The proof is deferred (follows from Lemma 1 below). I will use $\mathcal{O}(A)$ to refer to this measure.

Theorem. *Assume* A1 *and* A2, *and that* m_t *is a nondecreasing sequence of integers, diverging to* $+\infty$, *and satisfying*

$$m_t \leqslant k_3 \mathscr{I}(t)^{1-\delta}$$

for some positive constant k_3 *and* δ. *Then, as* $t \to \infty$,

$$\sup_{\alpha \in M_t} \int_{\lambda_1}^{\lambda_2} |\alpha(x) - g(x)|^2 \mathscr{O}(dx) \to 0 \quad \textit{almost surely.}$$

3. PROOF OF THE THEOREM

It will be convenient to assume that $x_0 \notin [\lambda_1, \lambda_2]$ (the modification for $x_0 \in [\lambda_1, \lambda_2]$ is trivial). Define two sequences of stopping times as follows:

$$e_1 = \inf\{t \geqslant 0\colon x_t \in [\lambda_1, \lambda_2]\}$$

Given $e_1, \ldots, e_k$,

$$l_k = \begin{cases} e_k + 1 & \text{if } x_{e_k+1} \notin [\lambda_1, \lambda_2], \\ \inf\{t \geqslant e_k + 1\colon x_t \notin [\lambda_1, \lambda_2]\} & \text{if } x_{e_k+1} \in [\lambda_1, \lambda_2]. \end{cases}$$

Given $l_1, \ldots, l_k$,

$$e_{k+1} = \inf\{t \geqslant l_k\colon x_t \in [\lambda_1, \lambda_2]\}.$$

Because of assumption A2, x_t is recurrent, and therefore these stopping times are well-defined. Notice that $\bigcup_{k=1}^{\infty} [e_k, l_k]$ includes all of the time that x_t spends on the interval $[\lambda_1, \lambda_2]$.*

Because of the strong Markov property of x_t, the discrete time process $x_{e_1}, x_{e_2}, \ldots$ is Markov with state space $\{\lambda_1, \lambda_2\}$ and stationary transition probabilities

$$p_{ij} = P(x_{e_2} = \lambda_j \mid x_{e_1} = \lambda_i) \qquad (i = 1, 2, \ j = 1, 2).$$

Obviously $p_{ij} > 0 \quad \forall i$ and j, and therefore there exists a unique stationary distribution $\{\pi_1, \pi_2\}$ on the states $\{\lambda_1, \lambda_2\}$. For any bounded measurable function α define

---

*After examining the proof the reader may wonder why I do not simply define e_{k+1} to be the first entrance of x_t into $[\lambda_1, \lambda_2]$ (upcrpssing of λ_1 or downcrossing of λ_2) after $e_k + 1$, and then take $l_k = e_{k+1}$. The difficulty here is that, without additional assumptions, it may happen that $E(l_k - e_k) = \infty$ (as when $g \equiv 0$). In this case the proof would no longer apply.

$$\tilde{E}[\alpha(x)] = \sum_{i=1}^{2} \sum_{j=1}^{2} \pi_i p_{ij} E\left[\int_{e_1}^{l_1} \alpha(x_s)\, ds \mid x_{e_1} = \lambda_i,\ x_{e_2} = \lambda_j\right].$$

A routine argument establishes the existence of a $\rho < 1$ such that

$$(3.1) \qquad P((l_1 - e_1) > p \mid x_{e_1} = \lambda_i,\ x_{e_2} = \lambda_j) \leqslant \rho^p$$

for all i and j, and all $p = 0, 1, \ldots$. Hence, the "expectation", $\tilde{E}[\alpha(x)]$, is finite for bounded α.

The key to the proof of the Theorem will be the following lemma:

Lemma 1. *Fix $\epsilon > 0$. There exists a constant k (which may depend on ϵ) such that for any constant $c \geqslant 1$, and any two measurable functions α and β uniformly bounded by c,*

$$P\left(\left|\frac{1}{n} \sum_{k=1}^{n} \left\{\int_{e_k}^{l_k} \alpha(x_s)\, ds + \int_{e_k}^{l_k} \beta(x_s)\, dw_s\right\} - \tilde{E}[\alpha(x)]\right| > \epsilon\right) \leqslant$$

$$\leqslant k\left(1 - \frac{1}{kc^2} + \frac{k}{n}\right)^{\frac{n}{k}} \quad \text{for all } n \geqslant 1.$$

If we assume Lemma 1 (it will be proved later) then we can establish the Proposition as follows: Take $\beta(x) = 0$ and $\alpha(x) = I_A(x)$, where A is any Borel set in $[\lambda_1, \lambda_2]$. By Lemma 1, and the Borel — Cantelli Lemma,

$$\frac{1}{n} \sum_{n=1}^{n} \int_{e_k}^{l_k} I_A(x_s)\, ds \to \tilde{E}[I_A(x)] \quad \text{almost surely,}$$

as $n \to \infty$ If $n_t = \sup\{k: l_k \leqslant t\}$ then

$$\frac{1}{n_t} \int_0^t I_A(x_s)\, ds = \frac{1}{n_t} \sum_{k=1}^{n_t} \int_{e_k}^{l_k} I_A(x_s)\, ds + \frac{1}{n_t} \int_{l_{n_t}}^t I_A(x_s)\, ds.$$

Obviously,

$$\frac{1}{n_t} \int_{l_{n_t}}^t I_A(x_s)\, ds \to 0 \quad \text{almost surely,}$$

as $t \to \infty$. Hence, as $t \to \infty$,

(3.2) $\dfrac{1}{n_t}\displaystyle\int_0^t I_A(x_s)\,ds \to \widetilde{E}[I_A(x)]$ almost surely

and, therefore,

(3.3) $\dfrac{\displaystyle\int_0^t I_A(x_s)\,ds}{\displaystyle\int_0^t I_{[\lambda_1,\lambda_2]}(x_s)\,ds} \to \dfrac{\widetilde{E}[I_A(x)]}{\widetilde{E}[I_{[\lambda_1,\lambda_2]}(x)]}$ almost surely.

Observe that:

1. This proves the Proposition since the right hand side of (3.3) is a probability measure.

2. To prove the Theorem, it will be enough to show

$$\sup_{\alpha \in M_t} \widetilde{E}\,|I_{[\lambda_1,\lambda_2]}(x_s)(\alpha(x)-g(x))|^2 \to 0 \quad \text{almost surely}$$

as $t \to \infty$, and

3. (3.2) implies that

(3.4) $\dfrac{\mathscr{I}(t)}{n_t} \to \widetilde{E}[I_{[\lambda_1,\lambda_2]}(x)]$ almost surely

as $t \to \infty$.

Another consequence of Lemma 1 is:

Lemma 2.

$$\varlimsup_{t\to\infty}\ \sup_{\alpha\in S_t}\ \left|\frac{1}{n_t}\log f_t(x_\cdot;\alpha)\ -\right.$$

$$\left. -\ \widetilde{E}\Big[I_{[\lambda_1,\lambda_2]}(x)\big(\alpha(x)g(x)-\tfrac{1}{2}\,\alpha(x)^2\big)\Big]\right| = 0 \quad \text{almost surely.}$$

Assume, for now, that this is proven too. Then, to prove the Theorem, choose for each $t \geqslant 0$ $\ \alpha_t \in S_t$ such that

$$\widetilde{E}\,|I_{[\lambda_1,\lambda_2]}(x)(\alpha_t(x)-g(x))|^2 \to 0$$

as $t \to \infty$. (The assumptions about $\{\psi_j(x)\}_{j=1}^{\infty}$ guarantee that this can be done.) And now reason that

$$\varliminf_{t \to \infty} \inf_{\alpha \in M_t} \widetilde{E}\left[I_{[\lambda_1,\lambda_2]}(x)\left(\alpha(x)g(x) - \tfrac{1}{2}\alpha(x)^2\right)\right] =$$

$$= \varliminf_{t \to \infty} \inf_{\alpha \in M_t} \frac{1}{n_t} \log f_t(x.;\alpha) \quad \text{almost surely} \geqslant$$

$$\geqslant \varliminf_{t \to \infty} \frac{1}{n_t} \log f_t(x.;\alpha_t) \quad \text{(from the definition of } M_t) =$$

$$= \varliminf_{t \to \infty} \widetilde{E}\left[I_{[\lambda_1,\lambda_2]}(x)\left(\alpha_t(x)g(x) - \tfrac{1}{2}\alpha_t(x)^2\right)\right] =$$

$$= \tfrac{1}{2}\widetilde{E}[I_{[\lambda_1,\lambda_2]}(x)g(x)^2]$$

$$\text{(since} \quad \widetilde{E}\,|\,I_{[\lambda_1,\lambda_2]}(x)(\alpha_t(x) - g(x))|^2 \to 0).$$

Hence

$$\varliminf_{t \to \infty} \inf_{\alpha \in M_t} \widetilde{E}\left[I_{[\lambda_1,\lambda_2]}(x) \times \right.$$

$$\left. \times \left(\alpha(x)g(x) - \tfrac{1}{2}\alpha(x)^2 - \tfrac{1}{2}g(x)^2\right)\right] = 0 \quad \text{almost surely,}$$

and then, finally,

$$\varlimsup_{t \to \infty} \sup_{\alpha \in M_t} \widetilde{E}\,|\,I_{[\lambda_1,\lambda_2]}(x)(\alpha(x) - g(x))|^2 = 0 \quad \text{almost surely.}$$

It remains to prove Lemmas 1 and 2.

Proof of Lemma 1. c_1, c_2, etc. will refer to constants which may depend on ϵ but are independent of α, β, c or n.

$$P\left(\left|\frac{1}{n}\sum_{k=1}^{n}\left\{\int_{e_k}^{l_k}\alpha(x_s)\,ds + \int_{e_k}^{l_k}\beta(x_s)\,dw_s\right\} - \widetilde{E}[\alpha(x)]\right| > \epsilon\right) \leqslant$$

$$\leqslant P\left(\left|\frac{1}{n}\sum_{k=1}^{n}\int_{e_k}^{l_k}\alpha(x_s)\,ds - \widetilde{E}[\alpha(x)]\right| > \tfrac{\epsilon}{2}\right) +$$

$$+ P\left(\left|\frac{1}{n}\sum_{k=1}^{n}\int_{e_k}^{l_k}\beta(x_s)\,dw_s\right| > \frac{\epsilon}{2}\right) \leqslant$$

$$\leqslant P\left(\left|\frac{1}{n}\sum_{k=1}^{n}\left(\int_{e_k}^{l_k}\alpha(x_s)\,ds - \right.\right.\right.$$

(3.5)

$$\left.\left.\left. - E\left[\int_{e_k}^{l_k}\alpha(x_s)\,ds\,\big|\,x_{e_k}, x_{e_{k+1}}\right]\right)\right| > \frac{\epsilon}{4}\right) +$$

(3.6)
$$+ P\left(\left|\frac{1}{n}\sum_{k=1}^{n}E\left[\int_{e_k}^{l_k}\alpha(x_s)\,ds\,\big|\,x_{e_k}, x_{e_{k+1}}\right] - \tilde{E}[\alpha(x)]\right| > \frac{\dot{\epsilon}}{4}\right) +$$

(3.7)
$$+ P\left(\left|\frac{1}{n}\sum_{k=1}^{n}\left(\int_{e_k}^{l_k}\beta(x_s)\,dw_s - \right.\right.\right.$$

$$\left.\left.\left. - E\left[\int_{e_k}^{l_k}\beta(x_s)\,dw_s\,\big|\,x_{e_k}, x_{e_{k+1}}\right]\right)\right| > \frac{\epsilon}{4}\right) +$$

(3.8)
$$+ P\left(\left|\frac{1}{n}\sum_{k=1}^{n}E\left[\int_{e_k}^{l_k}\beta(x_s)\,dw_s\,\big|\,x_{e_k}, x_{e_{k+1}}\right]\right| > \frac{\epsilon}{4}\right).$$

I will develop bounds for (3.7) and (3.8). The treatment of (3.5) and (3.6), being analogous and somewhat simpler, is omitted.

Begin with two preliminary bounds:

1. There exists a constant c_1 such that

(3.9)
$$E\left[\left|\int_{e_1}^{l_1}\beta(x_s)\,dw_s\right|^l\,\big|\,x_{e_1} = \lambda_i,\ x_{e_2} = \lambda_j\right] \leqslant c_1 c^l$$

for all $i = 1, 2$, $j = 1, 2$ and $l = 1, 2, 3, 4$.

2. There exists a constant c_2 such that

(3.10)
$$E\left[\exp\left\{t\left|\int_{e_1}^{l_1}\beta(x_s)\,dw_s\right|\right\}\,\big|\,x_{e_1} = \lambda_i,\ x_{e_2} = \lambda_j\right] \leqslant c_2$$

for all $t \leqslant \dfrac{1}{c_2 c}$, $i = 1, 2$ and $j = 1, 2$.

Observe that for any random variable z such that $E|z| < \infty$, and any $i = 1, 2$,

$$E[|z| \,| \, x_{e_1} = \lambda_i] = E[|z| \,| \, x_{e_1} = \lambda_i, \ x_{e_2} = \lambda_1] p_{i1} +$$
$$+ \, E[|z| \,| \, x_{e_1} = \lambda_i, \ x_{e_2} = \lambda_2] p_{i2}.$$

Recall that $p_{ij} > 0$ for all i and j. Hence with $c_3 = \max\limits_{i,j} \dfrac{1}{p_{ij}}$,

$$E[|z| \,| \, x_{e_1} = \lambda_i, \ x_{e_2} = \lambda_j] = c_3 E[|z| \,| \, x_{e_1} = \lambda_i]$$

for all i and j. Use this to establish (3.10):

$$(3.11) \qquad E\left[\exp\left\{ t \left| \int_{e_1}^{l_1} \beta(x_s) \, dw_s \right| \right\} \,| \, x_{e_1} = \lambda_i, \ x_{e_2} = \lambda_j \right] \leqslant$$

$$\leqslant c_3 E\left[\exp\left\{ t \left| \int_{e_1}^{l_1} \beta(x_s) \, dw_s \right| \right\} \,| \, x_{e_1} = \lambda_i \right] =$$

$$= c_3 \sum_{k=1}^{\infty} E\left[I_{[p-1,p)}(l_1 - e_1) \times \right.$$

$$\left. \times \exp\left\{ t \left| \int_{e_1}^{p} I_{[e_1, l_1]}(s) \beta(x_s) \, dw_s \right| \right\} \,| \, x_{e_1} = \lambda_i \right] \leqslant$$

$$\leqslant c_3 \sum_{p=1}^{\infty} \left\{ P((l_1 - e_1) \in [p-1, p) \,| \, x_{e_1} = \lambda_i)^{\frac{1}{2}} \times \right.$$

$$\left. \times E\left[\exp\left\{ 2t \left| \int_{e_1}^{p} I_{[e_1, l_1]}(s) \beta(x_s) \, dw_s \right| \right\} \,| \, x_{e_1} = \lambda_i \right]^{\frac{1}{2}} \right\}.$$

From (3.1)

$$(3.12) \qquad P((l_1 - e_1) \in [p-1, p) \,| \, x_{e_1} = \lambda_i) \leqslant \rho^{p-1}$$

where $\rho < 1$, and, from Lemma B, Section 5 of G e m a n and H w a n g [6]

$$(3.13) \qquad E\left[\exp\left\{ 2t \left| \int_{e_1}^{p} I_{[e_1, l_1]}(s) \beta(x_s) \, dw_s \right| \right\} \,| \, x_{e_1} = \lambda_i \right] \leqslant \sqrt{2} + e^{(4tc)^2 p}.$$

Use (3.12) and (3.13) in (3.11):

$$E\left[\exp\left\{t\left|\int_{e_1}^{l_1}\beta(x_s)\,dw_s\right|\right\}\,\middle|\,x_{e_1}=\lambda_i,\;x_{e_2}=\lambda_j\right]\leqslant$$

(3.14)

$$\leqslant c_3\sum_{p=1}^{\infty}\rho^{\frac{p-1}{2}}\left(\sqrt{2}+e^{(4tc)^2p}\right)^{\frac{1}{2}}.$$

Provided that c_2 is sufficiently large, the expression in (3.14) is bounded by c_2 whenever $tc\leqslant\dfrac{1}{c_2}$. This establishes (3.10). A similar argument can be used for (3.9).

Return now to (3.7). With

$$z_k=\int_{e_k}^{l_k}\beta(x_s)\,dw_s-E\left[\int_{e_k}^{l_k}\beta(x_s)\,dw_s\,\middle|\,x_{e_k},x_{e_{k+1}}\right]$$

(3.7) becomes

$$P\left(\left|\frac{1}{n}\sum_{k=1}^{n}z_k\right|>\frac{\epsilon}{4}\right),$$

and this is no bigger than

$$P\left(\frac{1}{n}\sum_{k=1}^{n}z_k>\frac{\epsilon}{4}\right)+P\left(\frac{1}{n}\sum_{k=1}^{n}z_k<-\frac{\epsilon}{4}\right).$$

$$P\left(\frac{1}{n}\sum_{k=1}^{n}z_k>\frac{\epsilon}{4}\right)=\text{(for any }t>0)\;P\left(t\sum_{k=1}^{n}\left(z_k-\frac{\epsilon}{4}\right)>0\right)\leqslant$$

(3.15)

$$\leqslant E\prod_{k=1}^{n}e^{t\left(z_k-\frac{\epsilon}{4}\right)}=$$

$$=E\left\{E\left[\prod_{k=1}^{n}e^{t\left(z_k-\frac{\epsilon}{4}\right)}\,\middle|\,x_{e_1},\dots,x_{e_{n+1}}\right]\right\}=$$

$$=\text{(strong Markov property)}$$

(3.16)

$$E\left\{\prod_{k=1}^{n}E\left[e^{t\left(z_k-\frac{\epsilon}{4}\right)}\,\middle|\,x_{e_k},x_{e_{k+1}}\right]\right\}.$$

Fix k. For any $i = 1, 2$ and $j = 1, 2$ define

$$\varphi_k(t) = E[e^{t(z_k - \frac{\epsilon}{4})} \mid x_{e_k} = \lambda_i, \ x_{e_{k+1}} = \lambda_j].$$

Then $\varphi_k(0) = 1$, $\dfrac{d}{dt} \varphi_k(0) = -\dfrac{\epsilon}{4}$, and

$$\frac{d^2}{dt^2} \varphi_k(t) = E\left[\left(z_k - \frac{\epsilon}{4}\right)^2 e^{t(z_k - \frac{\epsilon}{4})} \mid x_{e_k} = \lambda_i, \ x_{e_{k+1}} = \lambda_j\right] \leqslant$$

$$\leqslant E\left[\left(z_k - \frac{\epsilon}{4}\right)^4 \mid x_{e_k} = \lambda_i, \ x_{e_{k+1}} = \lambda_j\right]^{\frac{1}{2}} \times$$

$$\times E[e^{2t(z_k - \frac{\epsilon}{4})} \mid x_{e_k} = \lambda_i, \ x_{e_{k+1}} = \lambda_j]^{\frac{1}{2}} \leqslant$$

$$\leqslant \text{(use 3.9 and 3.10)} \ c_4 c^2$$

provided $t \leqslant \dfrac{1}{c_4 c}$, where c_4 is some sufficiently large constant. Integrate $\dfrac{d^2}{dt^2} \varphi_k(t)$ and then $\dfrac{d}{dt} \varphi_k(t)$:

$$\frac{d}{dt} \varphi_k(t) \leqslant -\frac{\epsilon}{4} + c_4 c^2 t \Rightarrow \varphi_k(t) \leqslant 1 - \frac{\epsilon}{4} t + \frac{1}{2} c_4 c^2 t^2$$

provided $t \leqslant \dfrac{1}{c_4 c}$. Set $t = \dfrac{1}{c_5 c^2}$ where $c_5 \geqslant c_4$ (recall that $c \geqslant 1$):

$$E[e^{t(z_k - \frac{\epsilon}{4})} \mid x_{e_k} = \lambda_i, \ x_{e_{k+1}} = \lambda_j] =$$

$$= \varphi_k(t) \leqslant 1 - \frac{1}{c_5 c^2} \left[\frac{\epsilon}{4} - \frac{c_4}{2c_5}\right] \leqslant 1 - \frac{\epsilon}{8 c_5 c^2}$$

for c_5 sufficiently large. Finally, put this back into (3.16):

$$P\left(\frac{1}{n} \sum_{k=1}^{n} z_k > \frac{\epsilon}{4}\right) \leqslant \left(1 - \frac{\epsilon}{8 c_5 c^2}\right)^n.$$

Since the same argument applies to the second term in (3.15), this establishes the required bound on (3.7).

For (3.8), define for each i and j a random variable n_{ij} by

$$n_{ij} = \text{number of times } (x_{e_k}, x_{e_{k+1}}) = (\lambda_i, \lambda_j)$$

$$(k = 1, 2, \ldots, n),$$

and write

$$\frac{1}{n} \sum_{k=1}^{n} \mathsf{E}\left[\int_{e_k}^{l_k} \beta(x_s)\, dw_s \mid x_{e_k}, x_{e_{k+1}}\right] =$$

$$= \sum_{i=1}^{2} \sum_{j=1}^{2} \frac{n_{ij}}{n} \mathsf{E}\left[\int_{e_1}^{l_1} \beta(x_s)\, dw_s \mid x_{e_1} = \lambda_i,\ x_{e_2} = \lambda_j\right].$$

Observe that for $i = 1$ or 2

$$\mathsf{E}\left[\int_{e_1}^{l_1} \beta(x_s)\, dw_s \mid x_{e_1} = \lambda_i,\ x_{e_2} = \lambda_1\right] p_{i1} +$$

$$+ \mathsf{E}\left[\int_{e_1}^{l_1} \beta(x_s)\, dw_s \mid x_{e_1} = \lambda_i,\ x_{e_2} = \lambda_2\right] p_{i2} =$$

$$= \mathsf{E}\left[\int_{e_1}^{l_1} \beta(x_s)\, dw_s \mid x_{e_1} = \lambda_i\right] = 0 \Rightarrow$$

$$\Rightarrow \frac{1}{n} \sum_{k=1}^{n} \mathsf{E}\left[\int_{e_k}^{l_k} \beta(x_s)\, dw_s \mid x_{e_k}, x_{e_{k+1}}\right] =$$

$$= \sum_{i=1}^{2} \sum_{j=1}^{2} \left(\frac{n_{ij}}{n} - \pi_i p_{ij}\right) \mathsf{E}\left[\int_{e_1}^{l_1} \beta(x_s)\, dw_s \mid x_{e_1} = \lambda_i,\ x_{e_2} = \lambda_j\right].$$

Using this, and (3.9),

$$(3.17) \quad \mathsf{P}\left(\left|\frac{1}{n} \sum_{k=1}^{n} \mathsf{E}\left[\int_{e_k}^{l_k} \beta(x_s)\, dw_s \mid x_{e_k}, x_{e_{k+1}}\right]\right| > \frac{\epsilon}{4}\right) \leqslant$$

$$\leqslant \mathsf{P}\left(\sum_{i=1}^{2} \sum_{j=1}^{2} \left|\frac{n_{ij}}{n} - \pi_i p_{ij}\right| > \frac{\epsilon}{4c_1 c}\right) \leqslant$$

$$\leqslant \sum_{i=1}^{2} \sum_{j=1}^{2} P\left(\left|\frac{n_{ij}}{n} - \pi_i p_{ij}\right| > \frac{\epsilon}{16c_1 c}\right).$$

Consider, for example,

$$P\left(\left|\frac{n_{11}}{n} - \pi_1 p_{11}\right| > \frac{\epsilon}{16c_1 c}\right).$$

Define for each i

$$n_i = \text{number of times } x_{e_k} = \lambda_i \qquad (k = 1, 2, \ldots, n).$$

Omitting details, a combinatorial-type argument demonstrates the existence of a constant c_6 sufficiently large that for all $\delta < 1$ and $n \geqslant 1$

(3.18)
$$\left\{\left|\frac{n_{11}}{n} - \pi_1 p_{11}\right| > \delta\right\} \Rightarrow$$

$$\Rightarrow \bigcup_{i=1}^{2} \left\{n_i > \frac{n}{c_6} \text{ and } \left|\frac{n_{ii}}{n_i} - p_{ii}\right| > \frac{\delta}{c_6} - \frac{c_6}{n}\right\}.$$

Take c_7 so large that $\frac{\epsilon}{16c_1} > \frac{1}{c_7}$. Then, using (3.18),

$$P\left(\left|\frac{n_{11}}{n} - \pi_1 p_{11}\right| > \frac{\epsilon}{16c_1 c}\right) \leqslant P\left(\left|\frac{n_{11}}{n} - \pi_1 p_{11}\right| > \frac{1}{c_7 c}\right) \leqslant$$

$$\leqslant P\left(n_1 > \frac{n}{c_6} \text{ and } \left|\frac{n_{11}}{n_1} - p_{11}\right| > \frac{1}{c_6 c_7 c} - \frac{c_6}{n}\right) +$$

(3.19)
$$+ P\left(n_2 > \frac{n}{c_6} \text{ and } \left|\frac{n_{22}}{n_2} - p_{22}\right| > \frac{1}{c_6 c_7 c} - \frac{c_6}{n}\right) \leqslant$$

$$\leqslant P\left(\left|\frac{b(r, p_{11})}{r} - p_{11}\right| > \frac{1}{c_6 c_7 c} - \frac{c_6}{n}\right) +$$

$$+ P\left(\left|\frac{b(r, p_{22})}{r} - p_{22}\right| > \frac{1}{c_6 c_7 c} - \frac{c_6}{n}\right)$$

where r is the smallest integer greater than or equal to $\frac{n}{c_6}$ and $b(r, p_{ii})$ is a binomial random variable with r "trials" and a p_{ii} "probability of

success" in each trial. Each of the probabilities in (3.19) is easily bounded
by

$$\left(1 - \frac{1}{c_8 c^2} + \frac{c_8}{n}\right)^{\frac{n}{c_6}}$$

for c_8 sufficiently large.

The proof of Lemma 1 can now be completed by the analogous treatment of the remaining terms in (3.17).

Proof of Lemma 2. Because of (3.4), and the assumed relation between m_t and $\mathcal{I}(t)$, there is a constant c_9 such that

$$\varlimsup_{t \to \infty} \frac{m_t}{[c_9 n^{1-\delta}]} < 1 \quad \text{almost surely},$$

where $[x]$ denotes the greatest integer less than or equal to x. Define $m_n = [c_9 n^{1-\delta}]$ and

$$\tilde{S}_m = \left\{ \sum_{j=1}^{m} a_j \psi_j(x): \sum_{j=1}^{m} |a_j| \leqslant k_1 (\log m)^{k_2} \right\}.$$

Then

$$\varlimsup_{t \to \infty} \sup_{\alpha \in S_t} \left| \frac{1}{n_t} \log f_t(x_.; \alpha) - \right.$$

$$\left. - \tilde{E}\left[I_{[\lambda_1, \lambda_2]}(x)\left(\alpha(x)g(x) - \tfrac{1}{2}\alpha(x)^2\right)\right] \right| =$$

$$= \varlimsup_{t \to \infty} \sup_{\alpha \in S_t} \left| \frac{1}{n_t} \log f_{l_{n_t}}(x_.; \alpha) - \right.$$

$$\left. - \tilde{E}\left[I_{[\lambda_1, \lambda_2]}(x)\left(\alpha(x)g(x) - \tfrac{1}{2}\alpha(x)^2\right)\right] \right| \quad \text{almost surely} \leqslant$$

$$\leqslant \varlimsup_{n \to \infty} \sup_{\alpha \in S_{m_n}} \left| \frac{1}{n} \log f_{l_n}(x_.; \alpha) - \right.$$

(3.20)

$$\left. - \tilde{E}\left[I_{[\lambda_1, \lambda_2]}(x)\left(\alpha(x)g(x) - \tfrac{1}{2}\alpha(x)^2\right)\right] \right| \quad \text{almost surely}.$$

For each n, let $\beta_{n1}, \beta_{n2}, \ldots, \beta_{nd_n}$ denote the collection of functions $\beta \in \tilde{S}_{m_n}$ of the form

$$\beta(x) = \sum_{j=1}^{m_n} b_j \psi_j(x)$$

where, for each j,

$$b_j = \frac{q_j}{m_n (\log m_n)^{k_2 + 1}}$$

for some integer q_j (positive, negative, or zero). Because

$$\sum_{k=1}^{m_n} |b_j| \leqslant k_1 (\log m_n)^{k_2}$$

for any $\beta \in \tilde{S}_{m_n}$, $d_n = O(m_n^{2m_n})$. Associated with each β_{ni} (say

$$\beta_{ni} = \sum_{j=1}^{m_n} a_j \psi_j(x)),$$

define a set of functions B_{ni}

$$B_{ni} = \left\{ \alpha(x) = \sum_{j=1}^{m_n} a_j \psi_j(x) : \alpha \in \tilde{S}_{m_n} \text{ and} \right.$$

$$\left. \sup_{1 \leqslant j \leqslant m_n} |a_j - b_j| \leqslant \frac{1}{m_n (\log m_n)^{k_2 + 1}} \right\}.$$

Then $\bigcup_{i=1}^{d_n} B_{ni}$ includes all of $\tilde{S}_{m_n}$, and since $|\psi_j(x)| \leqslant 1$ on $[\lambda_1, \lambda_2]$,

$$\sup_{x \in [\lambda_1, \lambda_2]} |\alpha(x) - \beta_{ni}(x)| \leqslant \frac{1}{(\log m_n)^{k_2 + 1}}$$

for all $\alpha \in B_{ni}$.

We can now rewrite (3.20):

$$\varlimsup_{n \to \infty} \sup_{i=1,\ldots,d_n} \sup_{\alpha \in B_{ni}} \left| \frac{1}{n} \log f_{l_n}(x\,;\alpha) - \right.$$

$$\left. - \mathsf{E}\left[I_{[\lambda_1,\lambda_2]}(x)\left(\alpha(x)g(x) - \frac{1}{2}\,\alpha(x)^2\right)\right]\right|$$

which is bounded by

(3.21)
$$\varlimsup_{n \to \infty} \sup_{i=1,\ldots,d_n} \left| \frac{1}{n} \log f_{l_n}(x\,;\beta_{ni}) - \right.$$

$$\left. - \widetilde{\mathsf{E}}\left[I_{[\lambda_1,\lambda_2]}(x)\left(\beta_{ni}(x)g(x) - \frac{1}{2}\,\beta_{ni}(x)^2\right)\right]\right| +$$

(3.22)
$$+ \varlimsup_{n \to \infty} \sup_{i=1,\ldots,d_n} \sup_{\alpha \in B_{ni}} \left| \frac{1}{n} \log f_{l_n}(x\,;\alpha) - \frac{1}{n} \log f_{l_n}(x\,;\beta_{ni})\right| +$$

$$+ \varlimsup_{n \to \infty} \sup_{i=1,\ldots,d_n} \sup_{\alpha \in B_{ni}} \left| \widetilde{\mathsf{E}}\left[I_{[\lambda_1,\lambda_2]}(x) \times \right. \right.$$

(3.23)
$$\left. \times \left(\alpha(x)g(x) - \frac{1}{2}\,\alpha(x)^2\right)\right] -$$

$$\left. - \widetilde{\mathsf{E}}\left[I_{[\lambda_1,\lambda_2]}(x)\left(\beta_{ni}(x)g(x) - \frac{1}{2}\,\beta_{ni}(x)^2\right)\right]\right|.$$

The expression in (3.23) equals zero, since

$$\sup_{\alpha \in B_{ni}} \sup_{x \in [\lambda_1,\lambda_2]} |\alpha(x) - \beta_{ni}(x)| \leqslant \frac{1}{(\log m_n)^{k_2+1}} \to 0$$

as $n \to \infty$.

Next, argue that the expression in (3.21) also equals zero (almost surely): fix an arbitrary $\epsilon > 0$. Notice that $\alpha \in S_{m_n} \Rightarrow \sup_{x \in [\lambda_1,\lambda_2]} |\alpha(x)| \leqslant$ $\leqslant k_1 (\log m_n)^{k_2}$. Apply Lemma 1 with

$$\alpha(x) = I_{[\lambda_1,\lambda_2]}(x)\left(\beta_{ni}(x)g(x) - \frac{1}{2}\,\beta_{ni}(x)^2\right)$$

$$\beta(x) = \sigma I_{[\lambda_1,\lambda_2]}(x)\beta_{ni}(x) \quad \text{and} \quad c = c_n = O((\log m_n)^{2k_2}):$$

$$P\left(\sup_{i=1,\ldots,d_n}\left|\frac{1}{n}\log f_{l_n}(x\,.;\beta_{ni})-\right.\right.$$

$$\left.\left.-\tilde{E}\left[I_{[\lambda_1,\lambda_2]}\left(\beta_{ni}(x)g(x)-\frac{1}{2}\beta_{ni}(x)^2\right)\right]\right|>\epsilon\right)\leqslant$$

$$\leqslant\sum_{i=1}^{d_n}P\left(\left|\frac{1}{n}\log f_{l_n}(x\,.;\beta_{ni})-\right.\right.$$

$$\left.\left.-\tilde{E}\left[I_{[\lambda_1,\lambda_2]}\left(\beta_{ni}(x)g(x)-\frac{1}{2}\beta_{ni}(x)^2\right)\right]\right|>\epsilon\right)\leqslant$$

$$\leqslant d_nk\left(1-\frac{1}{kc_n^2}+\frac{k}{n}\right)^{\frac{n}{k}}=$$

$$=O\left(n^{2c_9n^{1-\delta}}\left(1-\frac{1}{k(\log n)^{4k_2}}+\frac{k}{n}\right)^{\frac{n}{k}}\right).$$

Since

$$\sum_{n=1}^{\infty}n^{2c_9n^{1-\delta}}\left(1-\frac{1}{k(\log n)^{4k_2}}+\frac{k}{n}\right)^{\frac{n}{k}}<\infty,$$

the Borel – Cantelli Lemma implies that the expression in (3.21) is equal to zero with probability one.

All that remains is the expression in (3.22).

$$\sup_{i=1,\ldots,d_n}\sup_{\alpha\in B_{ni}}\left|\frac{1}{n}\log f_{l_n}(x\,.;\alpha)-\frac{1}{n}\log f_{l_n}(x\,.;\beta_{ni})\right|\leqslant$$

$$\leqslant\sup_{i=1,\ldots,d_n}\sup_{\alpha\in B_{ni}}\left|\frac{1}{n}\sum_{k=1}^{n}\int_{e_k}^{l_k}I_{[\lambda_1,\lambda_2]}(x_s)\times\right.$$

(3.24)
$$\times\left[g(x_s)(\alpha(x_s)-\beta_{ni}(x_s))-\right.$$

$$\left.\left.-\frac{1}{2}(\alpha(x_s)-\beta_{ni}(x_s))(\alpha(x_s)+\beta_{ni}(x_s))\right]ds\right|+$$

(3.25)
$$+\sup_{i=1,\ldots,d_n}\sup_{\alpha\in B_{ni}}\left|\frac{\sigma}{n}\sum_{k=1}^{n}\int_{e_k}^{l_k}I_{[\lambda_1,\lambda_2]}(x_s)\times\right.$$

$$\left.\times(\alpha(x_s)-\beta_{ni}(x_s))\,dw_s\right|.$$

If c_{10} is chosen large enough, then the expression in (3.24) is bounded by

$$(3.26) \qquad \frac{c_{10}}{\log m_n} \frac{1}{n} \sum_{k=1}^{n} (l_k - e_k).$$

Lemma 1 (with $\alpha(x) = 1$ and $\beta(x) = 0$) implies that

$$\frac{1}{n} \sum_{k=1}^{n} (l_k - e_k) \rightarrow \tilde{\mathsf{E}}[1] \quad \text{almost surely}$$

so ((3.26) (and therefore (3.24))) converges to zero almost surely as $n \rightarrow \infty$. In (3.25), write

$$\alpha(x) = \sum_{j=1}^{m_n} a_j \psi_j(x) \quad \text{and} \quad \beta_{ni}(x) = \sum_{j=1}^{m_n} b_j \psi_j(x):$$

$$\left| \frac{\sigma}{n} \sum_{k=1}^{n} \int_{e_k}^{l_k} I_{[\lambda_1, \lambda_2]}(x_s)(\alpha(x_s) - \beta_{ni}(x_s)) \, dw_s \right| =$$

$$= \left| \frac{\sigma}{n} \sum_{k=1}^{n} \sum_{j=1}^{m_n} (a_j - b_j) \int_{e_k}^{l_k} I_{[\lambda_1, \lambda_2]}(x_s) \psi_j(x_s) \, dw_s \right| \leq$$

$$\leq (\text{since } \alpha \in B_{ni})$$

$$\frac{\sigma}{(\log m_n)^{k_2 + 1}} \frac{1}{n} \sum_{k=1}^{n} \frac{1}{m_n} \sum_{j=1}^{m_n} \int_{e_k}^{l_k} I_{[\lambda_1, \lambda_2]}(x_s) \psi_j(x_s) \, dw_s \Big|.$$

Since this last expression does not involve i or α, it is a bound on (3.25). It will therefore be sufficient to show that

$$\frac{1}{n} \sum_{k=1}^{n} \frac{1}{m_n} \sum_{j=1}^{m_n} \left| \int_{e_k}^{l_k} I_{[\lambda_1, \lambda_2]}(x_s) \psi_j(x_s) \, dw_s \right|$$

is with probability one bounded for all n.

By using the fact that $|\psi_j(x)| \leq 1$ on $[\lambda_1, \lambda_2]$, and imitating the argument used in the proof of Lemma 1 for bounding (3.7), we can show that for fixed $\epsilon > 0$ there exists a constant $c_{11} > 0$ such that

$$P\Big(\Big|\frac{1}{n}\sum_{k=1}^{n}\Big\{\frac{1}{m_n}\sum_{j=1}^{m_n}\Big|\int_{e_k}^{l_k}I_{[\lambda_1,\lambda_2]}(x_s)\psi_j(x_s)\,dw_s\Big|-$$

$$(3.27)\qquad -\frac{1}{m_n}\sum_{j=1}^{m_n}E\Big[\Big|\int_{e_k}^{l_k}I_{[\lambda_1,\lambda_2]}(x_s)\psi_j(x_s)\,dw_s\Big|\,\Big|\,x_{e_k},x_{e_{k+1}}\Big]\Big\}\Big|>\epsilon\Big)\leqslant$$

$$\leqslant 2\Big(1-\frac{1}{c_{11}}\Big)^{n}$$

for all $n\geqslant 1$. (The only modification to the previous argument is in developing an analogue to (3.10). For this purpose, first observe that

$$E\Big[\exp\Big\{t\,\frac{1}{m_n}\sum_{j=1}^{m_n}\Big|\int_{e_k}^{l_k}I_{[\lambda_1,\lambda_2]}(x_s)\psi_j(x_s)\,dw_s\Big|\Big\}\,\Big|\,x_{e_k},x_{e_{k+1}}\Big]\leqslant$$

$$(3.28)\qquad \leqslant \prod_{j=1}^{m_n}E\Big[\exp\Big\{t\int_{e_k}^{l_k}I_{[\lambda_1,\lambda_2]}(x_s)\psi_j(x_s)\,dw_s\Big|\Big\}\,\Big|\,x_{e_k},x_{e_{k+1}}\Big]^{\frac{1}{m_n}},$$

and the apply (3.10) to conclude that the expression in (3.28) is no bigger than c_2 provided $t\leqslant\frac{1}{c_2}$.)

One consequence of (3.27) is that

$$\frac{1}{n}\sum_{k=1}^{n}\frac{1}{m_n}\sum_{j=1}^{m_n}\Big|\int_{e_k}^{l_k}I_{[\lambda_1,\lambda_2]}(x_s)\psi_j(x_s)\,dw_s\Big|-$$

$$-\frac{1}{n}\sum_{k=1}^{n}\frac{1}{m_n}\sum_{j=1}^{m_n}E\Big[\Big|\int_{e_k}^{l_k}I_{[\lambda_1,\lambda_2]}(x_s)\times$$

$$\times\psi_j(x_s)\,dw_s\Big|\,\Big|\,x_{e_k},x_{e_{k+1}}\Big]\to 0\quad\text{almost surely.}$$

Since, by (3.9)

$$\frac{1}{n}\sum_{k=1}^{n}\frac{1}{m_n}\sum_{j=1}^{m_n}E\Big[\Big|\int_{e_k}^{l_k}I_{[\lambda_1,\lambda_2]}(x_s)\times$$

$$\times\psi_j(x_s)\,dw_s\Big|\,\Big|\,x_{e_k},x_{e_{k+1}}\Big]\leqslant c_1,$$

this completes the proof of Lemma 2. ∎

REFERENCES

[1] G . B a n o n , Nonparametric identification for diffusion processes, *SIAM J. on Control and Optimization,* 16 (1978), 380-395.

[2] G . B a n o n – H . T . N g u y e n , Recursive estimation in diffusion model, Preliminary report.

[3] B . M . B r o w n – J . I . H e w i t t , Asymptotic likelihood theory for diffusion processes, *J. Appl. Prob.,* 12 (1975), 228-238.

[4] P . D . F e i g i n , Maximum likelihood estimation for continuous time stochastic processes, *Adv. Appl. Prob.,* 8 (1976), 712-736.

[5] A . F r i e d m a n , *Stochastic differential equations and applications,* Vol. 1, Academic Press, New York, 1975.

[6] S . G e m a n – C . - R . H w a n g , Nonparametric maximum likelihood estimation by the method of sieves, *Ann. of Stat.* (to appear).

[7] U . G r e n a n d e r , *Abstract inference,* John Wiley & Sons, New York, 1981, in preparation.

[8] T . S . L e e – F . K o z i n , Almost sure asymptotic likelihood theory for diffusion processes, *J. Appl. Prob.,* 14 (1977), 527-537.

[9] R . S . L i p t s e r – A . N . S h i r y a y e v , *Statistics of random processes* II, (Chapter 17), Springer-Verlag, New York, 1978.

[10] H . T . N g u y e n – T . D . P h a m , On the law of large numbers for continuous time martingales and applications to statistics, Preliminary report.

S . G e m a n

Division of Applied Mathematics, Brown University, Providence, Rhode Island 02912, USA.

COLLOQUIA MATHEMATICA SOCIETATIS JÁNOS BOLYAI

32. NONPARAMETRIC STATISTICAL INFERENCE,

BUDAPEST (HUNGARY), 1980.

LIL TYPE PROPERTIES (ON THE WHOLE LINE) OF THE PL AND THE BAYES ESTIMATORS

J.K. GHORAI[*] — L. REJTŐ[**] — V. SUSARLA[***]

1. INTRODUCTION AND SUMMARY

Let $X_1, \ldots, X_n$ and $Y_1, Y_2, \ldots, Y_n$ be i.i.d. random variables with distribution functions F and G respectively. Define $Z_i = \min\{X_i, Y_i\}$ and

$$\delta_i = \begin{cases} 1 & \text{if} \quad X_i \leqslant Y_i \\ 0 & \text{if} \quad X_i > Y_i \end{cases} \quad \text{for} \quad i = 1, \ldots, n.$$

Two most important estimators of $\bar{F} = 1 - F$ based on $(Z_1, \delta_1), \ldots, (Z_n, \delta_n)$ are the product limit estimator $\bar{F}_n^*$ [4] and the Bayes estimator $\bar{F}_n^\alpha$ [6]. Uniform strong consistency results with rates for $\bar{F}_n^*$ on the entire line were proved by Földes and Rejtő [2], Földes,

[*]The work of this author is partly supported by a grant from UWM Graduate School.

[**]The work of this author was done when she was visiting in the University of Wisconsin – Milwauke.

[***]The work of this author is partly supported by NIH Grant 1R01GM23129.

Rejtő and Winter [3] under various assumptions on F and G. Point-wise strong consistency for the PL estimator was also established by Phadia and Van Ryzin [5]. Pointwise strong consistency and uniform strong consistency results on $(-\infty, T]$ for the Bayes estimator $\bar{F}_n^\alpha$ were proved by Susarla and Van Ryzin [7], [8]. In their paper Földes and Rejtő [2] assumed the continuity of F and that $T_F < T_G$ (see Section 2 for definitions of T_F and T_G). Without the condition, $T_F < T_G$, the uniform strong consistency results with rates are not available for the product limit estimator $\bar{F}_n^*$. In this paper we let $T_F \leqslant T_G \leqslant +\infty$ and establish the uniform strong consistency of $\bar{F}_n^*$, with rate, on the entire line under the assumption that there exist constants $0 < k < 1$, $0 \leqslant \gamma \leqslant 1$ and $T < T_F$ such that $k\bar{F}^\gamma \leqslant \bar{G}$ on (T, T_F). This condition is automatically satisfied when $T_F < T_G$ by taking $k = \bar{G}(T_F^-)$ and $\gamma = 0$. The rate of convergence under this condition is $O\left(\left(\frac{\log\log n}{n}\right)^{\frac{1}{2(1+\gamma)}}\right)$, which reduces to $O\left(\frac{\sqrt{\log\log n}}{\sqrt{n}}\right)$ when $\gamma = 0$. The LIL type result of Földes and Rejtő [2] for $\bar{F}_n^*$ can be obtained from this with the additional assumption of continuity of G on (T, T_F).

The condition introduced here has an intuitive meaning. It ensures enough observations from F in order to be able to estimate F in the tail. In particular, for example, if $T_G < T_F$ then we will never be able to observe uncensored values beyond T_G and hence we will not be able to estimate F beyond T_G. The present condition precisely avoids these situations.

Uniform strong consistency for F_n^* with rate was proved by Földes and Rejtő [2] under the assumption of continuity of F and $T_F < T_G$. In this paper we provide uniform strong consistency results for the Bayes estimator, $\bar{F}_n^\alpha$, on the entire line when $T_F = T_G \leqslant +\infty$. This is achieved by first showing that the sup norm distance between $\bar{F}_n^*$ and $\bar{F}_n^\alpha$ is small for large n. In Lemma 4.1 we provide an equivalent form of the product limit estimator which is valid even when censored and uncensored observations are tied together. This allows us to handle the distance between $\bar{F}_n^*$ and $\bar{F}_n^\alpha$ under most general conditions. The expression (2.4) for the Bayes estimator is also new in the sense that only special forms of this expression have appeared in the literature [7]. Recently B u r k e , C s ö r g ő and H o r v á t h [1] proved a strong embedding theorem for $\sqrt{n}\,(\bar{F}_n^*(t) - \bar{F}(t))$ assuming F and G to be continuous. As a consequence of the embedding theorem they obtained the following result.

$$\sup_{-\infty < t \leqslant T} |\bar{F}_n^*(t) - \bar{F}(t)| = O\left(\frac{\sqrt{\log \log n}}{\sqrt{n}}\right) \quad \text{a.s.}$$

where $T < T_F$ and F and G are continuous.

The rest of the paper contains the following. In Section 2 all necessary notations are described. Section 3 deals with the strong consistency of the PL estimator. Section 4 contains the results concerning the distance between $\bar{F}_n^*$ and $\bar{F}_n^\alpha$ along with uniform strong consistency results for the Bayes estimator.

2. NOTATIONS AND ASSUMPTIONS

Let $X_1, \ldots, X_n$ denote i.i.d. observation with common distribution function F and let $Y_1, \ldots, Y_n$ denote i.i.d. observations with common distribution function G. Define, for $i = 1, \ldots, n$,

$$(2.1) \qquad Z_i = \min\{X_i, Y_i\}, \quad \delta_i = \begin{cases} 1 & \text{if } X_i \leqslant Y_i \\ 0 & \text{if } X_i > Y_i \end{cases}$$

$$F(u) = P(X \leqslant u), \quad \bar{F}(u) = 1 - F(u), \quad \tilde{F}(t) = P(Z \leqslant t, \ \delta = 1)$$

$$G(u) = P(Y \leqslant u), \quad \bar{G}(u) = 1 - G(u)$$

$$\bar{H}(u) = \bar{F}(u)\bar{G}(u), \quad H(u) = 1 - \bar{H}(u)$$

$$T_F = \sup\{t: \bar{F}(t) > 0\}, \quad T_G = \sup\{t: \bar{G}(t) > 0\}$$

$$[A] = \text{indicator of the set } A.$$

$$N^+(t) = \sum_{j=1}^{n} [Z_j > t], \quad N(t) = \sum_{j=1}^{n} [Z_j \geqslant t]$$

$$\lambda(u) = \sum_{j=1}^{n} [Z_j = u, \ \delta_j = 1], \quad \mu(u) = \sum_{j=1}^{n} [Z_j = u, \ \delta_j = 0]$$

$$\lambda_i = \lambda(Z_i), \quad \mu_i = \mu(Z_i)$$

$$n\bar{H}_n(t) = N^+(t)$$

$$n\tilde{G}_n(t) = \sum_{j=1}^{n} [Z_j \leqslant t, \ \delta_j = 0], \quad n\tilde{F}_n(t) = \sum_{j=1}^{n} [Z_j \leqslant t, \ \delta_j = 1].$$

Let $Z_{i_1} < Z_{i_2} < \ldots < Z_{i_m}$ denote the distinct ordered observations in the sample.

Product limit estimator of $\bar{F}$

$$(2.2) \qquad \bar{F}_n^*(t) = \begin{cases} \displaystyle\prod_{i=1}^{n} \left(\frac{N^+(Z_i) + \mu_i}{N^+(Z_i) + \mu_i + \lambda_i} \right)^{\frac{[Z_i \leqslant t, \, \delta_i = 1]}{\lambda_i}} & \text{for} \quad t \leqslant \max_i \{Z_i\}, \\[20pt] 0 & \text{for} \quad t > \max_i \{Z_i\}. \end{cases}$$

If F is assumed to be continuous then $\lambda_i = 1$ or 0 with probability one for all i and hence the expression (2.2) can be expressed as

$$(2.3) \qquad \bar{F}_n^*(t) = \begin{cases} \displaystyle\prod_{i=1}^{n} \left(\frac{N^+(Z_i)}{N^+(Z_i) + 1} \right)^{[Z_i \leqslant t, \, \delta_i = 1]} & \text{if} \quad t \leqslant \max_i \{Z_i\} \\[20pt] 0 & \text{if} \quad t > \max_i \{Z_i\}. \end{cases}$$

Bayes estimator of $\bar{F}$

Let $\alpha(\cdot)$ be a finite nonnull measure on the Borel σ-field of the real line R. The Bayes estimator of $\bar{F}(t)$ can be expressed as s

$$(2.4) \qquad \bar{F}_n^\alpha(t) =$$

$$= \frac{N^+(t) + \alpha(t, \infty)}{n + \alpha(R)} \prod_{i=1}^{n} \left(\frac{N^+(Z_i) + \alpha(Z_i, \infty) + \mu_i}{N^+(Z_i) + \alpha(Z_i, \infty)} \right)^{\frac{[Z_i \leqslant t, \, \delta_i = 0]}{\mu_i}}$$

whenever $F(\cdot)$ is a Dirichlet process with parameter measure $\alpha(\cdot)$ and the loss function is integrated squared error. If G is assumed to be continuous then $\mu_i = 1$ or 0 for all i and hence (2.4) can be expressed as

$$\bar{F}_n^\alpha(t) =$$

(2.5)

$$= \frac{N^+(t) + \alpha(t, \infty)}{n + \alpha(R)} \prod_{i=1}^{n} \left(\frac{N^+(Z_i) + \alpha(Z_i, \infty) + 1}{N^+(Z_i) + \alpha(Z_i, \infty)} \right)^{[Z_i \leqslant t,\, \delta_i = 0]}.$$

In the exponents of (2.2) and (2.4) $\dfrac{0}{0} = 1$.

3. UNIFORM STRONG CONSISTENCY OF PL ESTIMATOR

Theorem 3.1. *Suppose that F is continuous and G is arbitrary satisfying the following conditions*

(a) $T_F = T_G \leqslant + \infty$;

(b) *there exists T, γ and k such that $T < T_F$, $0 \leqslant \gamma \leqslant 1$, $0 < k \leqslant 1$, G is continuous for $x > T$ and $k\bar{F}^\gamma(x) \leqslant \bar{G}(x)$ for $x > T$.*

Then with probability one

$$\limsup_{n \to \infty} \left(\frac{n}{\log \log n} \right)^{\frac{1}{2(1+\gamma)}} \sup_{-\infty < u < \infty} |\bar{F}_n^*(u) - \bar{F}(u)| \leqslant$$

$$\leqslant \frac{C_1}{k} 2^{-\frac{1}{2(1+\gamma)}}$$

where $C_1 = 24.9$.

Proof. The proof of this theorem is essentially the same as the proof of the iterated logarithm type result of F ö l d e s and R e j t ő [2]. Thus only a sketch of the proof is given here. Let us introduce the following notations

$$T(u) = -\log \bar{F}(u) = \int_{-\infty}^{u} \frac{d\tilde{F}(t)}{\bar{H}(t)}$$

(3.1)

$$T_n(u) = -\int_{-\infty}^{u} \frac{d\tilde{F}_n(t)}{\bar{H}_n(t)}.$$

Furthermore we need the modified version of the product limit estimator.

$$1 - \hat{F}_n(u) = \bar{\hat{F}}_n(u) =$$

$$= \begin{cases} \displaystyle\prod_{j=1}^{n} \left(\frac{N(Z_j)}{N(Z_j) + 1}\right)^{[Z_j \leqslant u, \, \delta_j = 1]} & \text{if} \quad u \leqslant \max_i \{Z_i\} \\[3mm] 0 & \text{if} \quad u > \max_i \{Z_i\}. \end{cases}$$

Then

(3.2) $\qquad \bar{F}_n^*(u) - \bar{F}(u) = (\bar{F}_n^*(u) - \bar{\hat{F}}_n(u)) + (\bar{\hat{F}}_n(u) - \bar{F}(u))$

and using Taylor's expansion

$$\bar{\hat{F}}_n(u) - \bar{F}(u) = e^{\log \bar{\hat{F}}_n(u)} - e^{-T_n(u)} + e^{-T_n(u)} - e^{-T(u)} =$$

(3.3) $\qquad = e^{-T_n^*(u)}(\log \bar{\hat{F}}_n(u) + T_n(u)) + \bar{F}(u)(T(u) - T_n(u)) +$

$$+ \frac{1}{2} e^{-T_n^{**}(u)}(T(u) - T_n(u))^2$$

where

$$\min\{-\log \bar{\hat{F}}_n(u), T_n(u)\} \leqslant T_n^*(u) \leqslant \max\{-\log \bar{\hat{F}}_n(u), T_n(u)\},$$

$$\min\{T(u), T_n(u)\} \leqslant T_n^{**}(u) \leqslant \max\{T(u), T_n(u)\}.$$

Let us choose a sequence $\{u_n\}$ in the following way.

(3.4) $\qquad \bar{H}(u_n) = k_1 \dfrac{\sqrt{\log \log n}}{\sqrt{2n}}.$

The fact that $\bar{H}$ is continuous nonincreasing on (T, T_F) implies that there exists $n_0 = n_0(T)$ and u_n such that for $n > n_0(T)$, $u_n > T$ and u_n satisfies (3.4). Furthermore using the condition (b) of Theorem 3.1, it is possible to give upper and lower bounds for F and H, namely,

$$k\bar{F}^{(1+\gamma)}(u_n) \leqslant \bar{H}(u_n) \leqslant \bar{F}(u_n),$$

(3.5)

$$k_1 \frac{\sqrt{\log \log n}}{\sqrt{2n}} \leqslant \bar{F}(u_n) \leqslant \left(\frac{k_1^2}{k^2} \frac{\log \log n}{2n}\right)^{\frac{1}{2(1+\gamma)}}.$$

In [2] the sequence $\{u_n\}$ was chosen with respect to F. If u_n is chosen according to (3.4) then inequality (3.5) will be satisfied. To complete the proof of Theorem 3.1 we need the following lemmas whose proofs are similar to those in [2]. In all the lemmas we assume that the conditions of Theorem 3.1 hold.

Lemma 3.1. *If* $u \leqslant u_n$, *then with probability* 1

$$\sup_{u \leqslant u_n} \left| \frac{\bar{H}_n(u^-)}{\bar{H}(u^-)} \right| \leqslant \frac{3}{2}, \qquad \inf_{u \leqslant u_n} \left| \frac{\bar{H}_n(u^-)}{\bar{H}(u^-)} \right| \geqslant \frac{1}{2},$$

$$\sup_{u \leqslant u_n} \left| \frac{\bar{H}_n(u)}{\bar{H}(u)} \right| \leqslant \frac{3}{2}, \qquad \inf_{u \leqslant u_n} \left| \frac{\bar{H}_n(u)}{\bar{H}(u)} \right| \geqslant \frac{1}{2}.$$

Lemma 3.2. *Let* $d_1 \geqslant 0$, $d_2 \geqslant 0$ *be fixed numbers such that* $d_1 + d_2 = 2$. *Then with probability one*

$$\int_{-\infty}^{u} \frac{d\tilde{F}_n(t)}{\bar{H}_n^{d_1}(t) \bar{H}^{d_2}(t)} \leqslant \frac{\sqrt{2n}}{\sqrt{\log \log n}} \frac{2^{d_1}}{k_1} \left(\frac{2}{k_1} + 1 \right)$$

for $u \leqslant u_n$ *and* $n \geqslant n_2(\omega)$.

Lemma 3.3. *If* $n \geqslant n_2(\omega)$, *then the following inequalities are satisfied with probability* 1.

(i) $$\sup_{-\infty < u \leqslant u_n} | \bar{F}_n^*(u) - \bar{\tilde{F}}_n(u) | \leqslant \frac{1}{\sqrt{n \log \log n}} \frac{4\sqrt{2}}{k_1} \left(\frac{2}{k_1} + 1 \right)$$

(ii) $$\sup_{-\infty < u \leqslant u_n} | \log \bar{\tilde{F}}_n(u) + T_n(u) | \leqslant \frac{1}{\sqrt{n \log \log n}} \frac{6\sqrt{2}}{k_1} \left(\frac{2}{k_1} + 1 \right)$$

(iii) $$\sup_{-\infty < u \leqslant u_n} | T_n(u) - T(u) | \leqslant \frac{4}{k_1} \left(\frac{1}{k_1} + 1 \right)$$

(iv) $$\sup_{-\infty < u \leqslant u_n} \bar{F}(u) | T_n(u) - T(u) | \leqslant$$

$$\leqslant \frac{4}{k_1} \left(\frac{1}{k_1} + 1 \right) \left(\frac{\log \log n}{2n} \right)^{\frac{1}{2(1+\gamma)}} \left(\frac{k_1}{k} \right)^{\frac{1}{1+\gamma}}$$

(v) *if* $k_1 \geqslant 6$ *then for all* $u \leqslant u_n$

$$\frac{1}{2} e^{-T_n^{**}(u)} (T(u) - T_n(u))^2 \leqslant \bar{F}(u) | T(u) - T_n(u) |.$$

Coming back to the proof of the theorem, the results follow from (3.2), (3.3) and the three lemmas. The constant C_1 is smallest when $k_1 = 6$.

Corollary 3.1. *Suppose* F *is continuous and* $\bar{G}(x) = \bar{F}(x)$ *for* $x > T$; *then*

$$\limsup_{n \to \infty} \left(\frac{n}{\log \log n}\right)^{\frac{1}{4}} \sup_{-\infty < x < +\infty} |F_n^*(x) - F(x)| \leqslant C_2$$

where C_2 *is a constant independent of* F *and* G. ($C_2 = 27$.)

This corollary is a consequence of the Theorem 3.1 when $G = F$, $k = 1$ and $\gamma = 1$. It should be pointed out that Theorem 3.1 can also be stated when condition (a) is changed to

(a') $T_F \leqslant T_G \leqslant \infty$.

If $T_F < T_G$ and G is continuous on (T, T_F) then we get a rate $O\left(\frac{\sqrt{\log \log n}}{\sqrt{n}}\right)$. The same rate is proved in [2] for this case but there the condition is a little bit more general, namely, the continuity of G on (T, T_F) is not needed. Theorem 3.1 improves the results of P h a d i a and V a n R y z i n [5] and S u s a r l a and V a n R y z i n [7].

4. DISTANCE BETWEEN F_n^* AND F_n^α

The following lemma gives the equivalent expressions for F_n^* and F_n^α. These equivalent expressions are used in Lemma 4.2 to obtain an upper bound of the distance between F_n^* and F_n^α.

Lemma 4.1. *Let* $\bar{F}_n^*(t)$ *be as defined in* (2.2). *Then* $\bar{F}_n^*(t)$ *can also be expressed as*

(4.1)
$$\bar{F}_n^*(t) = \frac{N^+(t)}{n} \prod_{l=1}^{m} \left(\frac{N^+(Z_{i_l}) + \mu_{i_l}}{N^+(Z_{i_l})}\right)^{[Z_{i_l} \leqslant t]} =$$

$$= \frac{N^+(t)}{n} \prod_{j=1}^{n} \left(\frac{N^+(Z_j) + \mu_j}{N^+(Z_j)}\right)^{\frac{[Z_j \leqslant t, \, \delta_j = 0]}{\mu_j}}$$

where $Z_{i_1} < Z_{i_2} < \ldots < Z_{i_m}$ $(m \leqslant n)$ are the distinct observations in the sample.

Proof. It is easy to see that the last two expressions in (4.1) are equal. Thus it is enough to show that the middle part of (4.1) is equal to the right side of (2.2). Suppose $Z_{i_j} \leqslant t < Z_{i_{j+1}}$, $j < n$. Then

(4.2)
$$A_n(t) = \frac{N^+(t)}{n} \prod_{l=1}^{m} \left(\frac{N^+(Z_{i_l}) + \mu_{i_l}}{N^+(Z_{i_l})} \right)^{[Z_{i_l} \leqslant t]} =$$

$$= \frac{N^+(t)}{n} \prod_{l=1}^{j} \frac{N^+(Z_{i_l}) + \mu_{i_l}}{N^+(Z_{i_l})}.$$

Using the fact that $N^+(t) = N^+(Z_{i_j})$, $n = N^+(Z_{i_1}) + \lambda_{i_1} + \mu_{i_1}$ and $N(Z_{i_l}) = N^+(Z_{i_{l+1}}) + \mu_{i_{l+1}} + \lambda_{i_{l+1}}$ we get

RHS of (4.2) =

$$= \frac{N^+(Z_{i_j})}{N^+(Z_{i_1}) + \lambda_{i_1} + \mu_{i_1}} \prod_{l=1}^{j} \frac{N^+(Z_{i_l}) + \mu_{i_l}}{N^+(Z_{i_{l+1}}) + \mu_{i_{l+1}} + \lambda_{i_{l+1}}} =$$

$$= \prod_{l=1}^{j} \left(\frac{N^+(Z_{i_l}) + \mu_{i_l}}{N^+(Z_{i_l}) + \mu_{i_l} + \lambda_{i_l}} \right) =$$

$$= \prod_{l=1}^{m} \left(\frac{N^+(Z_{i_l}) + \mu_{i_l}}{N^+(Z_{i_l}) + \mu_{i_l} + \lambda_{i_l}} \right)^{[Z_{i_l} \leqslant t]} =$$

$$= \prod_{i=1}^{n} \left(\frac{N^+(Z_i) + \mu_i}{N^+(Z_i) + \mu_i + \lambda_i} \right)^{\frac{[Z_i \leqslant t, \delta_i = 1]}{\lambda_i}} = \bar{F}_n^*(t).$$

This completes the proof of the lemma.

The following lemma gives the sup norm distance between the product limit estimate and the Bayes estimate.

Lemma 4.2. Let $\bar{F}_n^*(t)$ and $\bar{F}_n^\alpha(t)$ be as defined in (2.2) and (2.4) respectively.

(a) *For fixed* $t < \max\{Z_1 \ldots Z_n\}$

$$|\bar{F}_n^*(t) - \bar{F}_n^\alpha(t)| \le$$

$$\le \frac{\alpha(R)(1 + \alpha(R))}{n} \sum_{j=1}^{n} \frac{[Z_j \le t, \delta_j = 0]}{n\bar{H}_n^2(Z_j)} + \frac{2\alpha(R)}{n\bar{H}_n(t)}.$$

(b) *For fixed* $T < T_F$ *and arbitrary* F *and* G

$$\sup_{-\infty < t \le T} |\bar{F}_n^*(t) - \bar{F}_n^\alpha(t)| = O\left(\frac{1}{n}\right) \quad a.s.$$

(c) *Let* u_n *be as defined in* (3.4). *If* G *is continuous and* F *is arbitrary then, with probability one,*

$$\sup_{-\infty < t \le u_n} |\bar{F}_n^*(t) - \bar{F}_n^\alpha(t)| \le \frac{c}{\sqrt{n \log \log n}}$$

where

$$c = \frac{16\sqrt{2}\,\alpha(R)(1 + \alpha(R))}{k_1}\left(\frac{8}{k_1} + 5\right)\left(= \frac{152\sqrt{2}}{9}\,\alpha(R)(1 + \alpha(R))\right.$$

$$\left. \text{with}\quad k_1 = 6\right).$$

Proof.

(a) Without loss of generality let $Z_{i_1} < \ldots < Z_{i_m}$ denote the distinct observations in the sample. Also for a fixed $t < T_F$ let $Z_{i_j} \le t < Z_{i_{j+1}}$.

Then expressing $\bar{F}_n^*(t)$ and $\bar{F}_n^\alpha(t)$ in terms of distinct observations we get from (4.1) and (2.4)

(4.3) $$|\bar{F}_n^\alpha(t) - \bar{F}_n^*(t)| =$$

$$= \left| \frac{N^+(t) + \alpha(t, \infty)}{n + \alpha(R)} \prod_{l=1}^{j} \left(\frac{N^+(Z_{i_l}) + \alpha(Z_{i_l}, \infty) + \mu_{i_l}}{N^+(Z_{i_l}) + \alpha(Z_{i_l}, \infty)} \right) - \right.$$

$$\left. - \frac{N^+(t)}{n} \prod_{l=1}^{j} \left(\frac{N^+(Z_{i_l}) + \mu_{i_l}}{N^+(Z_{i_l})} \right) \right| \le$$

$$\leqslant \left| \frac{N^+(t) + \alpha(t, \infty)}{n + \alpha(R)} - \frac{N^+(t)}{n} \right| \prod_{l=1}^{j} a_{i_l} +$$

$$+ \left| \frac{N^+(t)}{n} \left(\prod_{l=1}^{j} a_{i_l} - \prod_{l=1}^{j} b_{i_l} \right) \right|$$

where

$$a_{i_l} = \frac{N^+(Z_{i_l}) + \alpha(Z_{i_l}, \infty) + \mu_{i_l}}{N^+(Z_{i_l}) + \alpha(Z_{i_l}, \infty)}$$

and

$$b_{i_l} = \frac{N^+(Z_{i_l}) + \mu_{i_l}}{N^+(Z_{i_l})}.$$

The proof of part (a) will follow from the following inequalities.

$$(4.4) \qquad \left| \frac{N^+(t) + \alpha(t, \infty)}{n + \alpha(R)} - \frac{N^+(t)}{n} \right| \leqslant \frac{2\alpha(R)}{n + \alpha(R)}.$$

$$(4.5) \qquad \prod_{l=1}^{j} a_{i_l} \leqslant \frac{N^+(Z_{i_1}) + \alpha(Z_{i_1}, \infty) + \mu_{i_1}}{N^+(Z_{i_j}) + \alpha(Z_{i_j}, \infty)}.$$

Let $b_{i_0} = 1$. Then using the fact that $n = N^+(Z_{i_1}) + \lambda_{i_1} + \mu_{i_1}$, $N^+(t) = N^+(Z_{i_j})$ and

$$\frac{N^+(Z_{i_{l+1}}) + \mu_{i_{l+1}} + \lambda_{i_{l+1}} + \alpha(Z_{i_{l+1}}, \infty)}{N^+(Z_{i_{l-1}})} \leqslant 1 + \alpha(R)$$

we get

$$\frac{N^+(t)}{n} b_{i_0} b_{i_1} \dots b_{i_{l-1}} a_{i_{l+1}} \dots a_{i_j} =$$

$$= N^+(t) \prod_{s=1}^{l-1} \left(\frac{N^+(Z_{i_s}) + \mu_{i_s}}{N^+(Z_{i_s}) + \mu_{i_s} + \lambda_{i_s}} \right) \times$$

$$\times \left(\frac{N^+(Z_{i_{l+1}}) + \lambda_{i_{l+1}} + \mu_{i_{l+1}} + \alpha(Z_{i_{l+1}}, \infty)}{N^+(Z_{i_{l-1}})} \right) \times$$

$$\times \prod_{s=l+1}^{j} \frac{N^+(Z_{i_s}) + \mu_{i_s} + \alpha(Z_{i_s}, \infty)}{N^+(Z_{i_s}) + \mu_{i_s} + \lambda_{i_s} + \alpha(Z_{i_s}, \infty)} \times$$

$$\times \frac{1}{N^+(Z_{i_j}) + \alpha(Z_{i_j}, \infty)} \leqslant 1 + \alpha(R).$$

Hence

$$\left| \frac{N^+(t)}{n} \left(\prod_{l=1}^{j} a_{i_l} - \prod_{l=1}^{j} b_{i_l} \right) \right| =$$

$$= \left| \sum_{l=1}^{j} \frac{N^+(t)}{n} b_{i_0} \dots b_{i_{l-1}} (a_{i_l} - b_{i_l}) a_{i_{l+1}} \dots a_j \right| \leqslant$$

$$\leqslant \sum_{l=1}^{j} (1 + \alpha(R)) |a_{i_l} - b_{i_l}| \leqslant (1 + \alpha(R)) \sum_{l=1}^{j} \frac{\alpha(R) \mu_{i_l}}{(N^+(Z_{i_l}))^2} =$$

(4.6)

$$= \alpha(R)(1 + \alpha(R)) \sum_{l=1}^{m} \frac{[Z_{i_l} \leqslant t] \mu_{i_l}}{(N^+(Z_{i_l}))^2} \leqslant$$

$$\leqslant \alpha(R)(1 + \alpha(R)) \sum_{i=1}^{n} \frac{[Z_i \leqslant t, \delta_i = 0]}{(N^+(Z_i))^2} =$$

$$= \frac{\alpha(R)(1 + \alpha(R))}{n} \sum_{i=1}^{n} \frac{[Z_i \leqslant t, \delta_i = 0]}{n \bar{H}_n^2(Z_i)}.$$

Now (4.4), (4.5) and (4.6) together imply part (a) of the lemma.

Part (b) follows directly from (a) since for large n $\quad T <$ $< \max \{Z_1 \dots Z_n\}$ with probability one.

To show part (c) observe that, for $t \leqslant u_n$ (u_n is defined in (3.4))

(4.7) $\quad |\bar{F}_n^*(t) - \bar{F}_n^\alpha(t)| \leqslant$

$$\leqslant \frac{\alpha(R)(1 + \alpha(R))}{n} \left\{ \sum_{i=1}^{n} \frac{[Z_i \leqslant t, \delta_i = 0]}{n \bar{H}_n^2(Z_i)} \right\} + \frac{2\alpha(R)}{n \bar{H}_n(t)} \leqslant$$

$$\leqslant \frac{4\alpha(R)(1 + \alpha(R))}{n} \sum_{i=1}^{n} \frac{[Z_i \leqslant t, \delta_i = 0]}{n \bar{H}^2(Z_i)} + \frac{4\alpha(R)}{n \bar{H}(t)}$$

by Lemma 3.1 $=$

$$= \frac{4\alpha(R)(1 + \alpha(R))}{n} \int_{-\infty}^{t} \frac{d\tilde{G}_n(u)}{\bar{H}^2(u)} + \frac{4\alpha(R)}{n\bar{H}(u_n)} \leqslant$$

$$\leqslant \frac{4\alpha(R)(1 + \alpha(R))}{n} \frac{\sqrt{2n}}{\sqrt{\log\log n}} \frac{4}{k_1} \left(\frac{8}{k_1} + 5\right) \quad \text{by Lemma 3.2} =$$

$$= \frac{152\sqrt{2}}{9} \alpha(R)(1 + \alpha(R)) \sqrt{\frac{1}{n\log\log n}} \quad \text{for} \quad k_1 = 6, \ t \leqslant u_n.$$

Hence

(4.8)
$$\sup_{-\infty < t \leqslant u_n} |\bar{F}_n^*(t) - \bar{F}_n^\alpha(t)| \leqslant \frac{152\sqrt{2}}{9} \frac{\alpha(R)(1 + \alpha(R))}{\sqrt{n\log\log n}} \quad \text{a.s.}$$

This completes the proof of the lemma.

Theorem 4.1. *Let F and G be continuous. Then*

$$\limsup_{n \to \infty} \left(\frac{n}{\log\log n}\right)^{\frac{1}{2(1+\gamma)}} \sup_{-\infty < t < \infty} |F_n^\alpha(t) - F(t)| \leqslant$$

$$\leqslant \left(C_0 + \frac{C_1}{k} 2^{-\frac{1}{2(1+\gamma)}} + \frac{k_1}{k}\right)$$

where C_0, C_1, k and γ are as given in Theorem 3.1 and Lemma 4.2.

Proof. Observe that

(4.9)
$$\sup_{-\infty < t < \infty} |F_n^\alpha(t) - F(t)| = \sup_{-\infty < t < \infty} |\bar{F}_n^\alpha(t) - \bar{F}(t)| \leqslant$$

$$\leqslant \sup_{-\infty < t < \infty} |\bar{F}_n^\alpha(t) - \bar{F}_n^*(t)| + \sup_{-\infty < t < \infty} |\bar{F}_n^*(t) - \bar{F}(t)|.$$

Hence it is enough to consider the first term on the right hand side of (4.9). Choose u_n according to (3.4). Then using the following fact that

$$\sup_{t \geqslant u_n} |\bar{F}_n^\alpha(t) - \bar{F}(t)| \leqslant \max\{\bar{F}_n^\alpha(u_n), \bar{F}_n^*(u_n)\} \leqslant$$

$$\leqslant |\bar{F}_n^\alpha(u_n) - \bar{F}_n^*(u_n)| + |\bar{F}_n^*(u_n) - \bar{F}(u_n)| + \bar{F}(u_n) \leqslant$$

$$\leqslant \sup_{u \leqslant u_n} |\bar{F}_n^\alpha(u) - \bar{F}_n^*(u)| + \sup_{-\infty < t < \infty} |\bar{F}_n^*(t) - \bar{F}(t)| + \bar{F}(u_n)$$

we get

$$\sup_{-\infty < t < \infty} |\bar{F}_n^\alpha(t) - \bar{F}_n^*(t)| \leqslant$$

(4.10)
$$\leqslant \sup_{-\infty < t \leqslant u_n} |\bar{F}_n^\alpha(t) - \bar{F}_n^*(t)| + \sup_{t \geqslant u_n} |\bar{F}_n^\alpha(t) - \bar{F}_n^*(t)| \leqslant$$

$$\leqslant \sup_{-\infty < t \leqslant u_n} |\bar{F}_n^\alpha(t) - \bar{F}_n^*(t)| +$$

$$+ \sup_{-\infty < t < \infty} |\bar{F}_n^*(t) - \bar{F}(t)| + \bar{F}(u_n).$$

Hence Theorem 3.1, Lemma 4.2 (c), (3.5) and (4.9) together imply

(4.11)
$$\sup_{-\infty < t < \infty} |\bar{F}_n^\alpha(t) - \bar{F}(t)| \leqslant \frac{c}{\sqrt{n \log \log n}} +$$

$$+ \left(\frac{\log \log n}{n}\right)^{\frac{1}{2(1+\gamma)}} \left(C_0 + \frac{C_1}{k} 2^{-\frac{1}{2(1+\gamma)}}\right) +$$

$$+ \left(\frac{\log \log n}{n}\right)^{\frac{1}{2(1+\gamma)}} \frac{k_1}{k}.$$

Hence

$$\limsup_{n \to \infty} \left(\frac{n}{\log \log n}\right)^{\frac{1}{2(1+\gamma)}} \sup_{-\infty < t < \infty} |\bar{F}_n^\alpha(t) - \bar{F}(t)| \leqslant$$

$$\leqslant \left(C_0 + \frac{C_1}{k} 2^{-\frac{1}{2(1+\gamma)}} + \frac{k_1}{k}\right).$$

This completes the proof of the theorem.

Theorem 4.2. *Suppose F and G are arbitrary. Let $T < T_F$ be fixed. If $\bar{H}(T^-) < 1$ then*

$$\sup_{-\infty < t \leqslant T} |\bar{F}_n^\alpha(t) - \bar{F}(t)| = O\left(\left(\frac{\log n}{n}\right)^{\frac{1}{4}}\right) \quad \text{a.s.}$$

The proof of this theorem follows from Lemma 4.2 (b) and Theorem 2.2 in [3].

REFERENCES

[1] M.D. Burke – S. Csörgő – L. Horváth, Strong approximations of some biometric estimates under random censorship, *Z. Wahrscheinlichkeitstheorie verw. Geb.,* 56 (1981), 87-112.

[2] A. Földes – L. Rejtő, A LIL type result for the product limit estimator, *Z. Wahrscheinlichkeitstheorie verw. Geb.,* 56 (1981), 75-86.

[3] A. Földes – L. Rejtő – B.B. Winter, Strong consistency properties of nonparametric estimators for randomly censored data. The product limit estimator, estimation of density and failure rate. *Trans. of the eighth Prague conference on information theory, statistical decision functions, random processes,* Vol. C, (1979), 105-121.

[4] E.L. Kaplan – P. Meier, Nonparametric estimation from incomplete observations, *J. Amer. Statist. Assoc.,* 53 (1958), 457-481.

[5] E.G. Phadia – J. Van Ryzin, A note on convergence rates for the product limit estimator, *Ann. Statist.,* 8 (1980), 673-678.

[6] V. Susarla – J. Van Ryzin, Nonparametric estimation of survival curves from incomplete observations, *JASA,* 71 (356) (1976), 897-902.

[7] V. Susarla – J. Van Ryzin, Addendum to large sample theory for a Bayesian nonparametric survival curve estimator based on censored samples, *Ann. Statist.,* 8, (1980), 693.

[8] V. Susarla – J. Van Ryzin, Large sample theory for survival curve estimators under variable censoring. *Optimization methods in statistics,* Academic Press, Inc., (1979), 475-508.

J. Ghorai

Math. Dept., Univ. of Wisconsin--Milwauke, Milwauke, WI 53201, USA.

L. Rejtő

Math. Inst. Hung. Acad. Sci., H-1053 Budapest, Reáltanoda u. 13-15, Hungary.

V. Susarla

Dept. of Stat. and Probability, M.S.U., E. Lansing, MI 48824, USA.

COLLOQUIA MATHEMATICA SOCIETATIS JÁNOS BOLYAI

32. NONPARAMETRIC STATISTICAL INFERENCE,

BUDAPEST (HUNGARY), 1980.

NONPARAMETRIC ESTIMATION OF MEAN RESIDUAL LIFE TIME WITH CENSORED DATA*

J. GHORAI — A. SUSARLA — V. SUSARLA — J. VAN RYZIN

1. INTRODUCTION

In survival analysis, one is generally interested in the survival function $F(t)$ = probability of survival beyond t units of time, in $r(t)$ = conditional probability of survival in a small interval of time $(t, t + \Delta t)$ given survival to time t as $\Delta t \to 0$, in the mean survival time $\mu = \int_0^\infty F(t)\, dt$, and in $f(t) = \lim_{\Delta t \to 0} \dfrac{F(t) - F(t + \Delta t)}{\Delta t}$. Gehan [7] calls the function r the age-specific mortality rate function. Also, of particular interest in the survival analysis, and replacement versus renewal policies (Bryson and Siddiqui [3], Esary, Marshall and Proschan [4], Marshall and Proschan [15], Barlow and Proschan [1]) is the concept of residual lifetime, $e(t)$, given survival to at least time t, $t > 0$. In symbols,

*Research supported by NIH Grant R01 GM-28405.

(1.1) $\qquad e(t) = \dfrac{1}{F(t)} \int\limits_{t}^{\infty} F(u)\, du.$

If $e(t)$ is constant and F is continuous on $(0, \infty)$, it implies that $F(t) = e^{-vt}$ for some $v > 0$. In the statistical literature, the characteristic $e(t)$ being constant is known as "no aging". $e(t)$ is related to $r(t)$ in the same way as μ is related to F. $e(t) - t$, which gives the remaining average survival time given that the unit survived at least t units of time, is important to researchers in medicine, reliability, and system analysis.

If one has a set of n independent random survival times $X_1, \ldots, X_n$ from $1 - F$ corresponding to n units, it is easy to estimate $e(\cdot)$. An intuitive estimator of $e(\cdot)$ is given by $\hat{e}(\cdot) = \dfrac{1}{\hat{F}(\cdot)} \int \hat{F}(u)\, du$ where

$n\hat{F}(t) = \sum\limits_{i=1}^{n} [X_i > t] = \#$ of units of surviving at least t units of time. H a l l and W e l l n e r [9], and Y a n g [23] studied the asymptotic properties of $\hat{e}$. But in several survival analysis studies, especially in medical studies where the period of study is fixed, one does not get to observe the lifetimes $X_1, \ldots, X_n$; but only their right censored versions leading to the information

(1.2) $\qquad (\delta_1, Z_1), \ldots, (\delta_n, Z_n)$

where for $i = 1, \ldots, n$,

(1.3) $\qquad \delta_i = [X_i \leqslant Y_i] \quad \text{and} \quad Z_i = X_i \wedge Y_i$

with Y_i representing the i-th censoring random variable. Unless otherwise stated, it is assumed throughout that $Y_1, \ldots, Y_n$ are i.i.d. with $G(y) = P[Y > y] > 0$, and that G is continuous. In survival analysis studies, the information described in (1.2) and (1.3) comes about due to the fact that the patient might have been withdrawn alive or might have died for some reason other than the disease under consideration or might have been alive at the end of the period of study. (See, for example, G r o s s and C l a r k [8].)

The object of the present investigation is to present a class of estimators of $\hat{e}_\alpha(\cdot)$ of $e(\cdot)$ based on $(\delta_1, Z_1), \ldots, (\delta_n, Z_n)$, one for each measure α involved in the definition of $\hat{e}_\alpha(\cdot)$ and study their asymptotic properties like mean square consistency, and weak convergence of $\hat{e}(\cdot)$ on a fixed interval on $[0, T]$, $T < \infty$. The estimators $\hat{e}_\alpha(\cdot)$ have a Bayesian interpretation. We also provide a method for obtaining a confidence band for $e_M(\cdot) = \frac{1}{F(\cdot)} \int\limits_\cdot^M F(t)\, dt$ with $M = M(n) \to \infty$ as $n \to \infty$. We conclude the paper with a brief application. All the proofs are postponed to the Appendix.

2. DEFINITION AND ASYMPTOTIC PROPERTIES OF SOME ESTIMATORS OF $e(\cdot)$

Let $(\delta_1, Z_1), \ldots, (\delta_n, Z_n)$ be as defined in (1.2) and (1.3). For the sake of describing a Bayes estimator of F, when $1 - F$ is distributed according to a Dirichlet process with parameter measure α on the Borel σ-field in $(0, \infty)$ (for example, see F e r g u s o n [5]), let $Z_1, \ldots, Z_k$ denote the uncensored observations, and $Z_{(k+1)}, \ldots, Z_{(m)}$ denote the ordered distinct censored observations. Let λ_j denote the frequency of $Z_{(j)}$ $(j = k + 1, \ldots, m)$, among censored observations. Let $N^+(\cdot)$ denote the number of censored and uncensored observations greater than . If we assume squared loss function

(2.1) $L(\hat{e}(\cdot), e(\cdot)) = (\hat{e}(\cdot) - e(\cdot))^2$

and the Dirichlet process prior with parameter α for $1 - F$, then the Bayes estimator of $e(\cdot)$ is given by

(2.2) $\hat{e}_\alpha(\cdot) = \int\limits_\cdot^\infty \mathsf{E}\left[\frac{F(u)}{F(\cdot)} \mid (\delta_1, Z_1), \ldots, (\delta_n, Z_n) \right] du.$

The inside expectation can be evaluated using the fact that given the uncensored observations $(1, Z_1), \ldots, (1, Z_k)$, $1 - F$ is again a Dirichlet process with parameter measure $\alpha^*(\cdot) = \alpha(\cdot) + \sum\limits_{i=1}^k I_\cdot(Z_i)$ where

$$I_A(a) = \begin{cases} 1 & \text{if} \quad a \in A \\ 0 & \text{otherwise,} \end{cases}$$

for any measurable A and the following lemma where $E_{\alpha*}[\ ]$ stands for expectation with respect to the Dirichlet process with parameter measure α^*.

Lemma 2.1. *Let $T(F)$ be any nonnegative real valued function defined over the class of all survival functions. Then*

$$E_{\alpha*}[T(F) \mid (\lambda_j, Z_{(j)}), \ j = k+1, \ldots, m] =$$

$$= \frac{E_{\alpha*}\left[T(F) \prod_{j=k+1}^{m} F^{\lambda_j}(Z_{(j)})\right]}{E_{\alpha*}\left[\prod_{j=k+1}^{m} F^{\lambda_j}(Z_{(j)})\right]}.$$

Using the above lemma, one can show that

Lemma 2.2. *If $Z_{(l)} \leqslant x < Z_{(l+1)}$, $Z_{(l')} \leqslant u < Z_{(l'+1)}$, with $l \leqslant l'$ $(l, l' = k, \ldots, m)$, $Z_{(k)} = 0$, and $Z_{(m+1)} = \infty$, and $u \leqslant x$, then*

$$E_{\alpha*}\left[\frac{F(u)}{F(x)} \mid (\lambda_j, Z_{(j)}), \ j = k+1, \ldots, n\right] = \frac{\hat{F}_{\alpha}(u)}{\hat{F}_{\alpha}(x)}$$

where

$$\hat{F}_{\alpha}(t) = \frac{\alpha[t, \infty) + N^+(t)}{\alpha[0, \infty) + n} \times$$

(2.3)

$$\times \prod_{j=1}^{n} \left\{\frac{\alpha[Z_j, \infty) + N^+(Z_j) + \lambda_j}{\alpha[Z_j, \infty) + N^+(Z_j)}\right\}^{\frac{[\delta_j = 0, Z_j \leqslant t]}{\lambda_j}}$$

for $t = u$ or x.

Remark 2.1. It is worth noting here that the conditional expectation of the ratio $\frac{F(u)}{F(x)}$ given the data is same as the ratio of the conditional expectations (given the data) of $F(u)$ and $F(x)$, a property similar to the one shared by multivariate beta random variables.

Theorem 2.1. *The Bayes estimator of $e(\cdot)$ with squared error loss function and the Dirichlet process prior with parameter $\alpha(\cdot)$ for $1 - F$ is given by*

$$(2.4) \qquad \hat{e}_\alpha(\cdot) = \frac{1}{\hat{F}_\alpha(\cdot)} \int\limits^\infty \hat{F}_\alpha(u) \, du$$

where $\hat{F}_\alpha$ is given by (2.3).

Remark 2.2. If $\alpha[x, \infty) > 0$ for each $x > 0$, then the estimator $\hat{e}_\alpha(\cdot)$ is properly defined on $[0, \infty)$, and it utilizes all censored and un-censored observations. This is very important where heavy censoring occurs. It reduces to the estimator of Y a n g [22] if $\alpha([0, \infty)) \to 0$, and all the observations are uncensored. If $\alpha([0, \infty)) \to 0$, then it simply converges to the estimator obtained by replacing $\dfrac{\hat{F}_\alpha(u)}{\hat{F}_\alpha(\cdot)}$ by the ratio of K a p l a n − M e i e r [12] estimators for $F(u)$ and $\hat{F}(\cdot)$.

Remark 2.3. By taking $x = 0$, we get the Bayes estimator of the mean $\mu = \int\limits_0^\infty F(u) \, du$, namely $\hat{\mu} = \int\limits_0^\infty \hat{F}_\alpha(u)$, under squared error loss.

In the rest of the paper, we will study the asymptotic properties of the estimator $\hat{e}_\alpha(\cdot)$ or its modifications in a non-Bayesian setup. That is, F and G are fixed, $X_1, \ldots, X_n$ are i.i.d. $1 - F$ with $\mu = \int\limits_0^\infty F(u) \, du < \infty$, $Y_1, \ldots, Y_n$ are i.i.d. $1 - G$, and that F and G are continuous. Because of the explosive behavior of $\lim\limits_{n \to \infty} \mathrm{var} \, (\hat{F}_\alpha(t))$, $\lim\limits_{t \to \infty} \lim\limits_{n \to \infty} \mathrm{var} \, (\hat{F}_\alpha(t)) = \infty$ the asymptotic properties of $\hat{e}_\alpha(t)$ are difficult to study under the above general conditions, and moreover, from the large sample theory point of view, the parameter α does not play any role. Consequently we consider the estimator

$$(2.5) \qquad \bar{e}_n(x) = \int\limits_x^M \frac{\bar{F}_n(u)}{\bar{F}_n(x)}$$

where $M = M(n) \to \infty$, and

$$(2.6) \qquad \bar{F}_n(\cdot) = \frac{N^+(\cdot)}{n} \prod\limits_{j=1}^n \left\{ \frac{2 + N^+(Z_j)}{1 + N^+(Z_j)} \right\}^{[\delta_j = 0, Z_j \leqslant \cdot]}.$$

Before proceeding further, it is worth pointing out here that, while the large sample properties are not effected by the presence or absence of

α, it has been observed in some simulations that the Bayes estimator $\hat{F}_\alpha$ performs better than the Kaplar $-$ Meier estimator of F in L_1, L_2, and L_∞ norms. This better performance is considerable in the presence of heavy censoring. For details, see R a i, S u s a r l a and V a n R y z i n [16].

Theorem 2.2. *Let the following conditions hold*

(A1) $\dfrac{M}{n} \displaystyle\int_a^M \dfrac{du}{G^4(u)F^2(u)} \to 0$ *for an* $a > 0$.

(A2) $\dfrac{M^2}{n} \to 0$.

(A3) $F(T)G(T) > 0$.

Then

$$\sup_{x \leqslant T} \mathsf{E}[(\bar{e}_n(x) - e(x))^2] \to 0.$$

Remark 2.4. From the bounds obtained in the proof of the above theorem, it is possible to obtain a rate for the conclusion of the theorem.

Remark 2.5. A rate for the strong consistency of $\bar{e}_n$ can also be obtained using the methods of Section 4 of [21], but as noted there, this rate could be very slow. Also, the choice of $M (= M(n))$ which is discussed there applies here as well. Typically, a choice like $M = C (\log \log n)$ would suffice for large classes of choices for F and G including gamma densities for F and G.

To state the weak convergence result for

$$\left\{ \sqrt{n} \left(\bar{e}_n(x) - \frac{1}{F(x)} \int_x^M F(u) \, du \right) \mid 0 \leqslant x \leqslant T \right\},$$

we need the following notation: Define

(2.7) $\quad \tilde{\tilde{H}}(s) = \mathsf{P}[\delta_1 = 1, Z_1 \leqslant s]$,

(2.8) $\quad H(s) = F(s)G(s)$,

(2.9) $\qquad \mu(s) = \int\limits_s^\infty F(u)\,du,$

and

(2.10) $\qquad \sigma(x,y) = \dfrac{1}{F(x)F(y)} \int\limits_y^\infty \dfrac{\mu^2(v)}{H^2(v)}\,d\widetilde{H}(v)$

where $0 \leqslant x \leqslant y \leqslant T < \infty$.

Theorem 2.3. *Let* $\sigma(x,y) < \infty$ *for all* $0 \leqslant x \leqslant y \leqslant T < \infty$, *and let the following conditions hold.*

(A4) $\qquad \dfrac{1}{\sqrt{n}\,H^3(M)} \to 0.$

(A5) $\qquad \dfrac{1}{\sqrt{n}} \int\limits_0^M \dfrac{du}{H^4(u)G(u)} \to 0.$

Then

$$\left\{ \sqrt{n}\left(\bar{e}_n(x) - \dfrac{1}{F(x)} \int\limits_x^M F(u)\,du\right) \mid 0 \leqslant x \leqslant T \right\}$$

converges weakly to a mean zero Gaussian process with covariance function $\sigma(x,y)$.

Two important special cases of this theorem are given below.

Remark 2.6. If there is no censoring, then $G(u) \equiv 1$. In this case,

$$\sigma(x,y) = \dfrac{1}{F(x)F(y)} \int\limits_y^\infty u^2 d(1-F(u)) - \dfrac{1}{F(y)}\left(\int\limits_y^\infty u\,d(1-F(u))\right)^2.$$

The above expression is given by Y a n g ((3.2) in [22]).

Remark 2.7. If $x = y = 0$, then we obtain that $\bar{e}_n(0) = \int\limits_0^M \bar{F}_n(u)\,du$ as an estimator of μ. Specializing $\sigma(x,y)$ to this case, we obtain

$$\sigma(0,0) = \int\limits_0^\infty \dfrac{1}{H^2(s)}\left(\int\limits_s^\infty F(u)\,du\right)^2 d\widetilde{H}(s).$$

This expression is given by S u s a r l a and V a n R y z i n (Theorem 2.1 in [21]).

Remark 2.8. Even in the case of no censoring, that is, $G \equiv 1$, the conditions (A1) through (A5) are *not* automatically satisfied. $\sigma(x, y) < \infty$ for all $x \leqslant y \leqslant T < \infty$ is implied by the hypothesis that $\sigma(0, 0) < \infty$, and $F(T) > 0$. Hall and Wellner [9], assume in the case of no censoring, that the second moment is finite.

3. CONFIDENCE BANDS FOR $\int\limits_{.}^{M} F(u)\, du$

In this section, we will show how one can use Theorem 2.3 to obtain confidence bands for $\left\{ \int\limits_{x}^{M} F(u)\, du \mid 0 \leqslant x \leqslant T \right\}$, $T < \infty$. The technique, similar to the one given in Hall and Wellner [9], is modified to the censored data situation considered here.

Recall from the previous section (Theorem 2.3) that

$$(3.1) \qquad \left\{ \sqrt{n} \left(\bar{e}_n(x) - \frac{1}{F(x)} \int\limits_{x}^{M} F(u)\, du \right) \mid 0 \leqslant x \leqslant T \right\}$$

converges weakly ($\xrightarrow{w}$) to a mean zero Gaussian process with covariance function

$$(3.2) \qquad \sigma(x, y) = \frac{1}{F(x)F(y)} \int\limits_{y}^{\infty} \frac{\mu^2(u)}{H^2(u)}\, d\widetilde{H}(u) \qquad (0 \leqslant x \leqslant y \leqslant T).$$

Let

$$(3.3) \qquad Q(x) = \frac{F^2(x)\, \sigma(x, x)}{\sigma(0, 0)},$$

$$(3.4) \qquad Z(x) = \frac{\sqrt{\sigma(0, 0)}}{F(x)}\, B(Q(x)),$$

where $\{ B(t) \mid 0 \leqslant t < \infty \}$ is the standard Brownian motion on $[0, \infty)$, and

$$(3.5) \qquad W(x) = B(Q(x)).$$

Lemma 3.1.

(a) Q *is a survival function on* $[0, \infty)$. *(That is, Q is nonincreasing, $Q(0) = 1$, and $Q(\infty) = 0$.)*

(b) $\{Z(t) \mid 0 \leqslant t \leqslant T\}$ is a zero mean Gaussian process with co-variance function $\sigma(x, y)$ $(x \leqslant y)$.

(c) $\left\{\sqrt{n}\left(\bar{e}_n(x) - \dfrac{1}{F(x)} \int\limits_x^M F(u)\, du\right) \mid 0 \leqslant x \leqslant T\right\} \xrightarrow{\ w\ }$

$$\{Z(x) \mid 0 \leqslant x \leqslant T\}.$$

(d) $\left\{\dfrac{F(x)}{\sqrt{\sigma(0, 0)}}\, \sqrt{n}\left(\bar{e}_n(x) - \dfrac{1}{F(x)} \int\limits_x^M F(u)\, du\right) \mid 0 \leqslant x \leqslant T\right\} \xrightarrow{\ w\ }$

$$\{W(x) \mid 0 \leqslant x \leqslant T\}.$$

The next theorem provides conditions under which we can replace $F(x)$ and $\sigma(0, 0)$ in (d) of the above lemma by their estimators. We estimate $F(x)$ and $\sigma(0, 0)$ by $\bar{F}_n(x)$ of (2.6) and $\hat{\sigma}(0, 0)$ defined below.

(3.6) $\hat{\sigma}(0, 0) = \int\limits_0^N \dfrac{\left(\int\limits_u^N \bar{F}_n(v)\, dv\right)^2}{H_n^2(u)}\, d\widetilde{\widetilde{H}}_n(u)$

where $N \to \infty$, $\bar{F}_n$ is given by (2.6),

(3.7) $n H_n(\cdot) = \#$ of $Z_j > \cdot$ $(j = 1, \ldots, n)$,

and

(3.8) $n\widetilde{\widetilde{H}}_n(\cdot) = \#$ of (δ_j, Z_j) such that $\delta_j = 1$, $Z_j \leqslant \cdot$,

$$(j = 1, \ldots, n).$$

Theorem 3.1. *If* $\hat{\sigma}(0, 0) \xrightarrow{\ P\ } \sigma(0, 0)$, *then*

$$\left\| \dfrac{\bar{F}_n(x)}{\sqrt{\hat{\sigma}(0, 0)}}\, \sqrt{n}\left(\bar{e}_n(x) - \dfrac{1}{F(x)} \int\limits_x^M F(u)\, du\right) \right\|_0^T \xrightarrow{\ D\ } \| B(Q(t)) \|_0^T.$$

Theorem 3.1 can be used to obtain confidence bands for $\dfrac{1}{F(x)} \int\limits_x^M F(u)\, du$ with $0 \leqslant x \leqslant T$ with asymptotic confidence coefficient at least β as follows: Let $0 < \beta < 1$. Then find $a(\beta)$ such that

(3.9) $P[\| B(t) \|_0^1 \leqslant a(\beta)] \geqslant \beta$.

(For example, see page 79 of Billingsley [2].) By Theorem 3.1,

$$\mathsf{P}\Big[\Big\| \frac{\bar{F}_n(x)}{\sqrt{\hat{\sigma}(0,0)}} \sqrt{n}\Big\{\bar{e}_n(x) - \frac{1}{F(x)} \int_x^M F(u)\,du\Big\} \Big\|_0^T \leqslant a(\beta)\Big] \rightarrow$$

$$\rightarrow \mathsf{P}[\|B(Q(t))\|_0^T \leqslant a(\beta)] = \mathsf{P}[\|B(t)\|_1^{Q(T)} \leqslant a(\beta)] \geqslant$$

$$\geqslant \mathsf{P}[\|B(t)\|_0^1 \leqslant a(\beta)] = \beta.$$

Hence a confidence band for $\Big\{\frac{1}{F(x)} \int_x^M F(u)\,du \mid 0 \leqslant x \leqslant T\Big\}$ with (asymptotic) confidence coefficient at least β is given by

$$(3.10) \qquad \Big[\bar{e}_n(x) - \frac{\sqrt{\hat{\sigma}(0,0)}}{\bar{F}_n(x)} \frac{a(\beta)}{\sqrt{n}}, \ \bar{e}_n(x) + \frac{\sqrt{\hat{\sigma}(0,0)}}{\bar{F}_n(x)} \frac{a(\beta)}{\sqrt{n}}\Big].$$

Remark 3.1. The confidence band given by (3.10) is very wide in the right tail; slight improvement of this can be made if one can use the distribution of $\|B(t)\|^{Q(T)}$ where Q is *unknown*, but can be estimated by $\hat{Q}$ given by

$$\hat{Q}(\cdot) = \frac{1}{\hat{\sigma}(0,0)} \int_{\cdot}^N \frac{\Big(\int_u^N \bar{F}_n(v)\,dv\Big)^2}{H_n^2(v)}\, d\widetilde{\widetilde{H}}_n(v)$$

where $\hat{\sigma}(0,0)$ is given by (3.6). If $N \rightarrow \infty$ at a specified rate, then both $\hat{\sigma}(0,0)$ and $\hat{Q}(\cdot)$ are consistent estimators of $\sigma(0,0)$ and $Q(\cdot)$ respectively.

Remark 3.2. For $\hat{\sigma}(0,0) \xrightarrow{\mathsf{P}} \sigma(0,0)$, it is enough that the following condition holds:

$$\frac{N^2}{\sqrt{N}\,H^4(N)} \rightarrow 0.$$

This condition is stronger than is really necessary, but easy to verify along with the conditions of Theorems 2.2 or 2.3. It can be verified that $\|\bar{F}_n(x) - F(x)\|_0^T = O\Big(\frac{\sqrt{\log\log n}}{\sqrt{n}}\Big)$ whenever $F(T)G(T) > 0$ using the details given in Susarla and Van Ryzin [18], [20] or Földes and Rejtő [24].

4. A BRIEF APPLICATION

We consider the data given in the Appendix of Susarla and Van Ryzin [19]. From this set of data, we removed all the duplicates leaving us with 69 observations, which are listed below, corresponding to the survival times (in weaks) of 69 participants of a melanoma study conducted by the Central Oncology Group with headquarters at the University of Wisconsin – Madison.

Uncensored observations. 16, 44, 55, 67, 73, 76, 80, 81, 86, 93, 100, 108, 114, 120, 125, 129, 134, 140, 147, 148, 151, 152, 158, 181, 190, 193, 213, 215.

Censored observations. 13, 14, 19, 20, 21, 23, 25, 26, 27, 31, 32, 34, 37, 38, 40, 46, 50, 53, 54, 57, 57, 59, 60, 65, 66, 70, 85, 90, 98, 102, 103, 110, 118, 124, 130, 136, 138, 141, 194, 234.

For purposes of calculation, it is easier to work with $M = \infty$ (in practice, we can select M large enough even before the sample has been drawn so that most of the observations are less than or equal to M. For example, in the cancer data situation we have here, we can and do take $M \geqslant 20$ years.) Also, it is easier to work with

$$(4.1) \qquad \bar{F}_{n,1}(\cdot) = \frac{N^+(\cdot) + 1}{n + 1} \prod_{j=1}^{n} \left\{ \frac{2 + N^+(Z_j)}{1 + N^+(Z_j)} \right\}^{[\delta_j = 0, Z_j \leqslant \cdot]}$$

instead of $\bar{F}_n$ of (2.6). It is clear that both $\bar{F}_{n,1}(\cdot)$ and $\bar{F}_n(\cdot)$ are asymptotically equivalent and for purposes of calculation, it is easier to work with $\bar{F}_{n,1}(\cdot)$ than with $\bar{F}_n(\cdot)$ since $\bar{F}_{n,1}(\cdot)$ has jumps only at uncensored observations while $\bar{F}_n(\cdot)$ has jumps at all Z_j. Also,

$$(4.2) \qquad \begin{aligned} \text{jump of } & \bar{F}_{n,1}(\cdot) \text{ at } Z_l = \\ &= \frac{\delta_l}{n + 1} \prod_{j=1}^{n} \left\{ \frac{2 + N^+(Z_j)}{1 + N^+(Z_j)} \right\}^{[\delta_j = 0, Z_j \leqslant Z_l]} \end{aligned}$$

Let

$$(4.3) \qquad \bar{e}_{n,1}(x) = \frac{1}{\bar{F}_{n,1}(x)} \int_x^M \bar{F}_{n,1}(u) \, du,$$

be an estimator of $\frac{1}{F(x)} \int_x^M \bar{F}(u)\,du$. Then $\{\bar{e}_{n,1}(x) \mid 0 \leqslant x \leqslant T\}$ has the same asymptotic properties as $\{\bar{e}_n(x) \mid 0 \leqslant x \leqslant T\}$. $\bar{e}_{n,1}(x)$ can be rewritten as

$$\bar{e}_{n,1}(x) = x + \frac{1}{(n+1)\bar{F}_{n,1}(x)} \sum_{l=1}^n \delta_l Z_l \times$$

(4.4)

$$\times \prod_{j=1}^n \left\{ \frac{2 + N^+(Z_j)}{1 + N^+(Z_j)} \right\}^{[\delta_j = 0,\, Z_j \leqslant Z_l]},$$

where we have used (4.2) and the fact that all the observations are less than equal to $M = 20$ years. The table below provided values $\bar{e}_{n,1}(x) - x$ for the data described above. Observe that $\bar{e}_{n,1}(x) - x$ is right continuous.

Table 1. Estimator of mean residual life.

x in	$\bar{e}_{n,1}(x) - x$	x in	$\bar{e}_{n,1}(x) - x$
[0,16)	125.7495	[129,132)	143.5508
[16,44)	127.3841	[132,134)	151.6697
[44,55)	128.9454	[134,140)	144.5653
[55,67)	127.7829	[140,147)	144.7464
[67,73)	126.1251	[147,148)	144.3811
[73,76)	127.4602	[148,151)	143.8571
[76,80)	128.7854	[151,152)	142.9240
[80,81)	130.0726	[152,158)	141.6708
[81,86)	131.4031	[158,181)	139.3434
[86,93)	132.7018	[181,190)	133.0177
[93,100)	133.9004	[190,193)	123.0918
[100,108)	134.9821	[193,213)	108.5507
[108,120)	135.9234	[213,215)	72.7053
[120,125)	142.2135	[215,000)	0.0000
[125,129)	142.9469		

Some comments on the table are in order: A close look at the values of $\bar{e}_{n,1}(x) - x$ shows that the function is approximately constant, except for $x \geqslant 193$ where the remaining data is small. This type of observation seems to be in line with the conclusion arrived by K o u l and S u s a r l a [13] where a test procedure has been considered to test for the hypothesis that F is exponential against the alternative that F is in the new better than used class of distributions. In this testing problem the hypothesis of exponentiality of F has not been rejected at level 0.05. Our calculation of $\bar{e}_{n,1}(x) - x$ also points to this direction. It is perhaps not unreasonable to expect from the values of $\bar{e}_{n,1}(x) - x$ that probably F has a slowly varying failure rate over most of the interval and with a possibly decreasing failure rate on the rest of the interval in $(0, \infty)$, but that the variability is not large enough to reject the hypothesis of exponentiality at level 0.05.

From the values of $\bar{e}_n(x) - x$, and from the estimators of F and $\sigma(0, 0)$ as given by (4.1) and (3.6) respectively, it is not difficult to calculate the confidence band for $\left\{ \frac{1}{F(x)} \int_x^M F(u)\, du \mid 0 \leqslant x \leqslant T \right\}$ using the formula (3.10). As has been pointed out earlier, these bands would be very wide in the right tail. See also Remark 3.1 in this connection.

5. CONCLUDING REMARKS

Throughout the paper we assumed that F and G are continuous. It can however be shown that for all asymptotic results stated here, it is enough that just F be continuous. In reality, we encounter situations where many censored observations are equal (for example, the situations in which units are put on test at time zero, and some (more than one) of these units are still working at the termination of the period of study). To take care of the ties in the censored observations, we propose the estimator

$$(5.1) \qquad \hat{F}_n(u) = \prod_{j=1}^n \left\{ \frac{1 + N^+(Z_j)}{2 + N^+(Z_j)} \right\}^{[\delta_j = 1, Z_j \leqslant u]}$$

as an estimator of F. An estimator of $\frac{1}{F(\cdot)} \int\limits_{\cdot}^{M} F(u)\,du$ is then simply given by $\frac{1}{\hat{F}_n(\cdot)} \int\limits_{\cdot}^{M} \hat{F}_n(u)\,du$. It can be shown that these two estimators have the same asymptotic properties given by Theorems 2.2 and 2.3, but with just F continuous and other additional conditions as stated in these theorems. These results remain valid if the censoring random variables $Y_1, \ldots, Y_n$ are *independent*, but not identically distributed under certain conditions on F, and $n\bar{G}(\cdot) = \sum\limits_{i=1}^{n} P[Y_i > \cdot]$ (for example, see [20]).

Professor Lidia Rejtő pointed out in a personal communication that it is probably enough to have the condition that F and G do *not* have the same points of discontinuity and other conditions in Theorems 2.2 and 2.3 for the conclusions of these theorems to hold. Y a n g [22] considered this problem when both X_i and Y_i in our notation are bounded, and gives the limiting Gaussian process along with its covariance function. However, we consider the problem when X_i or Y_i need not be bounded, obtain Bayes estimators of $e(\cdot)$, and study their asymptotic behavior including mean square, almost sure and weak convergence results concerning $\bar{e}_n(\cdot)$ process. Also, our proofs use a different technique than the one used by Yang.

APPENDIX

(1) Proofs of Lemmas 2.1 and 2.2 are similar to those in [17].

(2) Proof of Theorem 2.1 follows directly from Lemmas 2.1 and 2.2 and the squared error loss assumption.

(3) **Proof of Theorem 2.2** (Need (A1), (A2) and (A3)).

$$E(\bar{e}_n(x) - e(x))^2 \leqslant 2[A_n + B_n]$$

where

$$A_n = E\left(\int_x^{M_n} \left(\frac{\bar{F}_n(u)}{\bar{F}_n(x)} - \frac{F(u)}{F(x)}\right) du\right)^2$$

and

$$B_n = \left(\int_M^\infty \frac{F(u)}{F(x)} du\right)^2 = \frac{1}{(F(x))^2} \left(\int_M^\infty F(u) du\right)^2.$$

Since $\mu = \int_0^\infty F(u) du < \infty$ by assumption, $B_n \to 0$ as $n \to \infty$. Also

$$A_n \leqslant M \int_x^M E\left(\frac{\bar{F}_n(u)}{\bar{F}_n(x)} - \frac{F(u)}{F(x)}\right)^2 du \leqslant$$

$$\leqslant C_1 M \int_x^M E[\bar{F}_n(u) - F(u)]^2 du + C_2 M^2 E[\bar{F}_n(x) - F(x)]^2 \leqslant$$

$$\leqslant \frac{C_1 M}{n} \int_x^M C_n(u) du + C_2 \frac{M^2}{n} C_n(x)$$

where C_1 and C_2 are constants and

$$C_n(u) = 2\left(\frac{n^2 H(u)(1 - H(u)) + nH^2(u)}{(\alpha(R^+) + n)^2}\right) + \frac{6}{H^4(u)G^3(u)} + 2$$

Since the first and third term in $C_n(u)$ are bounded by some constants, $A_n \to 0$ when the assumed condition hold. This completes the proof of Theorem 2. The expression given for $C_n(u)$ can be obtained by following the details of Section 2 of [18].

(4) **Proof of Theorem 2.3.** In Steps I and II, we show that the process $\sqrt{n}\left(\bar{e}_n(x) - \int_x^M \frac{F(u)}{F(x)} du\right)$ is equivalent to a process $\{R_n^*(x): 0 \leqslant x \leqslant T\}$. Then we provide three lemmas proving

    (i)   the tightness condition for the process R_n^*,

    (ii) covariance function of the limiting process,

    (iii) convergence of the finite dimensional distribution of R_n^* to corresponding multivariate normal distribution.

Let

$$R_n(x) = \sqrt{n} \int_x^M (\bar{F}_n(u) - F(u))\, du - \sqrt{n}\,(\bar{F}_n(x) - F(x)) e_M(x)$$

where

$$e_M(x) = \frac{1}{F(x)} \int_x^M F(u)\, du.$$

Step I. Define $\hat{F}_n(t)$ by

$$\frac{\hat{F}_n(t)}{\sqrt{n}} =$$

$$= \frac{H_n(t) - H(t)}{G(t)} + F(t) \left(\int_0^t \left(\frac{2}{H(t)} - \frac{H_n(u)}{H^2(u)} \right) du - \int_0^t \frac{dH(u)}{H(u)} \right).$$

It was shown in [21] that the asymptotic distributions of $R_n(0)$ and $R_n^*(0)$ (defined below) are the same. The same proof can be used to show that the asymptotic distribution of the process $\{R_n(x): 0 \leqslant x \leqslant T\}$ is the same as that of $\{R_n^*(x): 0 \leqslant x \leqslant T\}$. This shows that it is enough to show the weak convergence of the process

$$R_n^*(x) = \int_x^{M_n} \hat{F}_n(u)\, du - \hat{F}_n(x) e_M(x)$$

where $\hat{F}_n$ is as defined above.

Step II. Simple algebraic manipulations show that $\dfrac{R_n^*}{\sqrt{n}}$ can be alternatively expressed as follows

$$\frac{R_n^*(x)}{\sqrt{n}} = \int_x^M \frac{H_n(u)}{G(u)}\,du + \int_x^M \left(\frac{2}{H(u)} - \frac{H_n(u)}{H^2(u)}\right)\mu_M(u)\,d\tilde{H}_n(u) -$$

$$- \frac{e_M(x)}{G(x)}(H_n(x) - H(x)) - \left(\int_x^M \frac{H(u)}{G(u)}\,du + \int_x^M \mu_M(u)\,\frac{d\tilde{H}(u)}{H(u)}\right) =$$

$$= \frac{S_n(x)}{\sqrt{n}} - \frac{e_M(x)}{G(x)}\frac{T_n(x)}{\sqrt{n}} - \left(\int_x^M \frac{H(u)}{G(u)} + \int_x^M \mu_M(x)\,\frac{d\tilde{H}(u)}{H(u)}\right)$$

where

$$\mu_M(x) = \int_x^M F(u)\,du, \quad e_M(x) = \frac{\mu_M(x)}{F(x)},$$

$$\frac{S_n(x)}{\sqrt{n}} = \frac{1}{\binom{n}{2}} \sum_{1 \leqslant j < k \leqslant n} \Phi_x[(\delta_j, Z_j), (\delta_k, Z_k)],$$

$$2\Phi_x[(\delta_1, Z_1), (\delta_2, Z_2)] = \frac{2\mu_M(Z_1)}{H(Z_1)}[\delta_1 = 0,\ x \leqslant Z_1 \leqslant M] +$$

$$+ 2\frac{\mu_M(Z_2)}{H(Z_2)}[\delta_2 = 0,\ x \leqslant Z_2 \leqslant M] +$$

$$+ \int_x^{Z_1 \wedge M} \frac{du}{G(u)} + \int_x^{Z_2 \wedge M} \frac{du}{G(u)} -$$

$$- \left(\frac{n-1}{n}\right)\left\{[Z_1 > Z_2]\frac{\mu_M(Z_2)}{H^2(Z_2)}[\delta_2 = 0,\ x \leqslant Z_2 \leqslant M] +\right.$$

$$\left. + [Z_2 > Z_1]\frac{\mu_M(Z_1)}{H^2(Z_1)}[\delta_1 = 0,\ x \leqslant Z_1 \leqslant M]\right\}$$

and

$$\frac{T_n}{\sqrt{n}} = H_n(x) - H(x) = \frac{1}{n}\sum_{j=1}^n \{[Z_j > x] - H(x)\}.$$

Hence

$$\frac{S_n}{\sqrt{n}} - \frac{e_M(x)}{G(x)} \frac{T_n}{\sqrt{n}} = \frac{1}{\binom{n}{2}} \sum_{1 \leq j < k \leq n} \Phi_n^*[(\delta_j, Z_j), (\delta_k, Z_k)]$$

where

$$\Phi_x^*[(\delta_1, Z_1), (\delta_2, Z_2)] =$$

$$= \Phi_x[(\delta_1, Z_1), \delta_2, Z_2)] - \frac{e_M(x)}{G(x)} \{[Z_2 > x] - H(x)\} -$$

$$- \frac{e_M(x)}{G(x)} \{[Z_1 > x] - H(x)\}.$$

This shows that

$$\frac{R_n^*(x)}{\sqrt{n}} = \frac{1}{\binom{n}{2}} \sum_{1 \leq j < k \leq n} \Phi_x^*[(\delta_j, Z_j), (\delta_k, Z_k)] -$$

$$- \int_x^{M_n} \left(\frac{H(u)}{G(u)} \right) du + \int_x^{M_n} \mu_M(u) \, d\tilde{H}(u).$$

Lemma A.1. *Consider the process* $\{R_n^*(t): 0 \leq t \leq T\}$. *For* $0 \leq$ $\leq t_1 \leq t \leq t_2 \leq T$.

$$\mathsf{E}|R_n^*(t) - R_n^*(t_1)|^2 |R_n^*(t_2) - R_n^*(t)|^2 \leq C|t_2 - t_1|^2$$

where C *is some constant independent of* t_1, t *and* t_2.

Proof.

$$R_n^*(t) - R_n^*(t_1) = - \int_{t_1}^{t} \hat{F}_n(u) \, du - (\hat{F}_n(t) - \hat{F}_n(t_1)) e_M(t) -$$

$$- \hat{F}_n(t_1)(e_M(t) - e_M(t_1)) =$$

$$= - [A(t_1, t) + e_M(t) B(t_1, t) + \hat{F}_n(t_1) C(t_1, t)].$$

Similarly

$$R_n^*(t_2) - R_n^*(t) = - [A(t, t_2) + e_M(t_2) B(t, t_2) + \hat{F}_n(t) C(t, t_2)].$$

Hence

$$E \, | R_n^*(t_2) - R_n^*(t) |^2 \, | R_n^*(t) - R_n^*(t_1) |^2 \leqslant$$

$$\leqslant 9 E [A(t_1, t) + e_M(t) B(t_1, t) + \hat{F}_n(t_1) C(t_1, t)] \times$$

$$\times [A(t, t_2) + e_M(t_2) B(t, t_2) + \hat{F}_n(t) C(t, t_2)].$$

The rest of the proof consists in showing that the expectation of each term in the product is bounded by $C(t_2 - t_1)^2$ where C is a constant independent of t_1, t and t_2. Since the calculations are straightforward but tedious, we omit the details for brevity. (The details of the proof will be available from the authors upon request.)

Lemma A.2. *Let $\sigma^2(x, y)$ be as defined in (3.9). Then*

$$\lim_{n \to \infty} \, \mathrm{cov} \, [R_n^*(x), R_n^*(y)] = F(x) F(y) \sigma(x, y) = \tilde{\sigma}(x, y).$$

Proof. Observe that the second term in (A.1) is a constant depending only on n and the first term is a U-statistics based on independent observations $(\delta_1, Z_1), \ldots, (\delta_n, Z_n)$. Hence using (6.5) in [10], we get

$$\frac{1}{n} \, \mathrm{cov} \, [R_n^*(x), R_n^*(y)] = \frac{1}{\binom{n}{2}} \, [2(n - 2) \xi_1(x, y) + \xi_2(x, y)]$$

where

$$\xi_1(x, y) = E \{ [\psi_{x1} - E \psi_{x1}][\psi_{y1} - E \psi_{y1}] \},$$

$$\xi_2(x, y) = E \{ (\Phi_x^* - E \Phi_x^*)(\Phi_y^* - E \Phi_y^*) \}$$

and

$$\psi_{x1} = E \{ \Phi_x^*[(\delta_1, Z_1), (\delta_2, Z_2)] \, | \, (\delta_1, Z_1) \}.$$

It is not difficult to show that

$$2 \psi_{x1}(\delta_1, Z_1) =$$

$$= \int_x^{Z_1 \wedge M} \frac{du}{G(u)} + \frac{\mu_M(Z_1)}{H(Z_1)} \, [\delta_1 = 0, \, x \leqslant Z_1 \leqslant M] \, -$$

$$- \int_x^{Z_1 \wedge M} \frac{\mu_M(u)}{H^2(u)} \, d\tilde{H}(u) - \frac{e_M(x)}{G(x)} \{[Z_1 > x] - H(x)\} +$$

$$+ \frac{1}{n} \left\{ \frac{\mu_M(Z_1)}{H(Z_1)} [\delta_1 = 0, \; x \leqslant Z_1 \leqslant M] + \right.$$

$$+ \int_x^{Z_1 \wedge M} \frac{\mu_M(u)}{H^2(u)} \, d\tilde{H}(u) \right\} + \text{constant} =$$

$$= A(x) + B(x) - C(x) - Q(x) + \frac{D(x)}{n} + \text{constant}.$$

Hence

$$4 \, \text{cov} \, [\psi_{x1}, \psi_{y1}] = \text{cov} \, [A(x) + B(x) - C(x) - Q(x),$$

$$A(y) + B(y) - C(y) - Q(y)] +$$

$$+ \frac{1}{n} \{ \text{cov} \, [A(x) + B(x) - C(x) - Q(x), D(y)] +$$

$$+ \, \text{cov} \, [A(y) + B(y) - C(y) - Q(y), D(x)] \} +$$

$$+ \frac{1}{n^2} \, \text{cov} \, (D(x), D(y)).$$

Evaluating the individual covariance terms and simplifying, we get

$$4 \, \text{cov} \, [\psi_{x1}, \psi_{y1}] = 2 \int_y^M F(u) \int_y^u \frac{dv}{G(v)} \, du - \frac{\mu_M^2(y)}{H(y)} +$$

$$+ \mu_M(y) \int_x^y \frac{\mu_M(u)}{H(u)} \, dH(u) + \int_y^M \frac{\mu_M^2(v)}{H^2(v)} \, d\tilde{H}(v) + O\left(\frac{1}{n}\right) \to \tilde{\sigma}(x, y).$$

Also it can be shown that $\text{cov} \, [\Phi_x^*, \Phi_y^*] = O\left(\frac{1}{H^2(T)}\right)$. This shows that

$$\lim_{n \to \infty} \, \text{cov} \, [R_n^*(x), R_n^*(y)] = \frac{n}{\binom{n}{2}} [2(n-2)\xi_1(x, y) + \xi_2(x, y)] =$$

$$= \lim_{n \to \infty} \left\{ 4 \frac{n-2}{n-1} \, \text{cov} \, [\psi_{x1}, \psi_{y1}] + \frac{2}{n-1} \, \text{cov} \, (\Phi_x^*, \Phi_y^*) \right\} =$$

$$= \tilde{\sigma}(x, y).$$

Hence

$$\lim_{n \to \infty} \text{cov} \, [\sqrt{n} \, (\hat{e}(x) - e(x)), \, \sqrt{n} \, (\hat{e}(y) - e(y))] = \sigma(x, y).$$

[Follows from step I and definition of R_n^*.]

Lemma A.3. *Let* $0 \leqslant t_1 < t_2 < \ldots < t_k \leqslant T.$ *Then*

$$(\sqrt{n} \, (\hat{e}(t_1) - e_M(t_1)), \ldots, \sqrt{n} \, (\hat{e}(t_k) - e_M(t_k))) \xrightarrow{w} N(0, \Sigma_k)$$

where $\Sigma_k = ((\sigma_{ij})),$ $\sigma_{ij} = \sigma(t_i, t_j)$ *and* $e_M(t) = \int\limits_t^M \dfrac{\bar{F}(u) \, du}{\bar{F}(t)}.$

Proof. Follows from Lemma A.2, Theorem 7.1 in [2], and the Cramér – Wold device to prove the asymptotic normality of a sequence of random vectors.

(5) The proofs of the Lemma 3.1 and Theorem 3.1 are similar to the proofs given in H a l l and W e l l n e r [18]. That $\{\bar{F}_n(x) \,|\, 0 \leqslant x \leqslant T\}$ is a consistent estimator of $\{\bar{F}(x) \,|\, 0 \leqslant x \leqslant T\}$ follows either from the details of Section 2 of [18] or from the details of F ö l d e s and R e j t ő [6]. The proof that $\hat{\sigma}(0, 0)$ given by (3.6) is a consistent estimator of $\sigma(0, 0)$ follows by applying the triangle inequality, and using the moments of $(1 + B)^{-r}$ where B is a binomial random variable and $r > 0$. Upper bounds on such binomial moments are available in [18].

REFERENCES

[1] R.E. B a r l o w – F. P r o s c h a n , *Mathematical theory of reliability,* Wiley, New York, 1965.

[2] P. B i l l i n g s l e y , *Convergence of probability measures,* Wiley, New York, 1968.

[3] M.C. B r y s o n – M.M. S i d d i q u i , Some criteria for aging, *J. Amer. Stat. Assoc.,* 64 (1969), 1472-1483.

[4] J.D. E s a r y – A.W. M a r s h a l l – F. P r o s c h a n , Some reliability application of the hazard transfrom, *SIAM J. Appl. Math.,* 18 (1970), 849-860.

[5]   T.S. Ferguson, A Bayesian analysis of some nonparametric problems, *Ann. Stat.*, 1 (1973), 209-230.

[6]   A. Földes – L. Rejtő, Strong consistency of the Kaplan – Meier estimator under variable censoring, (to appear).

[7]   A.E. Gehan, Estimating survival function from the life tables, *J. Chron. Ds.*, 21 (1969), 629-644.

[8]   L. Gross – A. Clark, *Survival distributions: reliability applications in biomedical applications,* Wiley, New York, 1975.

[9]   W.J. Hall – J.A. Wellner, Confidence bands for a survival curve from censored data, *Biometrika,* 67 (1980), 133-143.

[10]   W. Hoeffding, A class of statistics with asymptotically normal distribution, *Ann. Math. Stat.*, 16 (1948), 293-325.

[11]   M. Hollander – F. Proschan, Tests for mean residual life, *Biometrika,* 62 (1975), 585-593.

[12]   E.L. Kaplan – P. Meier, Nonparametric estimation from incomplete observations, *J. Amer. Statist. Assoc.*, 53 (1958), 457-481.

[13]   H. Koul – V. Susarla, Testing for new better than used in expectation with incomplete data, to appear in *J. Amer. Stat. Assoc.*, 75 (1980), 952-956.

[14]   H. Koul – V. Susarla, Tests for mean residual life with incomplete data, to appear.

[15]   M. Marshall – F. Proschan, Classes of distributions applicable in replacement policy with renewal theory application, *Proc. 6th Berk. Symp. Math. Stat. and Prob.*, 1972.

[16]   K. Rai – V. Susarla – J. Van Ryzin, Shrinkage estimate in nonparametric Bayesian survival analysis: a simulation study, *Communications in statistics – simulation,* 9 (1980), 271-298.

[17] V. Susarla – J. Van Ryzin, Nonparametric Bayesian estimation of survival curves from incomplete observations, *J. Amer.
Stat. Assoc.,* 61 (1976), 897-902.

[18] V. Susarla – J. Van Ryzin, Large sample theory for a
Bayesian nonparametric survival curve estimator based on censored
samples, *Ann. Stat.,* 6 (1978), 755-768.

[19] V. Susarla – J. Van Ryzin, Empirical Bayes estimation
of a distribution (survival) function from right censored observations, *Ann. Stat.,* 6 (1978), 740-754.

[20] V. Susarla – J. Van Ryzin, Large sample theory of survival curve estimator under variable censoring, *Optimization
Methods in Statistics* (Ed. J.S. Rustagi), (1980), 475-508.

[21] V. Susarla – J. Van Ryzin, Large sample theory for an
estimator of the mean survival time from censored samples, *Ann.
Stat.,* 8 (1980), 1001-1016.

[22] G.L. Yang, Estimation of a biometric function, *Ann. Stat.,* 7
(1978), 112-116.

[23] G.L. Yang, Life espectancy under random censorship, *Stoch.
Processes and their Applications,* 6 (1977), 33-39.

J. Ghorai

Dept. of Math., Univ. of Wisconsin-Milwauke, Milwauke WI 53201, USA.

A. Susarla

Dept. of Internal Madicine, Hurley Medical Center, Flint, Michigan, USA.

V. Susarla

Dept. of Stat. and Prob., M.S.U., E. Lansing, MI 48824 and Suny, Binghampton, N.Y., USA.

J. Van Ryzin

Div. of Biostat. and Dept. of Math. Stat., Columbia Univ., New York, N.Y., USA.

COLLOQUIA MATHEMATICA SOCIETATIS JÁNOS BOLYAI

32. NONPARAMETRIC STATISTICAL INFERENCE,

BUDAPEST (HUNGARY), 1980.

BERRY — ESSEEN THEOREMS FOR U-STATISTICS IN THE NON I.I.D. CASE

M. GHOSH* — R. DASGUPTA

1. INTRODUCTION

Let $\{X_n, \ n \geqslant 1\}$ be a sequence of independent but not necessarily identically distributed random variables. Based on $X_1, \ldots, X_n$ $(n \geqslant r)$, a U-statistic with kernel φ and degree r is defined by

$$(1.1) \qquad U_n = \frac{1}{\binom{n}{r}} \sum_{1 \leqslant i_1 < \ldots < i_r \leqslant n} \varphi(X_{i_1}, \ldots, X_{i_r}),$$

where the kernel φ is symmetric in its arguments.

Central limit theorems for U_n were established by H o e f f d i n g [13]. Later, in the i.i.d. case, H o e f f d i n g [14] provided an elegant decomposition of U-statistics as a weighted average of martingales. He used this decomposition in proving a strong law of large numbers for U-statistics. Subsequently, B e r k [1] showed that U-statistics were

*Research supported by the Army Research Office Durham Grant Number DAAG 29-79-C-0083.

backward martingales and used this in providing an alternate proof of the strong law. Yet another use of this backward martingale property was made by Miller and Sen [18] in proving a weak invariance principle for U-statistics.

Recently, interest has been focussed on the rate of convergence to normality for U-statistics. Grams and Serfling [12] obtained the Berry — Esseen bound $O(n^{-\frac{r}{2r+1}})$ for centered U-statistics when φ admits a finite $(2r)$-th moment. For bounded φ, Bickel [2] obtained the order bound $O(n^{-\frac{1}{2}})$. This boundedness assumption on φ was later weakened by Chan and Wierman [5] to the finiteness of the fourth moment to get the same conclusion. The latter also obtained the error rate $O(n^{-\frac{1}{2}}(\log n)^{\frac{1}{3}})$ when only the finiteness of the third absolute moment of φ was assumed. Finally, Callaert and Janssen [4] obtained the sharpest order bound $O(n^{-\frac{1}{2}})$ assuming only the finiteness of the third absolute moment of φ.

In this note, in Section 2, we provide a Hoeffding [14] decomposition for U-statistics in the non-i.i.d. case, and use this in deriving certain Berry — Esseen bounds for U-statistics, assuming among other things finiteness of the $(2 + \delta)$-th absolute moment of the kernel φ for some $\delta > 0$. The same result assuming finiteness of the third absolute moment of the kernel φ was obtained independently by Janssen [15] in his Ph. D. thesis. Also applying a general result of Ghosh and Dasgupta [11] certain nonuniform rates of convergence to normality for U-statistics are established. In Section 3, these results are extended to random U-statistics. Finally, in Section 4, an application of the results of Section 3 is made in the i.i.d. case to the problem of fixed length interval estimation of means of U-statistics.

2. BERRY — ESSEEN THEOREMS FOR U-STATISTICS WITH FIXED SAMPLE SIZES

First, a few notations are introduced. Assume without loss of generality that $E\varphi(X_{i_1}, \ldots, X_{i_r}) = 0$ for all $1 \leqslant i_1 < \ldots < i_r \leqslant n$. To keep the notations simple, we show the decomposition only for $r = 2$, although it can very well be generalized. Let

$$\psi^{i_1}_{i_2}(x_{i_1}) = \varphi^{i_1}_{i_2}(x_{i_1}) = \mathsf{E}(\varphi(X_{i_1}, X_{i_2}) \mid X_{i_1} = x_{i_1})$$

(2.1)

$$(1 \leqslant i_1 \neq i_2 \leqslant n);$$

$$\psi_2(x_{i_1}, x_{i_2}) = \varphi(x_{i_1}, x_{i_2}) - \psi^{i_1}_{i_2}(x_{i_1}) - \psi^{i_2}_{i_1}(x_{i_2})$$

(2.2)

$$(1 \leqslant i_1 \neq i_2 \leqslant n).$$

Writing

$$\bar{\psi}^{(1)}_n(X_i) = \frac{1}{n-1} \sum_{\substack{j=1 \\ j \neq i}}^{n} \psi^i_j(X_i), \quad V^{(1)}_n = \frac{1}{n} \sum_{i=1}^{n} \bar{\psi}^{(1)}_n(X_i)$$

and

$$V^{(2)}_n = \frac{1}{\binom{n}{r}} \sum_{1 \leqslant i_1 < \ldots < i_r \leqslant n} \varphi(X_{i_1}, \ldots, X_{i_r}),$$

one has the representation

(2.3) $U_n = 2V^{(1)}_n + V^{(2)}_n.$

The following facts can be easily verified.

(2.4) $\mathsf{E}\,\psi^{i_1}_{i_2}(X_{i_1}) = 0, \quad \mathsf{E}(\psi_2(X_{i_1}, X_{i_2}) \mid X_{i_1} = x_{i_1}) = 0$ a.e.;

It follows from (2.3) and (2.4) that

$$\mathsf{E}(\psi^{i_1}_{i_2}(X_{i_1})\psi_2(X_{i_1}, X_{i_3})) = 0$$

(2.5)

for any $1 \leqslant i_1 \neq i_2 \leqslant n, \ 1 \leqslant i_1 \neq i_3 \leqslant n;$

(2.6) $\mathsf{E}(\psi_2(X_{i_1}, X_{i_2})\psi_2(X_{i_1}, X_{i_3})) = 0 \quad (1 \leqslant i_1 \neq i_2 \neq i_3 \leqslant n).$

Henceforth, unless otherwise mentioned, we work with U-statistics with kernel φ and degree 2. The generalization to a kernel of degree $r \ (> 2)$ is immediate. First the following lemma is proved.

Lemma 1. *If* $E\varphi(X_{i_1}, X_{i_2}) = 0$ $(1 \leqslant i_1 < i_2 \leqslant n)$ *and*

$$(2.7) \qquad \max_{1 \leqslant i_1 < i_2 \leqslant n} E|\varphi(X_{i_1}, X_{i_2})|^{2m} \leqslant K,$$

where in the above and in what follows, K *is a generic constant, positive and not dependent on* n, *then,*

$$(2.8) \qquad E(V_n^{(2)})^{2m} \leqslant \frac{K}{n^{2m}}.$$

Proof.

$$E(V_n^{(2)})^{2m} =$$

$$(2.9) \qquad = \frac{1}{\binom{n}{2}^{2m}} \sum_{1 \leqslant i_1 < j_1 \leqslant n} \cdots \sum_{1 \leqslant i_{2m} < j_{2m} \leqslant n} E(\psi_2(X_{i_1}, X_{j_1}) \cdots$$

$$\cdots \psi_2(X_{i_{2m}}, X_{j_{2m}})).$$

Note that if a pair of suffixes (i_k, j_k) occurs exactly once in $(\{i_1, j_1\}, \ldots \ldots \{i_{2m}, j_{2m}\})$, then in view of (2.4),

$$E(\psi_2(X_{i_1}, X_{j_1}) \cdots \psi_2(X_{i_{2m}}, X_{j_{2m}})) = 0.$$

Subject to the condition that each pair of suffixes (i_k, j_k) occurs at least twice, it follows from (2.9) that

$$E(V_n^{(2)})^{2m} =$$

$$(2.10) \qquad = \frac{1}{\binom{n}{2}^{2m}} \sum_{1 \leqslant i_1 < j_1 \leqslant n} \cdots \sum_{1 \leqslant i_m < j_m \leqslant n} E(\{\psi_2(X_{i_1}, X_{j_1})\}^2 \cdots$$

$$\cdots \{\psi_2(X_{i_m}, X_{j_m})\}^2).$$

Applying Hölder's inequality m times, one gets

$$E(\{\psi_2(X_{i_1}, X_{j_1})\}^2 \cdots \{\psi_2(X_{i_m}, X_{j_m})\}^2) \leqslant$$

$$(2.11)$$

$$\leqslant \prod_{k=1}^{m} E^{\frac{1}{m}} \{\psi_2(X_{i_k}, X_{j_k})\}^{2m}.$$

Applying Jensen's inequality and (2.7), it follows from (2.10) and (2.11) that

$$E(V_n^{(2)})^{2m} = O\left(\frac{1}{n^{4m}} n^{2m}\right) = O\left(\frac{1}{n^{2m}}\right).$$

Next define

(2.12)
$$s_n^2 = \sum_{i=1}^n E(\bar{\psi}_n^{(1)}(X_i))^2, \quad \sigma_n^2 = \frac{1}{n} s_n^2,$$

$$\bar{\nu}_n(\delta) = \frac{1}{n} \sum_{i=1}^n E|\bar{\psi}_n^{(1)}(X_i)|^{2+\delta} \quad (n \geqslant 1).$$

The main result of this section is as follows:

Theorem 1. *If*

(2.13) (i) $\displaystyle \max_{1 \leqslant i_1 < i_2 \leqslant n} E|\varphi(X_{i_1}, X_{i_2})|^{2+\delta} \leqslant K < \infty,$

for some $\delta > 0,$ *and*

(2.14) (ii) $\displaystyle \min_{1 \leqslant i \leqslant n} E(\bar{\psi}_n^{(1)}(X_i))^2 > 0,$

then

(2.15) $\displaystyle \left| P\left\{ \frac{\sqrt{n}\, U_n}{2\sigma_n} \leqslant x \right\} - \Phi(x) \right| \leqslant \frac{K\bar{\nu}_n(\delta')}{n^{\frac{1}{2}\delta'}\sigma_n^{2+\delta'}},$

where $\delta' = \min(\delta, 1)$ *and* $\Phi(x)$ *is the distribution function of a* $N(0, 1)$ *variable.*

Remark 1. When the X_i's are i.i.d, the conditions (2.13) and (2.14) simplify. Under these conditions with $\delta = 1$, Callaert and Janssen [3] proved (2.15). Our method of proof uses their argument as well as Esseen's classical lemma.

Proof of Theorem 2. First write

(2.16) $\displaystyle \frac{\sqrt{n}\, U_n}{2\sigma_n} = \frac{1}{s_n} \sum_{i=1}^n \bar{\psi}_n^{(1)}(X_i) + R_n,$

— 297 —

where $R_n = \dfrac{\sqrt{n}\, V_n^{(2)}}{\sigma_n}$. Following C h a n and W i e r m a n [5] or C a l l a e r t and J a n s s e n [3], write $R_n = \Delta_{n1} + \Delta_{n2}$, n large, where

(2.17)
$$\Delta_{n1} = \frac{\sqrt{n}}{\sigma_n \binom{n}{2}} \sum_{1 \leqslant i < j \leqslant c_n} \sum \psi_2(X_i, X_j),$$

$$\Delta_{n2} = \frac{\sqrt{n}}{\sigma_n \binom{n}{2}} \sum_{j=c_n+1}^{n} {}' \sum_{i=1}^{j-1} {}' \psi_2(X_i, X_j),$$

where $c_n = [n - 3\sqrt{n}\, \log n]$, the integer part of $[n - 3\sqrt{n}\, \log n]$.

Writing $\xi_j = \sum_{i=1}^{j-1} \psi_2(X_i, X_j)$ and $\mathscr{F}_j$ the σ-algebra generated by $\xi_2, \ldots, \xi_j$ $(j \geqslant 2)$, it follows that $\{\xi_j, \mathscr{F}_j, j \geqslant 2\}$ forms a martingale difference sequence. Hence, application of a martingale inequality of D h a r m a d h i k a r i, F a b i a n and J o g d e o [8] gives

(2.18)
$$\mathsf{E} \left| \sum_{j=c_n+1}^{n} {}' \sum_{i=1}^{j-1} {}' \psi_2(X_i, X_j) \right|^{2+\delta} \leqslant$$

$$\leqslant K(n - c_n)^{1+\frac{1}{2}\delta} \max_{c_n+1 \leqslant j \leqslant n} \mathsf{E} \left| \sum_{i=1}^{j-1} \psi_2(X_i, X_j) \right|^{2+\delta}.$$

Now for each fixed $j \geqslant c_n + 1$,

$$W_k = \sum_{i=1}^{k} \psi_2(X_i, X_j) \qquad (k = 1, 2, \ldots, j-1),$$

also forms a martingale sequence. Applying the same martingale inequality again, it follows that for $j \geqslant c_n + 1$,

(2.19)
$$\mathsf{E} \left| \sum_{i=1}^{j-1} \psi_2(X_i, X_j) \right|^{2+\delta} \leqslant$$

$$\leqslant (j-1)^{1+\frac{1}{2}\delta} \max_{1 \leqslant i \leqslant j-1} \mathsf{E} |\psi_2(X_i, X_j)|^{2+\delta}.$$

From (2.7), (2.13), (2.18) and (2.19), it follows that

$$\mathsf{E}\left|\sum_{j=c_n+1}^{n}\sum_{i=1}^{j-1}\psi_2(X_i,X_j)\right|^{2+\delta}\leqslant$$

(2.20)

$$\leqslant K(n-c_n)^{1+\frac{1}{2}\delta}(n-1)^{1+\frac{1}{2}\delta}\leqslant Kn^{\frac{3}{2}+\frac{3}{4}\delta}(\log n)^{1+\frac{1}{2}\delta}$$

Next, let $\dfrac{1}{T_n}=\dfrac{\bar{\nu}_n(\delta')}{n^{\frac{1}{2}\delta'}\sigma_n^{2+\delta'}}$. Using Markov's inequality, (2.13), (2.14), (2.17) and (2.20),

$$\mathsf{P}\left(|\Delta_{n1}|>\frac{1}{T_n}\right)\leqslant T_n^{2+\delta}\,\mathsf{E}\,|\Delta_{n2}|^{2+\delta}\leqslant$$

$$\leqslant KT_n^{2+\delta}\,n^{1+\frac{1}{2}\delta}\,\frac{1}{n^{2(2+\delta)}}\,n^{\frac{3}{2}+\frac{3}{4}\delta}(\log n)^{1+\frac{1}{2}\delta}=$$

(2.21)

$$=\frac{Kn^{\frac{1}{2}\delta'(2+\delta)}\sigma_n^{(2+\delta')(2+\delta)}(\log n)^{1+\frac{1}{2}\delta}}{\bar{\nu}_n^{2+\delta}(\delta')n^{\frac{3}{2}+\frac{3}{4}\delta}}\leqslant$$

$$\leqslant\frac{K(\log n)^{1+\frac{1}{2}\delta}}{n^{\frac{1}{2}+\frac{1}{4}\delta}}=o\left(\frac{1}{T_n}\right).$$

In view of (2.16), (2.17) and (2.21) it suffices to show that

(2.22) $\quad\left|\mathsf{P}\left\{\dfrac{1}{s_n}\sum_{i=1}^{n}\bar{\psi}_n^{(1)}(X_i)+\Delta_{n1}\leqslant x\right\}-\Phi(x)\right|\leqslant\dfrac{K\bar{\nu}_n(\delta')}{\sigma_n^{2+\delta'}n^{-\frac{1}{2}\delta'}}.$

But using Esseen's lemma and the notation $\eta_Z(t)$ for the characteristic function of a random variable Z, one gets,

left hand side of (2.22) $\leqslant$

(2.23)

$$\leqslant\int_{-T_n}^{T_n}\frac{1}{|t|}\left|\eta_{\frac{1}{s_n}\sum_{i=1}^{n}\bar{\psi}_n^{(1)}+\Delta_{n1}}(t)-\exp\left(-\frac{1}{2}t^2\right)\right|dt+\frac{K}{T_n}.$$

Using a result of E s s e e n [9],

$$(2.24) \quad \int_{-T_n}^{T_n} \frac{1}{|t|} \left| \eta_{\frac{1}{s_n} \sum_{i=1}^n \bar{\psi}_n^{(1)}(X_i)}(t) - \exp\left(-\frac{1}{2} t^2\right) \right| dt \leq \frac{K}{T_n}.$$

Hence, it suffices to show that

$$(2.25) \quad \int_0^{T_n} \frac{1}{t} \left| \eta_{\frac{1}{s_n} \sum_{i=1}^n \bar{\psi}_n^{(1)}(X_i)}(t) - \eta_{\frac{1}{s_n} \sum_{i=1}^n \bar{\psi}_n^{(1)}(X_i) + \Delta_{n1}}(t) \right| dt \leq \frac{K}{T_n}$$

as the other integral, namely the one over the range $(-T_n, 0)$ can be handled similarly. No choosing $\epsilon\,(>0)$ such that $\sqrt{T_n} > \epsilon$ for $n \geq n_0$ (say), proceed as in C h a n and W i e r m a n [5] to obtain the inequality

$$\int_0^{\epsilon\sqrt{T_n}} \frac{1}{t} \left| \eta_{\frac{1}{s_n} \sum_{i=1}^n \bar{\psi}_n^{(1)}(X_i)}(t) - \eta_{\frac{1}{s_n} \sum_{i=1}^n \bar{\psi}_n^{(1)}(X_i) + \Delta_{n1}}(t) \right| dt \leq$$

$$(2.26) \qquad \leq K \int_0^{\epsilon\sqrt{T_n}} \frac{1}{t} \left[\left| E\left\{ \exp\left(\frac{it}{s_n} \sum_{j=1}^n \bar{\psi}_n^{(1)}(X_j)\right) \Delta_{n1} \right\} \right| + \right.$$

$$\left. + \frac{1}{2} t^2 E\Delta_{n1}^2 \right] dt.$$

Now,

$$E\left[\exp\left(\frac{it}{s_n} \sum_{j=1}^n \bar{\psi}_n^{(1)}(X_j)\right) \Delta_{n1} \right] \leq$$

$$(2.27) \qquad \leq \frac{\sqrt{n}}{\sigma_n \binom{n}{2}} \sum_{1 \leq j < j' \leq c_n} \sum \left\{ \prod_{k \neq j \neq j'} \eta_{\bar{\psi}_n^{(1)}(X_k)}\left(\frac{t}{s_n}\right) \right\} \leq$$

$$\times E\left[\exp\left(\frac{it}{s_n} (\bar{\psi}_n^{(1)}(X_j) + \bar{\psi}_n^{(1)}(X_{j'}))\right) \psi_2(X_j, X_{j'}) \right].$$

But for $t \leq T_n$,

$$\frac{t}{s_n} \leq \frac{\epsilon n^{\frac{1}{2}\delta'} \sigma_n^{2+\delta'}}{\bar{v}_n(\delta')} \leq K\epsilon n^{\frac{1}{2}(\delta'-1)} \leq K\epsilon.$$

So

$$\prod_{k \neq j \neq j'} \eta_{\bar{\psi}_n^{(1)}(X_k)}\left(\frac{t}{s_n}\right) \leqslant \prod_{k \neq j \neq j'} \exp\left(-\frac{1}{2}\frac{t^2}{s_n^2} \, E(\bar{\psi}_n^{(1)}(X_k))^2\right) =$$

(2.28)
$$= \exp\left(-\frac{1}{2}t^2\right) \exp\left[\frac{1}{2}\frac{t^2}{s_n^2}\{E[\bar{\psi}_n^{(1)}(X_j)]^2 + E[\bar{\psi}_n^{(1)}(X_{j'})]^2\}\right] \leqslant$$

$$\leqslant K \exp\left(-\frac{1}{2}t^2\right).$$

Also, from (2.6), (2.13) and (2.17)

(2.29)
$$E\Delta_{n1}^2 \leqslant \frac{n}{\sigma_n^2\binom{n}{2}^2} \sum_{1 \leqslant i < j \leqslant c_n} E[\psi_2(X_i, X_j)]^2 = O\left(\frac{1}{n}\right).$$

Now, proceeding as C a l l a e r t and J a n s s e n [3], it follows from (2.26)-(2.29) that

$$\text{left hand side of } (2.26) \leqslant \frac{K\sqrt{n}}{\sigma_n s_n^2\binom{n}{2}} \times$$

$$\times \sum_{1 \leqslant i < j \leqslant c_n} E|\bar{\psi}_n^{(1)}(X_i)\bar{\psi}_n^{(1)}(X_j)\psi_2(X_i, X_j)| + O\left(\frac{T_n}{n}\right) \leqslant$$

(2.30)
$$\leqslant \frac{K\sqrt{n}}{\sigma_n s_n^2\binom{n}{2}} \{E[\bar{\psi}_n^{(1)}(X_i)]^2 E[\bar{\psi}_n^{(1)}(X_j)]^2 \times$$

$$\times E[\psi_2^2(X_i, X_j)]\}^{\frac{1}{2}} + O\left(\frac{T_n}{n}\right) \leqslant$$

$$\leqslant \frac{K\sqrt{n}\,\nu_n(\delta')}{\sigma_n^{2+\delta'}} + O\left(\frac{n^{\frac{1}{2}\delta'-1}\sigma_n^{2+\delta'}}{\bar{\nu}_n(\delta')}\right) = O\left(\frac{1}{T_n}\right).$$

Finally, repeating the steps of C a l l a e r t and J a n s s e n [3], it can be shown that

(2.31)
$$\int_{\epsilon\sqrt{T_n}}^{T_n} \frac{1}{|t|}\left|\eta_{\frac{1}{s_n}\sum_{i=1}^n \bar{\psi}_n^{(1)}(X_i)}(t) - \eta_{\frac{1}{s_n}\sum_{i=1}^n \bar{\psi}_n^{(1)}(X_i) + \Delta_{n1}}(t)\right| dt \leqslant$$

$$\leqslant \frac{K}{T_n}.$$

Remark 2. For a kernel φ with degree $r\,(> 2)$ one has a representation

$$(2.32)\qquad U_n = \sum_{c=1}^{r} \binom{r}{c} V_n^{(c)} \qquad (n \geqslant r),$$

similar to (2.3) where the $V_n^{(c)}$'s are appropriately defined. To obtain Berry — Esseen bounds in this case, one uses Markov's inequality on $V_n^{(3)}, \ldots, V_n^{(r)}$, while $V_n^{(2)}$ is decomposed in a similar fashion as (2.17).

Remark 3. Using Lemma 1 of this paper and a general theory of Ghosh and Dasgupta [11] (see in particular, their (3.18)) one gets the nonuniform bound

$$(2.33)\qquad \left| P\left\{ \frac{\sqrt{n}\,U_n}{2\sigma_n} \leqslant x \right\} - \Phi(x) \right| \leqslant \frac{K\,(\log n)^{g(\delta')}}{n^{\frac{1}{2}\delta'}(1 + |x|^{2+\delta})}.$$

3. BERRY — ESSEEN BOUNDS FOR RANDOM U-STATISTICS

Recall the definition of a U-statistic with kernel φ and degree r in (1.1). Define

$$\psi_{i_2,\ldots,i_r}^{i_1}(x_{i_1}) =$$
$$= E[\varphi(X_{i_1}, \ldots, X_{i_r}) \mid X_{i_2} = x_{i_2}, \ldots, X_{i_r} = x_{i_r}]$$

Let

$$\bar{\psi}_n(x_i) = \binom{n-1}{r}^{-1} \sum_{1 \leqslant i_2 < \ldots < i_r \leqslant n} \cdots \sum \psi_{i_2,\ldots,i_r}^{i}(x_i)$$

$$(i = 1, \ldots, n;\; i_j \neq i\; (j = 2, \ldots, r)),$$

$$s_n^2 = \sum_{1}^{n} V[\bar{\psi}_n(X_j)], \qquad \sigma_n^2 = \frac{1}{2} s_n^2.$$

Then Hoeffding [13] showed that if

$$(i)\; \sup_{n \leqslant r}\; \max_{1 \leqslant i_1 < \ldots < i_r \leqslant n}\; E|\varphi(X_{i_1}, \ldots, X_{i_r})|^{2+\delta} < \infty$$

and

$$\inf_{n \geqslant r} \sigma_n^2 > 0,$$

then

$$\frac{U_n - \mathrm{E}U_n}{r\sigma_n} \xrightarrow{\mathscr{L}} N(0, 1).$$

S p r o u l e [22] proved a central limit theorem for random U-statistics in the i.i.d. case. He showed that if $\frac{N_\nu}{n_\nu} \xrightarrow{\mathrm{P}} \tau \, (> 0)$ as $\nu \to \infty$ where $\{N_\nu, \, \nu \geqslant 1\}$ is a sequence of positive integer valued random variables and $\{n_\nu, \, \nu \geqslant 1\}$ is a nondecreasing sequence of positive integers converging to ∞ as $\nu \to \infty$, and τ is a positive constant, then

$$\sqrt{N_\nu}\,(U_{N_\nu} - \mathrm{E}U_{N_\nu}) \xrightarrow{\mathscr{L}} N(0, r^2 \zeta_1), \quad \zeta_1 = \mathrm{V}[\psi_1(X_1)],$$

$$\psi_1(X_1) = \mathrm{E}[\varphi(X_1, \ldots, X_r) \mid X_2, \ldots, X_r = x_r].$$

We address the question of rate of convergence to normality of $\sqrt{N_\nu}\,(U_{N_\nu} - \mathrm{E}U_{N_\nu})$ in this section for U-statistics based on independent but not necessarily identically distributed random variables. For simplicity of exposition, we first consider the case $r = 2$. The following well-known lemma is also needed.

Lemma 2. *Let* $\{X_n, \, n \geqslant 1\}$ *and* $\{Y_n, \, n \geqslant 1\}$ *be two sequences of random variables. Assume that*

$$\sup_x |\mathrm{P}(Y_n \leqslant x) - \Phi(x)| = O(a_n);$$

$$\mathrm{P}(|Z_n| > a_n) = O(a_n).$$

Then,

$$\sup_x |\mathrm{P}(Y_n + Z_n \leqslant x) - \Phi(x)| = O(a_n).$$

The main result of this section is now stated as follows:

Theorem 2. *Assume*

(i) $\quad \sup\limits_{n \geqslant 2} \max\limits_{1 \leqslant i_1 < i_2 \leqslant n} E|\varphi(X_{i_1}, X_{i_2})|^{2m} < \infty$

for some positive integer $m \geqslant 2$. *Let* $\{N_\nu, \nu \geqslant 1\}$ *be a sequence of positive integers valued random variables assuming the values* $r, r+1, \ldots$ *and satisfying*

$$(3.1) \qquad P\left(\left|\frac{N_\nu}{\tau\nu} - 1\right| > \epsilon_\nu\right) = O(\sqrt{\epsilon_\nu}),$$

where $1 > \epsilon_\nu \geqslant \dfrac{1}{\nu^{\frac{2m}{2m+1}}}$ *and* $\epsilon_\nu \to 0$ *as* $\nu \to \infty$. *Then,*

$$(3.2) \qquad \sup_x \left| P\left\{\frac{\sqrt{\tau_\nu}\,(U_{N_\nu} - E U_{N_\nu})}{2\sigma_{N_\nu}} \leqslant x\right\} - \Phi(x)\right| = O(\sqrt{\epsilon_\nu});$$

$$(3.3) \qquad \sup_x \left| P\left\{\frac{\sqrt{N_\nu}\,(U_{N_\nu} - E U_{N_\nu})}{2\sigma_{N_\nu}} \leqslant x\right\} - \Phi(x)\right| = O(\sqrt{\epsilon_\nu}).$$

Proof. We prove only (3.2). An argument similar to L a n d e r s and R o g g e 's [16] then proves (3.3).

Without any loss of generality assume $E\varphi(X_{i_1}, X_{i_2}) = 0$ for all $1 \leqslant i_1 < i_2 \leqslant n, \ n \geqslant 2$.

Note that using c_δ and Jensen's inequalities, and condition (i) of the theorem,

$$(3.4) \qquad \sup_{n \geqslant 2} \max_{1 \leqslant i \leqslant n} E|\bar{\psi}_n(X_i)|^{2+\delta} \leqslant K.$$

Using (3.1) and (3.2) it follows from a theorem of L a n d e r s and R o g g e [16] that

$$(3.5) \qquad \sup_x \left| P\left(\frac{\sqrt{\tau\nu}\, V_{N_\nu}^{(1)}}{\sigma_{N_\nu}} \leqslant x\right) - \Phi(x)\right| = O(\sqrt{\epsilon_\nu}).$$

In view of (2.3), (3.3) and Lemma 2, we complete the proof of the theorem by showing that

(3.6) $\quad P(\sqrt{\tau\nu}\,|\,V^{(2)}_{N_\nu}\,|> \sqrt{\epsilon_\nu}) = O(\sqrt{\epsilon_\nu})$.

Using (3.1) once again it follows that

$$P(\sqrt{\tau\nu}\,|\,V^{(2)}_{N_\nu}\,|> \sqrt{\epsilon_\nu}) \leqslant P\Big(|\,V^{(2)}_{N_\nu}\,|> \frac{\sqrt{\epsilon_\nu}}{\sqrt{\tau\nu}},$$

(3.7) $\qquad\qquad\qquad\qquad \tau_\nu(1-\epsilon_\nu) \leqslant N_\nu \leqslant \tau_\nu(1+\epsilon_\nu)\big) + O(\sqrt{\epsilon_\nu}) \leqslant$

$$\leqslant P\Big(\sup_{u_1 \leqslant n \leqslant u_2} |\,V^{(2)}_N\,| > \frac{\epsilon_\nu}{\sqrt{\tau\nu}}\Big) + O(\sqrt{\epsilon_\nu}),$$

where $u_1 = \max\,(r, [\tau\nu(1-\epsilon_\nu)])$ and $u_2 = \max\,(r, [\tau\nu(1+\epsilon_\nu)])$.

Observe next that by writing $W_n = \sum\sum_{1 \leqslant i < j \leqslant n} \psi_2(X_i, X_j)$ and F_n the σ-algebra generated by $X_1, \ldots, X_n$, it follows from (2.4) that $\{W_n, F_n; n \geqslant 2\}$ is a martingale sequence. Using now the maximal inequality for martingales, Markov's inequality and Lemma 1, it follows that

$$P\Big(\sup_{u_1 \leqslant n \leqslant u_2} |\,V^{(2)}_n\,| > \frac{\sqrt{\epsilon_\nu}}{\sqrt{\tau\nu}}\Big) \leqslant P\Big(|\,W_{u_2}\,| > \binom{u_1}{2}\frac{\epsilon_\nu}{\sqrt{\tau\nu}}\Big) \leqslant$$

(3.8)

$$\leqslant \frac{E\,|\,W_{u_2}\,|^{2m}}{\{\binom{u_1}{2}\frac{\sqrt{\epsilon_\nu}}{\sqrt{\tau\nu}}\}^{2m}} \leqslant \frac{Ku_2^{2m}\nu^m}{\epsilon_\nu^m u_1^{4m}} \leqslant \frac{K}{\nu^m \epsilon_\nu^m} \leqslant K\sqrt{\epsilon_\nu}.$$

Since $\epsilon_\nu \geqslant \dfrac{1}{\nu^{\frac{2m}{2m+1}}}$; (3.6) follows from (3.7) and (3.8).

Remark 4. In the *i.i.d.* case, under the assumption that $E\,|\varphi(X_1, \ldots, X_r)|^3$, C s e n k i [6] has obtained the order bound $O(\epsilon_\nu^{\frac{1}{3}})$ with $\epsilon_\nu \geqslant \dfrac{1}{\nu}$ in Theorem 2. His method of proof involves the Skorohod embedding principle. Although, the moment assumption of Csenki is slightly weaker than ours, his conclusion is also weaker. The special case of Berry — Esseen theorems for random means with emphasis on sequential applications are treated in G h o s h [10], C s e n k i [7] and C a l l a e r t and J a n s s e n [4].

Remark 5. If we consider U-statistics with kernels of degree $r \, (> 2)$, then the martingale argument used in Theorem 2 does not hold. It is not clear in this case whether (3.2) and (3.3) hold under the assumptions

$$\sup_{n \geqslant r} \max_{1 \leqslant i_1 < \ldots < i_r \leqslant n} E|\varphi(X_{i_1}, \ldots, X_{i_r})|^{2m} < \infty$$

for some positive integer m and (3.1) with $1 > \epsilon_\nu \geqslant \dfrac{1}{\nu^{\frac{2m}{2m+1}}}$, $\epsilon_\nu \to 0$

as $\nu \to \infty$. However, if one assumes the stronger condition

$$\sup_{n \geqslant r} \max_{1 \leqslant i_1 < \ldots < i_r \leqslant n} E|\varphi(X_{i_1}, \ldots, X_{i_r})|^u < \infty$$

for some $u > 0$, then if (3.1) holds for $1 > \epsilon_\nu \geqslant \nu^{-1+\delta}$, $0 < \delta < 1$, $\epsilon_\nu \to 0$ as $\nu \to \infty$, then conclusions (3.2) and (3.3) hold for U-statistics with kernels of arbitrary degree r. However in the i.i.d. case, the martingale argument continues to hold and Theorem 2 holds for U-statistics with kernels of arbitrary degree r.

In the situation $N_\nu = \nu$ with probability 1, the problem reduces to a fixed sample size problem. There, C a l l a e r t and J a n s s e n [3] have the sharper order bound $\left(\dfrac{1}{\sqrt{\nu}}\right)$. There may be situations, however, where the order $\nu^{-\frac{1-\delta}{2}}$ for some $0 < \delta < 1$ cannot be improved in (3.1) and there are bounds (3.2) and (3.3) are expected to be quite sharp in view of the results of L a n d e r s and R o g g e [16], and the technique used in our proof.

Random central limit theorems for sample sums and U-statistics are proved under the weaker condition $\dfrac{N}{\nu} \xrightarrow{P} \tau$ as $\nu \to \infty$, where τ is a positive random variable (see M o g y o r ó d i [19] and S p o u l e [22] rather than its special case (as considered by us) where τ is a positive constant. An example given by L a n d e r s and R o g g e [17] shows however that in the former case, no general approximation orders for normalized sample sums is possible. The same remark applies to our Theorem 2.

4. BERRY – ESSEEN BOUNDS IN FIXED LENGTH INTERVAL ESTIMATION OF MEANS OF U-STATISTICS

Consider a sequence $X_1, X_2, \ldots$ of i.i.d. r.v.'s. Recall the definition of a U-statistic in (1.1).

Let $\varphi_1(x_1) = E[\varphi(X_1, \ldots, X_r) | X_1 = x_1]$. Further following Sproule [21], (p. 15) some more notations are introduced.

For each $i = 1, \ldots, n,$ define a U-statistic based on $X_1, \ldots, X_{i-1}, X_{i+1}, \ldots, X_n$ by

$$(4.1) \qquad U_{(i)n} = \frac{1}{\binom{n-1}{r}} \sum^{(n-1,r)} \varphi(X_{\alpha_1}, \ldots, X_{\alpha_r}),$$

where the summation is over all combinations $(\alpha_1, \ldots, \alpha_r)$ formed from $(1, \ldots, i-1, i+1, \ldots, n)$. Define now the statistics

$$(4.2) \qquad W_{in} = nU_n - (n-r)U_{(i)n} \qquad (i = 1, 2, \ldots, n),$$

and $\bar{W}_n = \frac{1}{n} \sum_{i=1}^{n} W_{in}$. Sen [20] proposed the estimator

$$(4.3) \qquad s_{Wn}^2 = \frac{1}{n} \sum_{i=1}^{n} (W_{in} - \bar{W}_n)^2$$

for $r^2 \zeta_1$. For obtaining a confidence interval of fixed length $2d\ (> 0)$ $\theta = E\varphi(X_1, \ldots, X_r)$ with confidence coefficient $1 - \alpha$ $(0 < \alpha < 1)$, Sproule [21] suggested the following sequential procedure

$$(4.4) \qquad n \geq \frac{\tau_{\frac{\alpha}{2}}^2}{d^2} \left(s_{Wn}^2 + \frac{1}{n} \right),$$

where $\tau_{\frac{\alpha}{2}}$ is the upper $100\frac{\alpha}{2}\%$ point of a $N(0, 1)$ distribution. Propose the confidence interval $[U_N - d, U_N + d]$ for θ. Sproule [21] also has the representation of s_{Wn}^2 as a linear combination of U-statistics. With the end, for each $c = 0, 1, \ldots, r,$ let

$$(4.5) \qquad q^{(c)}(x_1, \ldots, x_{2r-c}) =$$

$$= \frac{1}{\binom{2r-c}{r}\binom{r}{c}} \overset{(c)}{\sum} \varphi(x_{\alpha_1}, \ldots, x_{\alpha_r})\varphi(x_{\beta_1}, \ldots, x_{\beta_r}).$$

where the summation $\overset{(c)}{\sum'}$ is over all combinations $(\alpha_1, \ldots, \alpha_r)$ and $(\beta_1, \ldots, \beta_r)$ each formed from $\{1, 2, \ldots, 2r-c\}$ such that there are exactly c integers in common between $(\alpha_1, \ldots, \alpha_r)$ and $(\beta_1, \ldots, \beta_r)$. Define

(4.6)
$$U_n^{(c)} = \frac{1}{\binom{n}{2r-c}} \overset{(n, 2r-c)}{\sum} q^{(c)}(X_{\alpha_1}, \ldots, X_{\alpha_{2r-c}})$$

$$(c = 0, 1, \ldots, r).$$

Then,

(4.7)
$$s_{Wn}^2 = r^2(U_n^{(1)} - U_n^{(0)}) + \sum_{c=0}^{r} \alpha_n^{(c)} U_n^{(c)},$$

where for each $c = 0, 1, \ldots, r$, $\alpha_n^{(c)} = O_e(n^{-1})$, where O_e denotes the exact order. It is easy to check that $E[U_n^{(1)} - U_n^{(0)}] = \zeta_1$. The main theorem of this section is now stated as follows.

Theorem 3. *For the sequential procedure described in (4.4) if*

(4.8)
$$E|\varphi(X_1, \ldots, X_r)|^m < \infty \quad \text{for all} \quad m > 0,$$

then,

(4.9)
$$\sup_x \left| P\left(\frac{\sqrt{N_d}(U_{N_d} - \theta)}{r\sqrt{\zeta_1}} \leqslant x \right) - \Phi(x) \right| = O(d^{\frac{1}{2}-\gamma}),$$

where $0 < \gamma < \frac{1}{2}$; γ can be arbitrarily small.

Proof. Write $\zeta_d = \frac{\tau_\alpha^2(r^2\zeta_1)}{d^2}$. In view of Theorem 3, all we need prove is

(4.10)
$$P\left\{ \left| \left(\frac{N_d}{\zeta_d}\right) - 1 \right| > \zeta_d^{-\frac{1}{2}+\gamma} \right\} = O(d^{\frac{1}{2}-\gamma}) \qquad 0 < \gamma < \frac{1}{2}.$$

First observe that if $t = [\xi_d + \xi_d^{\frac{1}{2}+\gamma}]$,

$$P(N_d > \xi_d + \xi_d^{\frac{1}{2}+\gamma}) \leqslant P(N > t) \leqslant P\left(t < \frac{\tau_\alpha^2}{d^2}\left(s_{Wt}^2 + \frac{1}{t}\right)\right) =$$

(4.11)

$$= P\left(s_{Wt}^2 - r^2\varsigma_1 > r^2\varsigma_1\left(\frac{t}{\xi_d} - 1\right) - \frac{1}{t}\right).$$

But

$$r^2\varsigma_1\left(\frac{t}{\xi_d} - 1\right) - \frac{1}{t} \geqslant r^2\varsigma_1\left(\xi_d^{-\frac{1}{2}+\gamma} - \frac{1}{\xi_d}\right) - \frac{1}{t} \geqslant K\xi_d^{-\frac{1}{2}+\gamma}$$

for small d, say $d \leqslant d_1$. Hence, for $d \leqslant d_1$, one gets from (4.11),

(4.12) $$P(N_d > \xi_d + \xi_d^{\frac{1}{2}+\gamma}) \leqslant P(s_{Wt}^2 - r^2\varsigma_1 > K\xi_d^{-\frac{1}{2}+\gamma}),$$

where K is a generic constant, positive but not depending on d. Using Lemma 1, the C_δ and Jensen inequalities, one gets

(4.13) $$E(s_{Wn}^2 - r^2\varsigma_1)^{2m} \leqslant \frac{K}{n^m},$$

where K does not depend on n. Hence, using Markov's inequality, it follows from (4.12) and (4.13) that for $d \leqslant d_1$,

(4.14) $$P(N_d > \xi_d + \xi_d^{\frac{1}{2}+\gamma}) = O(t^{-m}\xi_d^{m-m\gamma}) = O(d^{2m\gamma}).$$

Choosing $m > \frac{1}{4\gamma} - \frac{1}{2}$, it follows from (4.14) that for $d \leqslant d_1$,

(4.15) $$P(N_d > \xi_d + \xi_d^{\frac{1}{2}+\gamma}) = O(d^{\frac{1}{2}-\gamma}).$$

Next observe that from (4.4), $N \geqslant \frac{\tau_\alpha^2}{d}$. Write $a_1 = \left[\frac{\tau_\alpha^2}{d}\right]$, $a_2 = \left[\frac{1}{2}\xi_d\right]$, $a_3 = [\xi_d - \xi_d^{\frac{1}{2}+\gamma}] =$. Then for small $d\ (> 0)$, say $d \leqslant d_2$, $a_1 < a_2 < a_3$. Hence, for $d \leqslant d_2$,

$$P(N_d < \xi_d - \xi_d^{\frac{1}{2}+\gamma}) \leqslant$$

$$\leqslant \sum_{k=a_1}^{a_2} P(N = k) + \sum_{k=a_2+1}^{a_3} P\left(s_{Wk}^2 \leqslant kd^2\tau_{\frac{\alpha}{2}}^2 - \frac{1}{k}\right) \leqslant$$

$$\leqslant \sum_{k=a_1}^{a_2} P\left(s_{Wk}^2 \leqslant kd^2\tau_{\frac{\alpha}{2}}^2 - \frac{1}{k}\right) +$$

$$(4.16) \qquad\qquad + \sum_{k=a_2+1}^{a_3} P\left(s_{Wk}^2 \leqslant kd^2\tau_{\frac{\alpha}{2}}^2 - \frac{1}{k}\right) \leqslant$$

$$\leqslant \sum_{k=a_1}^{a_2} P(s_{Wk}^2 \leqslant kd^2\tau_{\frac{\alpha}{2}}^2) + \sum_{k=a_2+1}^{a_3} P(s_{Wk}^2 \leqslant kd^2\tau_{\frac{\alpha}{2}}^2) \leqslant$$

$$\leqslant \sum_{k=a_1}^{a_2} P\left(s_{Wk}^2 - r^2\varsigma_1 \leqslant -\frac{1}{2}r^2\varsigma_1\right) +$$

$$+ \sum_{k=a_2+1}^{a_3} P(s_{Wk}^2 - r^2\varsigma_1 \leqslant -K\xi_d^{-\frac{1}{2}+\gamma}).$$

Using Markov's inequality and (4.13), it follows from (4.16) that

$$P(N_d < \xi_d - \xi_d^{\frac{1}{2}+\gamma}) \leqslant$$

$$(4.17) \qquad \leqslant K\left\{ \sum_{k=a_1}^{a_2} \frac{1}{k^m} + \sum_{k=a_2+1}^{a_3} (\xi_d^{-\frac{1}{2}+\gamma})^{-2mk-m} \right\} =$$

$$= O(d^{m-1}) + o(d^{4m\gamma-2}).$$

Hence, choosing $m > \frac{5}{8\gamma} - \frac{1}{4}$ so that $m > \frac{3}{2} - \gamma$ for $0 < \gamma < \frac{1}{2}$ one gets from (4.17) that for $d \leqslant d_2$,

$$(4.18) \qquad P(N_d < \xi_d - \xi_d^{\frac{1}{2}+\gamma}) = O(d^{\frac{1}{2}-\gamma}).$$

Thus, from (4.15) and (4.18), (4.10) holds for $d \leqslant \min(d_1, d_2) = d_0$ (say); (4.10) trivially holds for $d > d_0$. This completes the proof of the theorem.

Added in proof. Since the submission of the paper, a related paper by Ibrahim A. Ahmad entitled "On the Berry – Esseen Theorem for random U-statistics" has appeared in the *Annals of Statistics*, (1980), V8, #6, pp. 1395-1398, where it is assumed that N is *independent* of the $\{X_i\}$ sequence.

REFERENCES

[1] R.H. Berk, Limiting behavior of posterior distributions when the model is incorrect, *Ann. Math. Statist.*, 37 (1966), 51-58.

[2] P.J. Bickel, Edgeworth expansions in nonparametric statistics, *Ann. Statist.*, 2 (1974), 1-20.

[3] H. Callaert – P. Janssen, The Berry – Esseen theorem for U-statistics, *Ann. Statist.*, 6 (1978), 417-421.

[4] H. Callaert – P. Janssen, The convergence rate of fixed width sequential confidence intervals for the mean, Submitted in *Sankhyā*.

[5] Y.K. Chan – J. Wierman, On the Berry – Esseen theorem for U-statistics, *Ann. Prob.*, 4 (1977), 136-139.

[6] A. Csenki, A theorem on the departure of randomly indexed U-statistics from normality with an application in fixed width sequential interval estimation, (unpublished paper).

[7] A. Csenki, On the convergence rate of fixed width sequential confidence intervals, *Scand. Actuarial J.*, (1980), 107-111.

[8] S.W. Dharmadhikari – V. Fabian – K. Jogdeo, Bounds on the moments of martingales, *Ann. Math. Statist.*, 39 (1968), 1719-1723.

[9] C.G. Esseen, Fourier analysis of distribution functions. A mathematical study of the Laplace – Gaussian law, *Acta. Math.*, 77 (1945), 1-125.

[10]   M . G h o s h ,   Rate of convergence to normality for random
means: Applications to sequential estimation.   *Sankhyā A, 42*
(1980), 231-240.

[11]   M . G h o s h – R . D a s g u p t a ,   On some nonuniform rates
of convergence to normality,   *Sankhyā A.,* 40 (1978), 347-368.

[12]   W . F . G r a m s – R . J . S e r f l i n g ,   Convergence rates for
U-statistics and related statistics,   *Ann. Statist.,* 1 (1973), 153-160.

[13]   W . H o e f f d i n g ,   A class of statistics with asymptotically normal
distribution, *Ann. Math. Statist.,* 19 (1948), 293-325.

[14]   W . H o e f f d i n g ,   The strong law of large numbers for   *U*-sta-
tistics,   *Inst. of Stat. Mimeo Series No. 302,* UNC, Chapel Hill.

[15]   P . L . J a n s s e n ,   De Berry – Esseen Stelling en een Asympto-
tische voor   *U*-Statistieken.   Ph. D. thesis, Department of Mathematics, Limburgs Universitair Centrum, Diepenbeek, Belgium.

[16]   D . L a n d e r s – L . R o g g e ,   The exact approximation order
in the central limit theorem for random summation, *Z. Wahrschein-*
lichkeitstheorie und verw. Geb., 36 (1976), 269-283.

[17]   D . L a n d e r s – L . R o g g e ,   A counterexample to the approximation theory of random summation,   *Ann. Prob.,* 5 (1977),
1018-1023.

[18]   R . G . M i l l e r – P . K . S e n ,   Weak convergence of   *U*-sta-
tistics and Von Mises' differentiable statistical functions,   *Ann.*
Math. Statist., 43 (1972), 31-41.

[19]   J . M o g y o r ó d i ,   Limit distributions for sequences of random
variables with random indices,   *Trans. 4th Prague Conf. Infor. Th.*
Statist. Dec. Fns. Random Proc., (1967), 463-470.

[20]   P . K . S e n ,   On some convergence properties of   *U*-statistics,
Cal. Statist. Assocn. Bull., 10 (1960), 1-18.

[21]   R.N. Sproule, A sequential fixed width confidence interval for the mean of a *U*-statistic, Ph. D. dissertation, UNC, Chapel Hill.

[22]   R.N. Sproule, Asymptotic properties of *U*-statistics, *Trans. Amer. Math. Soc.,* 199 (1974), 55-64.

M. Ghosh

Iowa State University, Ames, Iowa 50011, USA.

R. Dasgupta

Indian Statistical Institute, 203 B.T. Road, Calcutta-700035, India.

COLLOQUIA MATHEMATICA SOCIETATIS JÁNOS BOLYAI

32. NONPARAMETRIC STATISTICAL INFERENCE,

BUDAPEST (HUNGARY), 1980.

ASYMPTOTIC NORMALITY OF LINEAR COMBINATIONS OF FUNCTIONS OF ORDER STATISTICS IN ONE AND SEVERAL SAMPLES* **

Z. GOVINDARAJULU

1. INTRODUCTION

Linear combinations of order statistics naturally arise in best linear unbiased estimation of location and scale parameters of a distribution, based on order statistics (see B l o m [4]). J u n g [12] considered linear combinations of order statistics with continuous weight functions for estimating location and scale parameters and studied their asymptotic efficiencies. Trimmed and Windsorized means of samples are also considered in robust estimation of parameters. Hence, it is of interest and importance to study the large-sample properties of such statistics. Asymptotic normality of sample quantiles has been studied by many authors.

*This research was supported in part by the Mathematics Division of the Air Force Office of Scientific Research under Grant No. AF-AFOSR-741-65. This research is also supported by the National Science Foundation Grant NSF-GP-5664.

**Earlier versions of the results were obtained while the author was at the University of California, Berkeley and at Case Western Reserve University.

(For instance, see H a j e k [11]). If the distribution function has a con-
continuous density function which is strictly positive on the range of variation, using R é n y i ' s [17] representation of order statistics in random
samples drawn from continuous populations, one can obtain simple sufficient conditions for the asymptotic normality of linear combinations of
order statistics (LCOS's) based on trimmed samples, for example, sample
quantile, α-trimmed mean, Winsorized mean etc. (for definitions of these
see T u k e y [25]). However, when the tails of the ordered sample are
included in the linear combination, the study of its asymptotic normality
becomes quite nontrivial. The asymptotic distribution of linear combinations of functions of order statistics for the one-sample case has been
studied by C h e r n o f f et al [5], imposing smoothness condition on the
tails of the distribution. Also, the conditions of H á j e k [11] are somewhat too restrictive. Hence, it is of much interest to make a systematic
study of the asymptotic normality of LCFOS's in one- and c-samples
without the smoothness condition on the tails of the distribution. The
results of this paper constitute strengthening of the author's [6] earlier
results.

The organization of the paper is as follows. Assumptions and notation
are introduced in Section 2. Certain properties of functions (g, J) are
given in Section 3. Section 4 deals with certain properties of empirical
distribution functions. Some propositions that are used later are provided in
Section 5. The negligibility of the tails of the statistic is shown in Section
6. The asymptotic behavior of the statistic is considered in Section 7. The
asymptotic normality of the statistic when the sample size is random is
considered in Section 8. The asymptotic results for the one-sample case
are extended to the c-sample situation in Section 9. Some applications
are provided in Section 10. Recent work on the topic is briefly reviewed
in the concluding remarks.

2. ASSUMPTIONS AND NOTATION

Let $X_1, X_2, \ldots, X_N$ be a random sample of size N drawn from a
population having the continuous cumulative distribution function (c.d.f.)
$F(x)$. Also, let $F_N(x)$ denote the empirical distribution function (e.d.f.)

based on the random sample $X_1, X_2, \ldots, X_N$. $F(x)$ may depend on N, although this will not be stated explicitly. Define the class of statistics by

$$(2.1) \qquad N\bar{T}_N = \sum_{i=1}^{N} E_{N,i} g^*(X_{i,N})$$

where $g^*(x)$ is an arbitrary continuous function, $E_{N,i}$ are some given constants and $X_{i,N}$ are the ordered values of the X's. When $g^*(x) = x$, $\bar{T}_N$ generates the systematic statistics. Alternatively, one can write

$$(2.2) \qquad \bar{T}_N = \int_{-\infty}^{\infty} g(F(x)) J_N \left(\frac{N}{N+1} F_N(x) \right) dF_N(x)$$

where $J_N \left(\frac{i}{N+1} \right) = E_{N,i}$ and $g(F(x)) = g^*(x)$. Although J_N need be defined only at $\frac{1}{N+1}, \frac{2}{N+1} \cdots, \frac{N}{N+1}$, one can conveniently extend its domain of definition to $(0, 1)$ by adopting some convention such as J is constant on $\left(\frac{i}{N+1}, \frac{i+1}{N+1} \right)$ $(i = 0, 1, \ldots, N)$. $J_N(u)$ might depend on N, although the subscript N in $J_N(u)$ will be suppressed whenever convenient.

The assumption of continuity of F is somewhat too restrictive because if F is discontinuous having a denumerable number of jump points, then F can be made continuous by the following continuization process described by Govindarajulu, LeCam and Raghavachari [9] (which will be abbreviated as GLR [9]). If F has a jump of size α at a point t, remove the point t from the real line and insert in its place a closed interval of length α. Distribute the probability mass α uniformly over this interval. The new cumulative distribution F^* so obtained is continuous. The relative order relations among X's have the same probability distribution as if the sample was drawn from F.

As a further reduction, without loss of generality one can assume that F is uniform on $[0, 1]$. To see this, note that if after removal of discontinuities, the function F^* is constant over certain intervals, no observations will occur in these intervals. Thus, these intervals can be deleted from the real line without affecting the order of the observations. Then we are left with a continuous strictly increasing cumulative distribution function which can be transformed to the uniform cumulative $F(x) \equiv x$ by a strictly increasing continuous transformation.

Let us define classes of functions for which the asymptotic results hold.

Definition 2.1. A function h, $h \geqslant 1$ defined on $(0, 1)$ is said to belong to the class $\mathscr{U}$ if there exists an $\alpha \in (0, 1)$ such that h is monotone decreasing in $(0, \alpha]$ and monotone increasing in $[\alpha, 0)$.

Definition 2.2. A function $h \in \mathscr{U}$ will be said to belong to the class $\mathscr{U}_1$ $(\mathscr{U}_2)$ if it is integrable (square integrable) for the Lebesgue measure.

Let $h \in \mathscr{U}$ and $h_i \in \mathscr{U}_i$ $(i = 1, 2)$. Notice that in particular $h_1(x) = K[x(1-x)]^{-1+\delta}$ and $h_2(x) = K[x(1-x)]^{-\frac{1}{2}+\delta}$ for some $0 < \delta < \frac{1}{2}$ and $0 < K < \infty$. Let the function J be defined by an integral of the form $J(x) = \int_{\frac{1}{2}}^{x} J'(\xi)\, d\xi$. One could also consider the function J which differs

from the integral $\int_{\frac{1}{2}}^{x} J'(\xi)\, d\xi$ by a constant. However, this will introduce a normalized sum of random variables into the differences studied here. Thus, treatment of functions of the form $a + \int_{\frac{1}{2}}^{x} J'(\xi)\, d\xi$ is analogous.

We say that $(g, J) \in \mathscr{S}$ if g is continuous with $|g| \leqslant h^* \in \mathscr{U}$, J is absolutely continuous with $|J'| \leqslant h \in \mathscr{U}$ and $hh^* \leqslant h_1 h_2$. We also say that $(g, J) \in \mathscr{S}_0$ if $|g(u)| \leqslant K[u(1-u)]^{-\alpha+\delta}$ and $|J'(u)| \leqslant K[u(1-u)]^{-\frac{3}{2}+\alpha+\delta}$ for some $0 < \delta < \frac{1}{4}$ and $\alpha \leqslant \frac{3}{2}$ where K is a finite and positive generic constant. Notice that $\mathscr{S}_0 \subset \mathscr{S}$ since, in particular, we can have $h^*(u) = K[u(1-u)]^{-\alpha+\delta}$ and $h(u) = K[u(1-u)]^{-\frac{3}{2}+\alpha+\delta}$.

3. PROPERTIES OF FUNCTIONS OF (g, J)

Let us derive certain boundedness and integrability properties of functions of (g, J).

Lemma 3.1. *For $u \in (0, 1)$ let*

(3.1) $\widetilde{B}(u) = g(u)J(u)$ *and* $L'(u) = g(u)J'(u)$.

Then, there exists a number b_0 such that

(3.2) $\sup_{(g,J) \in \mathscr{S}} \left\{ \int \widetilde{B}^2(u) \, du \right\} < b_0$.

Furthermore for any $\epsilon > 0$ there exists a number b such that

(3.3) $\int_{|\widetilde{B}(u)| > b} \widetilde{B}^2(u) \, du < \epsilon$ *for every $(g, J) \in \mathscr{S}$.*

Analogous result holds for $L'(u)$.

Proof. One can easily show that both $\widetilde{B}(u)$ and $L'(u)$ are absolutely bounded by $h_1(u)h_2(u)$. Now, from Lemma 2 of G L R [9] we assert the desired result.

For small α, let

(3.4) $c^{*2}(\alpha) = \max \left[\int_0^\alpha h_2^2(u) \, du, \left\{ \int_0^\alpha h_1(u) \, du \right\}^2 \right]$.

Lemma 3.2. *If $(g, J) \in \mathscr{S}$, then*

$$\lim_{u \to 0} u |g(u)J(u)| = \lim_{u \to 1} (1 - u)|g(u)J(u)| = 0.$$

Proof. Since g and J' are bounded by u-shaped functions, it follows that J and hence gJ will be bounded by u-shaped functions. Furthermore, because of Lemma 3.1, gJ is uniformly square integrable, $gJ \leqslant K[u(1 - u)]^{-\frac{1}{2} + \delta}$. Thus it follows that

$$\lim_{u \to 0 \text{ or } 1} u(1 - u)[g(u)J(u)] = 0.$$

– 319 –

Consider the random variable B_N given by

$$(3.5) \quad B_N = \sqrt{N} \int g(F)J(F)d(F_N - F) + \sqrt{N} \int g(F)J'(F)(F_N - F) \, dF =$$

$$= \frac{1}{\sqrt{N}} \sum_{i=1}^{N} B(X_i)$$

where

$$(3.6) \quad B(X) = \int g(F)J(F)d(F_1 - F) + \int g(F)J'(F)(F_1 - F) \, dF =$$

$$= B_1 + B_2$$

(say) and F_1 denotes the empirical distribution function based on a single observation. Variance of B_N is given by

$$\text{Var } B = \text{Var } B_1 + \text{Var } B_2 + 2 \text{ Cov } (B_1, B_2),$$

the variance of $B_1 = \text{Var} \{g(U)J(U)\}$ where U is uniform on $(0, 1)$, is finite because of Lemma 3.1 and the variance of B_2 is finite on account of Lemma 3.2, because

$$(3.7) \quad \text{Var } B_2 = 2 \iint_{u < v} u(1 - v)L'(u)L'(v) \, dudv =$$

$$= \int_0^1 L^2(u) \, du - \left(\int_0^1 L(u) \, du \right)^2,$$

where $L'(u) = g(u)J'(u)$. Also

$$\text{Cov } (B_1, B_2) =$$

$$(3.8) \quad = \mathsf{E}\left[\left\{ \int g(F)J(F)dF_1 \right\} \left\{ \int g(F)J'(F)(F_1 - F)dF \right\} \right] =$$

$$= \mathsf{E}\left[g(U)J(U) \left\{ \int_U^1 (1 - v)g(v)J'(v) \, dv - \int_0^U vg(v)J'(v) \, dv \right\} \right].$$

The covariance is also finite because $\text{Cov}^2(B_1, B_2) \leqslant (\text{Var } B_1)(\text{Var } B_2)$. Thus $\text{Var } B$ is finite for every $(g, J) \in \mathscr{S}$.

Also, one can easily see that $\text{Var } B$ is positive provided $J \not\equiv 0$. When $J \equiv 1$, B_N denotes the normalized sample mean of $g(F(X_i))$ $(i = 1, \ldots, N)$. Now, performing integration by parts once in the first term of B_N, we obtain

$$(3.9) \qquad B_N = - \sqrt{N} \int J(F)(F_N - F) \, dg(F),$$

since by Lemma 3.2, $g(F)J(F)(F_N - F) \to 0$ as $F \to 0$ or 1. Now the variance of B_N given by (3.9) is

$$(3.10) \qquad \operatorname{Var} B_N = \sigma_N^2(F, J, g) = 2 \iint_{0 < u < v < 1} u(1 - v) J(u) J(v) \, dg(u) dg(v).$$

When $g(F(x)) = x$, the form of the variance of B_N as given in (3.10) agrees with Jung's [12] expression for the asymptotic variance of B_N.

Further, if $g(u)$ is differentiable, (3.10) takes the form of

$$(3.11) \qquad \operatorname{Var} B_N = \int_0^1 M^2(u) \, du - \left[\int_0^1 M(u) \, du \right]^2$$

where

$$M(u) = \int_{\frac{1}{2}}^{u} J(x) g'(x) \, dx$$

since

$$\operatorname{Var} B_N = 2 \iint_{x < y} \left[\iint_{x < u < v < y} J(u) J(v) g'(u) g'(v) \, du dv \right] dxdy =$$

$$= \iint_{0 < x < y < 1} \left[\int_x^y J(u) g'(u) \, du \right]^2 dxdy.$$

It is worthwhile to remark that if $|J(u) g'(u)| \leqslant h_1 h_2$ with $h_1 \in \mathscr{U}_1$ and $h_2 \in \mathscr{U}_2$, then $\operatorname{Var} B_N$ as given in (3.10) is finite. In order to see this consider

$$\operatorname{Var} B_N \leqslant$$

$$\leqslant 2 \iint_{u < v} \sqrt{u(1 - u) v(1 - v)} \, h_1(u) h_1(v) h_2(u) h_2(v) \, du dv =$$

$$= \left[\int_0^1 \sqrt{u(1 - u)} \, h_1(u) h_2(u) \, du \right]^2.$$

However, since

$$\xi h_2^2(\xi) \leqslant \int_0^\xi h_2^2(y)\, dy,$$

$$\int_0^\xi \sqrt{u}\, h_1(u) h_2(u)\, du \leqslant \int_0^\xi h_1(u) \left[\int_0^\xi h_2^2(v)\, dv \right]^{\frac{1}{2}} du \leqslant c^{*2}(\xi)$$

where $c^{*2}(\xi)$ is given by (3.4).

Then, as a consequence of Lemmas 3.1 and 3.2, we have the following proposition which will be used in Section 7.

Proposition 3.1. *Let* P_N *be the distribution of* B_N *and* Q_N *be the normal distribution having zero expectation and variance* $\sigma_N^2(F, J, g)$ *given by (3.10). Then, for every* $\epsilon > 0$ *there exists an* $N(\epsilon)$ *such that* $N \geqslant N(\epsilon)$ *implies*

$$|P_N\{(-\infty, y]\} - Q_N\{(-\infty, y]\}| \leqslant \epsilon$$

for every y, *every* $(g, J) \in \mathscr{S}$, *and every* F. *In addition, there is an* $N(\epsilon, a)$ *such that* $N \geqslant N(\epsilon, a)$ *and* $\sigma_N^2(F, J, g) \geqslant a$ *imply*

$$\sup |P_N\{(-\infty, y]\} - Q_N\{(-\infty, y]\}| < \epsilon$$

for every $(g, J) \in \mathscr{S}$, *and every* F.

Proof. The first part of the statement follows immediately from the usual central limit theorem. The second statement follows from the first by considering $\dfrac{B_N}{\sigma_N}$ instead B_N.

4. CERTAIN PROPERTIES OF EMPIRICAL DISTRIBUTION FUNCTIONS

In this section we shall state some inequalities and limit theorem pertaining to the empirical distribution function developed by G L R [9]. These constitute either generalizations or sharpened versions of results known up to that time. These results would be used in Sections 5 and 6 not only to bound the tails of the statistics T_N in probability, but also to show that the higher order random terms in the expansion of $\sqrt{N}\,T_N$ tend to zero in probability. For simplicity of notation, the results will be given for the uniform distribution on $[0, 1]$; however, they can easily be reworded so as to apply to arbitrary continuous distributions.

Let U_N be the empirical distribution function based on N independent observations from the uniform distribution on $[0, 1]$. Then, we have the following results.

Lemma 4.1. *For every $\epsilon > 0$ there exists $c(\epsilon)$ such that*

(4.1) $P\{\sup\limits_{t} \sqrt{N}\,h_2(t)\,|\,U_N(t) - t\,| \geqslant c(\epsilon)\} \leqslant \epsilon, \qquad h_2 \in \mathcal{U}_2.$

Proof. See G L R [9], Lemma 7.

Lemma 4.2. *For every $\epsilon > 0$ there exists a $\beta > 0$ such that*

(4.2) $P\left(\sup\limits_{t}\left\{\dfrac{U_N(t)}{t}\right\} > \dfrac{1}{\beta}\right) < \epsilon$

and

(4.3) $P(U_N(t) \geqslant \beta t \ \text{for every} \ t \ \text{such that} \ U_N(t) > 0) > 1 - \epsilon.$

Proof. See G L R [9], Lemma 8.

Define the process Z_N by $Z_N(t) = \sqrt{N}\{F_N(t) - F(t)\}$, where F_N is the empirical distribution function based on a sample of size N drawn from the distribution F. Also, for each positive integer N' let $K_{N'}$ be a random process defined on the interval $[0, 1]$. Let $Z_N^*(x) = Z_N[K_{N'}(x)]$.

Lemma 4.3. *If for every $\epsilon > 0$, $P(\sup\limits_{t}|K_{N'}(t) - t| > \epsilon) \to 0$, then*

(4.4) $P(\sup\limits_{x}|Z_N^*(x)| > \epsilon) \to 0.$

Proof. See G L R [9], Lemma 9.

In particular, we can have $K_N(x) = \inf(t: F_N(t) \geqslant x)$. If $F(t) = t$ for $t \in [0, 1]$, then $\sup\limits_{x}|K_N(x) - x|$ must tend to zero in probability since $\sup|F_N(t) - t|$ tends to zero in probability. Consequently, from Lemma 4.3 we have $\sup_x |Z_N(K_N(x)) - Z_N(x)|$ tends to zero in probability. For values of the form $x = \frac{j}{N'}$, the variable $Z_N[K_{N'}(x)]$ is equal to $\sqrt{N}[F_N(\xi_j) - F(\xi_j)]$ where ξ_j is the order statistic of rank j in a sample of size N' drawn from the uniform on $[0, 1]$. In other words, $NF_N[K_{N'}(\frac{j}{N'})]$ is the number of X_j's whose ranks are less than or equal to j.

5. CERTAIN PROPOSITIONS

In this section, we shall bound the higher order random terms of T_N. We need the following further notation. Let $\mathcal{M}$ denote the space of all finite signed measures μ on the interval $(0, 1)$ and their indefinite integrals. For $\mu \in \mathcal{M}$, let $J_\mu(x) = \mu\{(0, x)\}$ and let $\|\mu\|$ be the total mass of μ. The functions J_μ are simply those functions of bounded variation on $[0, 1]$ which are right continuous and vanish at zero. Further, let g_μ denote those functions that are continuous and bounded on $[0, 1]$.

Proposition 5.1. *For every* $\epsilon > 0$ *there exists an* $N(\epsilon)$ *such that* $N \geqslant N(\epsilon)$ *implies*

$$(5.1) \qquad \mathsf{P}\left\{\sqrt{N}\left|\int g_\mu(F)[J_\mu(F_N) - J_\mu(F)]\, d(F_N - F)\right| > \epsilon\|\mu\|\right\} < \epsilon$$

for every $\mu \in \mathcal{M}$ *and every* F.

Proof. Without loss of generality let F be uniform on $[0, 1]$ and let $Z_N(\xi) = \sqrt{N}[F_N(\xi) - \xi]$. Also, let $K_N(\xi) = \inf\{u: F_N(u) \geqslant \xi\}$. Now, integrating by parts once, we obtain

$$I_N = \sqrt{N}\int g_\mu(\xi)[J_\mu(F_N) - J_\mu(\xi)]\, d(F_N - \xi) =$$

$$= -\int Z_N(\xi) g_\mu(\xi)\, d[J_\mu(F_N) - J_\mu(\xi)] -$$

$$-\int Z_N(\xi)[J_\mu(F_N) - J_\mu(\xi)]\, dg_\mu(\xi) = -I_{1,N} - I_{2,N}$$

(say) where

$$|I_{1,N}| = \left| \int [Z_N(\xi - 0)g_\mu(\xi) - \right.$$

$$\left. - Z_N(K_N(\xi) - 0)g_\mu(K_N(\xi) - 0)]\mu(d\xi) \right| \leqslant$$

(5.2)
$$\leqslant \left| \int g_\mu(\xi)[Z_N(\xi - 0) - Z_N(K_N(\xi) - 0)]\mu(d\xi) \right| +$$

$$+ \left| \int Z_N(K_N(\xi) - 0)[g_\mu(\xi) - g_\mu(K_N(\xi) - 0)]\mu(d\xi) \right| \leqslant$$

$$\leqslant \|\mu\| \sup_x |Z_N(x) - Z_N(K_N(x))| +$$

$$+ \|\mu\| c(\epsilon) \sup |g_\mu(x) - g_\mu(K_N(x))|,$$

since Z_N is bounded in probability by $c(\epsilon)$. Now, noting that $K_N(\xi)$ tends to ξ in probability and g_μ is continuous on $[0, 1]$, one can assert that the second term tends to zero in probability. Because of Lemma 4.3, the first term on the right side also tends to zero. Next, consider

(5.3)
$$|I_{2,N}| \leqslant c(\epsilon) \int |J_\mu(F_N) - J_\mu(\xi)| |dg_\mu(\xi)|.$$

Since, $F_N(\xi) \to \xi$ in probability J is a continuous function, it follows from the boundedness of $g_\mu(\xi)$ that $I_{2,N}$ tends to zero in probability. This completes the proof of Proposition 5.1.

Remark 5.1.1. In Proposition 5.1, one could replace $J_\mu(F_N)$ by $J_\mu\left(\frac{N}{N+1} F_N\right)$ since

$$\sqrt{N} \int g_\mu(F)\left[J_\mu\left(\frac{N}{N+1} F\right) - J_\mu(F)\right] dF$$

is of the order $\frac{1}{\sqrt{N}}$ because

$$\sqrt{N} \int g_\mu(F)\left[J_\mu\left(\frac{N}{N+1} F\right) - J_\mu(F)\right] dF \leqslant$$

$$\leqslant \sqrt{N} \int |g_\mu(F)| \left(\int_{\frac{NF}{N+1}}^{F} |J'_\mu(u)| du \right) dF \leqslant$$

$$\leqslant \sqrt{N} \int |J'_\mu(u)| \left[\int_{u}^{\frac{(N+1)u}{N}} |g_\mu(F)| dF \right] du \leqslant$$

$$\leqslant \frac{K}{\sqrt{N}} \int |J_\mu'(u)| \, du \leqslant \frac{K}{\sqrt{N}}.$$

Corollary 5.1.1. *Proposition 5.1 holds if* $(g, J) \in \mathscr{S}$ *and the integral in* (5.1) *is taken over* $[\tau, 1 - \tau]$ *for some* $0 < 2\tau < 1$ *(in which case* g *and* J' *will both be bounded and hence integrable).*

Corollary 5.1.2. *Let*

$$(5.4) \qquad T_N = \sqrt{N} \left[\int g_\mu(F) J_\mu \left(\frac{N}{N+1} F_N \right) dF_N - \int g_\mu(F) J_\mu(F) \, dF \right]$$

and

$$T_N^* = \sqrt{N} \int g_\mu(F) \left[J_\mu \left(\frac{N}{N+1} F_N \right) - J_\mu(F) \right] dF +$$

$$(5.5)$$

$$+ \sqrt{N} \int g_\mu(F) J_\mu(F) \, d(F_N - F).$$

Then, it follows from Proposition 5.1 or Corollary 5.1.1 that the difference $T_N - T_N^*$ *converges to zero in probability, uniformly for* $\|\mu\|$ *bounded.*

For the following proposition we assume that $J \in \tilde{L}_1$, that is, J is absolutely continuous with $\int |J'(x)| \, dx \leqslant b$, and $g(F)$ is continuous and bounded on $(0, 1)$.

Proposition 5.2. *For every* $\epsilon > 0$ *there exists an* $N_1(\epsilon)$ *such that* $N \geqslant N_1(\epsilon)$ *implies that*

$$(5.6) \qquad \mathsf{P} \left[\sqrt{N} \left| \int g(F) [J(F_N) - J(F) - (F_N - F) J'(F)] \, dF \right| > \epsilon \right] < \epsilon$$

for every continuous and bounded $g(F)$, *every* $J \in \tilde{L}_1$ *and every* F.

Proof. First note that given $\epsilon > 0$ there is a $c < \infty$ such that

$$\mathsf{P}\{\sqrt{N} \|F_N - F\| \geqslant c\} < \epsilon.$$

Let

$$\tilde{F}_N = \begin{cases} F_N & \text{if} \quad \sqrt{N} \|F_N - F\| < c \\ F & \text{otherwise,} \end{cases}$$

where $\|F_N - F\| = \sup_x |F_N - F|$. Also, let

$$
(5.7) \quad
\begin{aligned}
\tilde{R}_{N,1} &= \sqrt{N} \int g(F)[J(\tilde{F}_N) - J(F)] \, dF = \\[2mm]
&= \sqrt{N} \int g(F) \left[\int_F^{\tilde{F}_N} J'(\xi) \, d\xi \right] dF
\end{aligned}
$$

and

$$
(5.8) \quad \tilde{R}_{N,2} = \int \sqrt{N} [\tilde{F}_N - F] g(F) J'(F) \, dF.
$$

If $\tilde{R}_N[F, J'] = \tilde{R}_{N,1} - \tilde{R}_{N,2}$, then

$$
(5.9) \quad |\tilde{R}_N[F, J']| \leqslant Kc \|J'\|, \qquad \|J'\| = \int |J'(\xi)| \, d\xi.
$$

So, (5.6) will follow, if we can show that for a fixed J', the term $\tilde{R}_N[F, J']$ converges to zero uniformly in F. Let

$$
\rho_N(x) = \sup_{|\xi| \leqslant c} \sqrt{N} \left| J\left(x + \frac{\xi}{\sqrt{N}}\right) - J(x) - \frac{\xi}{\sqrt{N}} J'(x) \right|.
$$

By definition of the derivative, this converges to zero for almost all x. In addition, because of (5.9) $\int \rho_N(x) \, dx$ converges to zero.

Next let $\{F^{(\nu)}\}$ be a sequence of distribution functions converging to F. Then, (5.6) holds for all F since g is bounded and the integral can be bounded by one which is free of F.

Corollary 5.2.1. *For every $\epsilon > 0$ there exists an $N_1(\epsilon)$ such that $N \geqslant N_1(\epsilon)$ implies (5.6) for every $(g, J) \in \mathscr{S}$ since in the interval $[\tau, 1 - \tau]$, g and J' are bounded and continuous and hence integrable.*

Remark 5.2.1. Proposition 5.2 and Corollary 5.2.1 are valid if $J(F_N)$ is replaced by $J\left(\frac{N}{N+1} F_N\right)$. That is,

$$
(5.10) \quad
\begin{aligned}
P\Big[\sqrt{N} \Big| \int g(F) \Big[J\Big(\frac{N}{N+1} F_N\Big) - \\
- J(F) - (F_N - F) J'(F) \Big] \, dF \Big| > \epsilon \Big] < \epsilon
\end{aligned}
$$

since $\sqrt{N} \int g(F) \left[\dfrac{F_N J'(F)}{N+1} \right] dF$ is of the order $\dfrac{1}{\sqrt{N}}$.

Corollary 5.2.2. *Let $\mathcal{M}$, J_μ and g_μ be as defined for Proposition 5.1. Also, let T_N be given by (5.4). Let*

$$T_N^{**} = \sqrt{N} \int g_\mu(F) J_\mu'(F)(F_N - F)\, dF +$$

(5.11)

$$+ \sqrt{N} \int g_\mu(F) J_\mu(F)\, d(F_N - F).$$

Then, it follows from Corollary 5.1.2 and Proposition 5.2 that $T_N - T_N^{**}$ converges to zero in probability uniformly for $\|\mu\|$ bounded.

6. BOUNDS FOR THE TAILS OF THE STATISTICS T_N

Let J be an indefinite integral of the form $J(x) = \int\limits_{\frac{1}{2}}^{x} J'(\xi)\, d\xi$ and let $(g, J) \in \mathcal{S}$. For τ such that $0 < 2\tau < 1$, let

(6.1) $\quad \Delta_N^*(J, g, \tau) = \sqrt{N} \int\limits_A g(F) \left[J\left(\dfrac{N}{N+1} F_N \right) - J(F) \right] dF_N$

where $A = (0, \tau] \cup [1 - \tau, 1)$. Then we have the following proposition.

Proposition 6.1. *For every $\epsilon > 0$ there exists a number τ_0 such that*

(6.2) $\quad P[\sup [|\Delta_N^*(J, g, \tau)|:\ 0 < \tau \leqslant \tau_0,\ J \in \mathcal{S}] > \epsilon] < \epsilon$

for every F and N.

Proof. If x is changed to $1 - x$, the order of observations is reversed, and the part of $\Delta_N^*(J, g, \tau)$ arising from the integral over $[1 - \tau, 1)$ is transformed into an analogous integral over $(0, \tau]$ for different g, J functions. Thus, it suffices to bound the part of $\Delta_N^*(J, g, \tau)$ relative to $(0, \tau]$. Let δ be such that $0 < 8\delta < \epsilon$, and h_1 and h_2 be monotone decreasing in $(0, \delta]$. Throughout, one can assume that $\tau \leqslant \delta$. Also, without loss of generality let F be uniform on $(0, 1]$. Let $\{U_j\}$ $(j = 1, \ldots, N)$, be independently and uniformly distributed on $[0, 1]$ and let R_j denote the rank of U_j $(j = 1, \ldots, N)$. Then it will be sufficient to bound

$$(6.3) \qquad S_N(J, g, \tau) = \frac{1}{N} \sum_{U_i \leqslant \tau} \left| g(U_i) \left[J\left(\frac{R_i}{N+1}\right) - J(U_i) \right] \right|.$$

According to Lemma 4.2, there exists a β $(\beta > 0)$ such that

$$(6.4) \qquad P\left[\beta U_i \leqslant \frac{R_i}{N+1} \leqslant \frac{U_i}{\beta} \quad \text{for all } i \right] > 1 - \frac{\epsilon}{4}.$$

After choosing β, we infer from Lemma 4.2 that there exists a number c_1 such that

$$(6.5) \qquad P\left\{ \sup_i \left[\sqrt{N} h_2(\beta U_i) \left| \frac{R_i}{N} - U_i \right|; \; U_i \leqslant \delta \right] \geqslant c_1 \right\} < \frac{\epsilon}{4}.$$

In (6.5) one can replace $\dfrac{R_i}{N}$ by $\dfrac{R_i}{N+1}$ because when (6.4) holds

$$(6.6) \qquad \begin{aligned} &\sup_i \left\{ \sqrt{N} h_2(\beta U_i) \left(\frac{R_i}{N} - \frac{R_i}{N+1} \right) \right\} = \\ &= \max_i \left\{ \frac{1}{\sqrt{N}} h_2(\beta U_i) \frac{R_i}{N+1} \right\} \leqslant \frac{1}{\beta \sqrt{N}} \max_i U_i h_2(\beta U_i) \end{aligned}$$

If the maximum is restricted to those values of i for which $U_i \leqslant \tau$, this last term does not exceed

$$(6.7) \qquad \frac{1}{\beta \sqrt{N}} \sup_{x \leqslant \tau} x h_2(\beta x)$$

and $x h_2(x) \to 0$ as $x \to 0$ and consequently the expression in (6.7) goes to zero as $N \to \infty$.

Let $W = \{(U_i, R_i)\}$. Then for given $\epsilon > 0$ there exists a τ_1 and a c such that $P(W \in \mathcal{R}) > 1 - \epsilon$ where $\mathcal{R}$ is the set of W such that

(i) $(N+1) \beta U_i \leqslant R_i \leqslant \dfrac{(N+1) U_i}{\beta}$ and

(ii) $\sup_i \left\{ \sqrt{N} h_2(\beta U_i) \left| \dfrac{R_i}{N+1} - U_i \right|; \; U_i \leqslant \tau_1 \right\} \leqslant c,$

when $W \in \mathcal{R}$, we have, since $(g, J) \in \mathcal{S}$,

$$(6.8) \qquad |g(U_i) J'(u)| \leqslant h_1(\beta U_i) h_2(\beta U_i)$$

for every u lying between U_i and $\dfrac{R_i}{N+1}$. Hence, with probability exceeding $1 - \epsilon$, $S_N(J, g, \tau)$ given by (6.3) satisfies

$$(6.9) \qquad S_N(J, g, \tau) \leqslant \frac{c}{N} \sum_{U_i \leqslant \tau} h_1(\beta U_i) = \frac{c}{N} \sum_{i=1}^{N} h_1(\beta U_i) I(U_i \leqslant \tau)$$

where $I(a \leqslant b) = 1$ when $a \leqslant b$ and zero elsewhere. Now, applying Markov inequality we obtain

$$\text{P}(S_N(J, g, \tau) > \epsilon) \leqslant \frac{c}{\epsilon} [\text{E} h_1(\beta U) I(U \leqslant \tau)] =$$

(6.10)

$$= \frac{c}{\epsilon} \int_0^\tau h_1(\beta u)\, du = \frac{c}{\beta \epsilon} \int_0^{\frac{\tau}{\beta}} h_1(v)\, dv$$

and there exists a $\tau_0 \leqslant \tau_1$ such that $c \int_0^{\frac{\tau}{\beta}} h_1(v)\, dv \leqslant \epsilon^2 \beta$. ∎

The tails of the statistic T_N is not equal to $\Delta_N^*(J, g, \tau)$; however it is given by

$$\Delta_N(J, g, \tau) = \sqrt{N} \left\{ \int_A g(F) J\left(\frac{N}{N+1} F_N\right) - \int_A g(F) J(F)\, dF \right\} =$$

(6.11)

$$= \Delta_N^* + \sqrt{N} \int_A g(F) J(F)\, d(F_N - F).$$

Thus $\Delta_N - \Delta_N^*$ is a normalized sum having zero expectation and a variance bounded by $[c^*(\tau)]^4$ given by (3.4). Thus, for every $\epsilon > 0$ there exists a τ_0 such that $\text{P}(|\Delta_N - \Delta_N^*| > \epsilon) < \epsilon$ for every $(g, J) \in \mathscr{S}$ and $\tau \leqslant \tau_0$. In other words, we have the following corollary.

Corollary 6.1.1. *Let* $A = (0, \tau] \cup [1 - \tau, 1)$ *and let* $\Delta_N(J, g, \tau)$ *be given by* (6.11). *Then, for every* $\epsilon > 0$ *there is a number* $\tau_0 > 0$ *such that* $\tau \leqslant \tau_0$ *and* $(g, J) \in \mathscr{S}$ *imply that*

$$(6.12) \qquad \text{P}(|\Delta_N(J, g, \tau)| > \epsilon) < \epsilon$$

for every N *and every* F.

7. THE ASYMPTOTIC BEHAVIOR OF THE STATISTIC T_N

Consider the pair of functions $(g, J) \in \mathcal{S}$. We consider functions J defined by the indefinite integrals of the form $J(x) = \int_{\frac{1}{2}}^{x} J'(\xi) \, d\xi$. Then, we have the following theorem.

Theorem 7.1. *Let* $(g, J) \in \mathcal{S}$ *and let*

$$(7.1) \qquad T_N = \sqrt{N} \left\{ \int g(F) J \left(\frac{N}{N+1} F_N \right) dF_N - \int g(F) J(F) \, dF \right\}.$$

Let P_N *be the distribution of* T_N *and* Q_N *be the normal distribution having zero expectation and variance* $\sigma_N^2(F, J, g)$. *Then, for every* $\epsilon > 0$ *there is an* $N(\epsilon)$ *such that* $N \geqslant N(\epsilon)$ *implies that*

$$|P_N\{(-\infty, y)\} - Q_N\{(-\infty, y)\}| < \epsilon$$

for every y, *every* $(g, J) \in \mathcal{S}$ *and every* F. *On sets where* σ_N^2 *is bounded away from zero,*

$$|P_N\{(-\infty, y]\} - Q_N\{(-\infty, y]\}|$$

may be replaced by

$$\sup_{y} |P_N\{(-\infty, y]\} - Q_N\{(-\infty, y]\}|.$$

Proof. Due to Proposition 6.1 and especially Corollary 6.1.1, for given $\epsilon > 0$ there is a number $\tau > 0$ and an $N_0(\epsilon)$ such that if $A = (0, \tau] \cup [1 - \tau, 1)$ and if T_N^τ is the expression

$$(7.2) \qquad T_N^\tau = \sqrt{N} \left\{ \int_A g(F) J \left(\frac{N}{N+1} F_N \right) dF_N - \int_A g(F) J(F) \, dF \right\}$$

then $P\{|T_N^\tau| > \epsilon\} < \epsilon$ for every $(g, J) \in \mathcal{S}$, and every F. Hence, it would suffice to prove the theorem for the truncated T_N on the interval $[\tau, 1 - \tau]$. Towards this one can expand T_N as

$$(7.3) \qquad T_N = B_N + C_{1,N} + C_{2,N}$$

where

(7.4) $\quad B_N = \sqrt{N} \int g(F)J(F)d(F_N - F) + \sqrt{N} \int g(F)J'(F)(F_N - F)\, dF,$

(7.5) $\quad C_{1,N} = \sqrt{N} \int g(F)\Big[J\Big(\dfrac{N}{N+1} F_N\Big) - J(F)\Big] d(F_N - F)$

and

(7.6) $\quad C_{2,N} = \sqrt{N} \int g(F)\Big[J\Big(\dfrac{N}{N+1} F_N\Big) - J(F) - (F_N - F)J'(F)\Big] dF.$

B_N is precisely the term studied in Proposition 3.1. Thus, it is sufficient to show that $C_{1,N}$ and $C_{2,N}$ tend to zero in probability for every $(g, J) \in \mathscr{S}$. However, when $(g, J) \in \mathscr{S}$, $C_{1,N}$ and $C_{2,N}$ tend to zero in probability because of Propositions 5.1 and 5.2.

Remark 7.1.1. Theorem 7.1 holds for every $(g, J) \in \mathscr{S}_0$ since $\mathscr{S}_0 \subset \mathscr{S}$.

Remark 7.1.2. In the following case, the uniformity asserted in Theorem 7.1 may be of interest. Suppose that for each N, $F = \Psi\Big[\dfrac{x - \theta}{\beta}\Big]$ where Ψ admits a density, with $\beta > 0$. If $\theta_\nu \to \theta_0$ and $\beta_\nu \to \beta_0 \neq 0$, the density of $\Psi\Big[\dfrac{x - \theta_\nu}{\beta_\nu}\Big]$ converges in measure to that of $\Psi\Big[\dfrac{x - \theta_0}{\beta_0}\Big]$. The convergence of $|P_N - Q_N|$ to 0 asserted in Theorem 7.1 is, therefore, uniform for $(g, J) \in \mathscr{S}$, and every bounded set of values $[\theta, \ln \beta]$. Besides, if for each integer N the value of (θ, β) is (θ_N, β_N), and $\theta_N \to 0$ and $\beta_N \to 1$, then the Kolmogorov distance $\sup |P_N - Q_N|$ also tends to zero uniformly for any set $S \subset \mathscr{S}$, such that $(g, J) \in S$ implies that $\infty > \lim_{N \to \infty} \sigma_N^2 \geq \alpha > 0$.

In some cases, the weights J_N may be expectations of suitable order statistics. In this connection, we introduce the additional notation and state some theorems of interest.

Let $\xi_{N,k}$ denote the k-th smallest order statistic in a random sample of size N from the uniform distribution on $[0, 1]$. For any function J let $\bar{J}_N$ be the function defined on $(0, 1)$ as follows. If $y = \dfrac{k}{N+1}$ $(k = 1, 2, \ldots, N)$, let

$$(7.7) \qquad \bar{J}_N(y) = \mathsf{E}J(\xi_{N,k}) = \int_0^1 J(x)\beta_N(x, k)\, dx$$

where

$$(7.8) \qquad \beta_N(x, k) = \frac{N!}{(k-1)!(N-k)!} x^{k-1}(1-x)^{N-k} \qquad (0 < x < 1),$$

is the density of $\xi_{N,k}$. Complete the definition of $\bar{J}_N$ by interpolating linearly between successive values, $\{\frac{k}{N+1}, \frac{k+1}{N+1}\}$ and leaving $\bar{J}_N$ constant below $\frac{1}{N+1}$ and above $\frac{N}{N+1}$. Then, we have

Lemma 7.1. *Let there be a constant K and a δ, $0 < \delta < \frac{1}{2}$ such that $|J'(x)| \leq K[x(1-x)]^{-\alpha+\delta}$ for some $\alpha \leq \frac{3}{2}$. Then, there exists a constant K_1 and an N_0 such that $N \geq N_0$ implies*

$$(7.9) \qquad |\bar{J}_N'(x)| \leq K_1 [x(1-x)]^{-\alpha+\delta}.$$

Furthermore, if $\{J_\nu'\}$ is a sequence such that J_ν' converges to J' in Lebesgue measure and $|J_\nu'| \leq K[x(1-x)]^{-\alpha+\delta}$, for all ν, then $\bar{J}_{\nu,N}' - J_\nu'$ converges to zero in Lebesgue measure as $N \to \infty$ uniformly in the index ν.

Proof. Replace the exponent $\frac{3}{2}$ by α in Lemma 1 of G L R [9]. The entire proof is valid verbatum for any α in the place of the exponent $\frac{3}{2}$.

Next, for each integer N let J_N' be a nonnegative function such that

$$0 \leq J_N'(x) \leq K[x(1-x)]^{-\frac{3}{2}+\alpha+\delta}$$

for some $0 < \delta < \frac{1}{4}$, $\alpha \leq \frac{3}{2}$ and fixed $K < \infty$. Also, let $|g(x)| \leq K[x(1-x)]^{-\alpha+\delta}$. Let J_N be an integral of J_N' and let

$$\bar{J}_N\left(\frac{i}{N+1}\right) = \mathsf{E}[J_N(\xi_{N,i})].$$

Then, we have the following theorem.

Theorem 7.2. *Let*

(7.10) $|g(x)| \leqslant K[x(1-x)]^{-\alpha+\delta}$ *and* $|J_N'| \leqslant K[x(1-x)]^{-\frac{3}{2}+\alpha+\delta}$

for some $0 < \delta < \frac{1}{4}$, $\alpha \leqslant \frac{3}{2}$ *and fixed* $K < \infty$. *Let* J_N' *converge in Lebesgue measure to a limit* J'. *Also, let* T_N *be the expression.*

(7.11) $T_N = \sqrt{N} \left\{ \int g(F) \bar{J}_N \left(\frac{N}{N+1} F_N \right) dF_N - \int g(F) J_N(F) dF \right\}.$

Let P_N *be the distribution of* T_N *and let* Q_N *be the noremal approximation of Theorem 7.1. Then, the conclusion of Theorem 7.1 holds uniformly in* F *as* $N \to \infty$.

Proof. Let

$$\tilde{T}_N = \sqrt{N} \left\{ \int g(F) \bar{J}_N \left(\frac{N}{N+1} F_N \right) dF_N - \int g(F) \bar{J}_N(F) dF \right\}.$$

From Lemma 7.1 we have that (g, J_N) belongs to $\mathcal{S}_0$, thus satisfying the conditions of Theorem 7.1. Besides J_N' converges in measure to a limit J', hence, $\bar{J}_N'$ converges to J' because of Lemma 7.1. From this, it is clear that the proof of Theorem 7.2 would be complete if T_N is replaced by $\tilde{T}_N$. Hence, it will be sufficient to bound the difference

$$T_N - \tilde{T}_N = \sqrt{N} \left\{ \int g(F) [\bar{J}_N(F) - J_N(F)] dF \right\}$$

which is smaller than

$$\int \sqrt{N} |g(x)| |[\bar{J}_N(x) - J_N(x)]| dx$$

which is again smaller than

$$K \int \sqrt{N} [x(1-x)]^{-\alpha+\delta} |\bar{J}_N(x) - J_N(x)| dx.$$

In the last integral, the terms in the absolute value sign are linear functions of J_N' and J_N' can be split into positive and negative parts and each of these into parts supported by $\left[0, \frac{1}{2}\right]$ and $\left[\frac{1}{2}, 1\right]$ respectively. Thus, it will be sufficient to prove the result under the assumption:

$J_N(x) = \int_x^1 h_N(\xi)\, d\xi$ where $0 < h_N(\xi) \leqslant x^{-\frac{3}{2} + \alpha + \delta}$, and $h_N(x) = 0$ for $x > \frac{1}{2}$. The parts relative to the interval $\left[\frac{1}{2}, 1\right]$ are handled by changing x to $1 - x$. With this definition, we have

(7.12) $$\sqrt{N} \int_0^x \xi^{-\alpha + \delta} J_N(\xi)\, d\xi \leqslant K\sqrt{N}\, x^{-\frac{1}{2} + \delta},$$

a similar inequality holding for $\bar{J}_N$. Thus, it will be sufficient to prove that

$$\sqrt{N} \int_{\epsilon_N}^1 [x(1 - x)]^{-\alpha + \delta} |\bar{J}_N(x) - J_N(x)|\, dx$$

tends to zero for $N\ \epsilon_N = N^{\frac{\delta}{2}}$. Now following the steps in the proof of Theorem 2 of GLR [9] with the modification that $J_N'(x) \leqslant Kx^{-\frac{3}{2} + \alpha + \delta}$, one can show that

$$\sqrt{N}[x(1 - x)]^{-\alpha + \delta} |\bar{J}_N(x) - J_N(x)|$$

is bounded by an integrable function. That it tends to zero in measure follows from Lemma 7.1. This completes the proof of Theorem 7.2.

As an application, let us mention the following corollary.

Corollary 7.2.1. *Let k be a fixed integer and let a_j $(j = 1, 2, \ldots, k)$ be bounded constants. Let*

(7.13) $$J_N\left(\frac{i}{N + 1}\right) = \sum_{j=1}^k a_j\, \mathrm{E}(W_{N,i}^j)$$

where $W_{N,i}$ is the i-th snallest order statistic in a random sample of size N drawn from a population whose distribution function is the inverse of a function V. If (g, V^j) satisfies (7.10) for $j = 1, 2, \ldots, k$, then the functions J_N given by (7.13) satisfy the conditions of Theorem 7.2.

Proof. The proof follows from the linearity of the transformation $J \to \bar{J}_N$ which is used for defining the functions occurring in Theorem 7.2.

8. CASE OF RANDOM SAMPLE SIZE

One is faced with random sample sizes in sequential estimation, in dealing with truncated samples, or when the sample size is determined by an independent random process. Hence, it is of interest to study the asymptotic normality of T_N when N is random. Anscombe [1] has shown that

(i) if a sequence of statistics $\{Y_N\}$ is asymptotically normal when suitably standardized and when N is nonstochastic and large,

(ii) Y_N is uniformly continuous in probability

then Y_N is also asymptotically normal when N is random. Anscombe [1] exhibits this property of uniform continuity in probability when Y_N is a sum of i.i.d. random variables. Rényi [18] provides a direct and elegant proof of the asymptotic normality of a sum of a random number of i.i.d. random variables. However, when a statistic is not a sum of i.i.d. random variables, it is not easy to establish the uniform continuity in probability of the statistic. Recently under a different set of regularity assumptions on g and J, Govindarajulu and Mason [10] have shown that if a statistic can be represented as a standardized sum of i.i.d. variables plus a remainder term which goes to zero almost surely, then it is uniformly continuous in probability. However, such a strong representation for T_N may require further regularity assumptions on (g, J). In the following we shall establish the asymptotic normality of T_N when N is random without additional assumptions on (g, J). Let N^* be a nonstochastic integer tending to infinity. We assume that $N = N(N^*)$ and $\dfrac{N}{N^*} \to 1$ in probability as $N^* \to \infty$.

Then, we have the following lemmas.

Lemma 8.1. Let $h_2(u) \in \mathcal{U}_2$. Then

$$\sqrt{\overline{N}}\, h_2(u)[F_N(F^{-1}(u)) - u]$$

is bounded in probability as $N^* \to \infty$.

Proof. Notice that Lemma 8.1 is an extension of Lemma 4.1 to the random sample size case. Lemma 8.1 follows from Theorem 1 of P y k e and S h o r a c k [15] who prove that

$$\sqrt{N}\, h_2(u)[F_N(F^{-1}(u) - u)]$$

converges weakly to $h_2(u)W(u)$ where $W(u)$ denotes the tied down Wiener process on $(0, 1)$ since

$$\{\sup_u \sqrt{N^*}\, h_2(u)[F_{N^*}(F^{-1}(u)) - u\}$$

is bounded in probability.

Lemma 8.2. *If* $\{Z_i\}$ *denotes a sequence of i.i.d. random variables having finite expectation* θ, *then* $\dfrac{1}{N}\sum\limits_{i=1}^{N} Z_i$ *converges in probability to* θ *as* $N^* \to \infty$.

Proof. This is an extension of Khintchine's weak law of large numbers to the random sample size case. For a proof of this, see for instance R é v é s z ([16], Theorem 10.2) or G o v i n d a r a j u l u ([8], Theorem 4.1).

Now, we are ready to state and prove the main theorem of this section.

Theorem 8.1. *With the above notation and assumptions we have*

$$\lim_{N^* \to \infty}\ P[T_N \le x] = \Phi(x)$$

for every x, *for every* F *and every* $(g, J) \in \mathscr{S}$ *where* T_N *is given by equation* (7.1).

Proof. The proof for the negligability of the tails of the statistic given in Section 6 can be repeated verbatum upon using Lemmas 8.1 and 8.2. The higher order random term C_{1N} tends to zero in probability because

$$Z_N(x) - Z_N(x)) \to 0$$

in probability after applying Lemma 8.1. Furthermore, the term C_{2N}

$$- 337 -$$

tends to zero in probability since

$$\sqrt{N}\,(F_N(F^{-1}(u)) - u)$$

is also bounded in probability as $N^* \to \infty$.

Remark 8.1. For the purpose of this section, the c.d.f. $F(x)$ and the function J may depend on $N.^*$. The conclusion of Theorem 8.1 is still valid. The analogues of Theorem 7.2 and Corollary 7.2.1 also hold when N is random. Remark 7.1.2 is applicable to Theorem 8.1 and the analogues of Theorem 7.2 and Corollary 7.2.1.

9. *c*-SAMPLE CASE

Let a rank sum statistic be a linear combination of expected values of order statistics in a random sample, the coefficients being random. If in the rank sum statistic, the expected values of order statistics are replaced by the order statistics themselves, the resultant statistic, called the randomized statistic, under mild assumptions, will be asymptotically equivalent to the rank sum statistic (for instance, see B e l l and D o k s u m [2], Theorem 2.5). Hence, it is of sufficient interest to extend the results of Section 7 to *c*-sample situations. We need the following additional notation. Let $X_{j,k}$ $(k = 1, 2, \ldots, n_j)$ denote the random sample drawn from a population having for its c.d.f. $F^{(j)}(x)$ $(j = 1, 2, \ldots, c)$. We assume that $F^{(1)}, F^{(2)}, \ldots, F^{(c)}$ have no common points of discontinuity. Also, let $F_{n_j}^{(j)}(x)$ denote the empirical c.d.f. based on the random sample $X_{j,1}, X_{j,2}, \ldots, X_{j,n_j}$ $(j = 1, 2, \ldots, c)$. Further, let

$$(9.1) \qquad m = n_1 + n_2 + \ldots + n_c,$$

$$(9.2) \qquad H(x) = \sum_{j=1}^{c} \lambda_j F^{(j)}(x)$$

$$(9.3) \qquad H_m(x) = \sum_{j=1}^{c} \lambda_j F_{n_j}^{(j)}(x)$$

where

(9.4) $\lambda_j = \dfrac{n_j}{m}$ and $0 < \lambda_0 \leqslant \lambda_j \leqslant 1 - \lambda_0 < 1$ $(j = 1, 2, \ldots, c)$

for some fixed $\lambda_0 \leqslant \dfrac{1}{c}$. Thus, $H_m(x)$ denotes the combined sample c.d.f. and $H(x)$ denotes the combined population c.d.f. Although $F^{(j)}(x)$ may depend on m, this will be suppressed for the sake of notational ease. Now define the class of statistics of interest by

(9.5) $n_j T_{m,j} = \sum\limits_{i=1}^{m} E_{m,i,j} g_j^*(W_{m,i}) Z_{m,i,j}$ $(j = 1, 2, \ldots, c)$,

where $E_{m,i,j}$ are some given constants, $Z_{m,i,j} = 1$ if the i-th smallest observation in the combined ordered sample belongs to j-th sample, and zero otherwise, $W_{m,1} < W_{m,2} < \ldots < W_{m,m}$ denotes the combined ordered sample and $g_j^*(x)$ are some continuous functions. Alternatively, one can write (9.5) as

(9.6) $T_{m,j} = \int\limits_{-\infty}^{\infty} g_j(H) J_{m,j}\left(\dfrac{m}{m+1} H_m\right) dF_{n_j}^{(j)}(x)$ $(j = 1, 2, \ldots, c)$,

where $J_{m,j}\left(\dfrac{i}{m+1}\right) = E_{m,i,j}$ and $g_j(H) = g_j^*(x)$. Again, as in Section 2 we can conveniently extend the domain of definition of $J_{m,j}$ to $(0, 1)$. It should be noted that unless otherwise stated, n_j $(j = 1, 2, \ldots, c)$ are nonrandom. Also, for the sake of simplicity, let us drop the subscript m in $J_{m,j}$.

Consider the random variable

(9.7)
$$B_m^{(j)} = \sqrt{m} \int g_j(H) J_j(H) d(F_{n_j}^{(j)}(x) - F^{(j)}(x)) +$$
$$+ \sqrt{m} \int g_j(H) J_j'(H)(H_m - H) dF^{(j)}(x).$$

Since

$$H_m - H = \sum_{k=1}^{c} \lambda_k (F_{n_k}^{(k)} - F^{(k)}),$$

one can rewrite $B_m^{(j)}$ as a linear combination of independent sums of independent and identically distributed random variables having expectation zero. The variance $B_m^{(i)}$ is denoted by $\sigma_{m,j}^2[F^{(j)}, J_j, g_j]$ where

$$\sigma_{m,j}^2 = \frac{1}{\lambda_j}\left[\int g_j^2(H)J_j^2(H)dF^{(j)}(x) - \left\{\int g_j(H)J_j(H)dF^{(j)}(x)\right\}^2\right] +$$

$$+ 2\iint\limits_{x<y} H(x)[1 - H(y)]g_j(H(x))g_j(H(y)) \times$$

$$\times J_j'(H(x))J_j'(H(y))dF^{(j)}(x)dF^{(j)}(y) +$$

(9.8)

$$+ 2\iint\limits_{x<y} [1 - F^{(j)}(y)]g_j(H(x))g_j(H(y)) \times$$

$$\times J_j(H(x))J_j'(H(y))dF^{(j)}(x)dF^{(j)}(y) +$$

$$- 2\iint\limits_{x>y} F^{(j)}(y)g_j(H(x))g_j(H(y)) \times$$

$$\times J_j(H(x))J_j'(H(y))dF^{(j)}(x)dF^{(j)}(y).$$

Since $F^{(j)}(x) \leqslant \dfrac{H(x)}{\lambda_0}$, it follows from Lemmas 3.1 and 3.2 that $\sigma_{m,j}^2$ is also finite for $j = 1, 2, \ldots, c$. Also, when

$$F^{(1)} = \ldots = F^{(c)}, \quad g_1 = \ldots = g_c,$$

$\sigma_{m,j}^2$ takes a simpler form.

Further, the covariance of $B_m^{(i)}$ and $B_m^{(j)}$ $(i \neq j = 1, 2, \ldots, c)$ is denoted by $\sigma_{m,i,j}[F^{(i)}, F^{(j)}, J_i, J_j, g_i, g_j]$ where straightforward computations yield

(9.9)

$$\sigma_{m,i,j} = \iint\limits_{x<y} [1 - F^{(i)}(y)]g_i(H(x))g_j(H(y)) \times$$

$$\times J_i(H(x))J_j'(H(y))dF^{(i)}(x)dF^{(j)}(y) -$$

$$- \iint\limits_{x>y} F^{(i)}(y)g_i(H(x))g_j(H(y)) \times$$

$$\times J_i(H(x))J_j'(H(y))dF^{(i)}(x)dF^{(j)}(y) +$$

$$+ \iint_{x>y} [1 - F^{(j)}(x)] g_i(H(x)) g_j(H(y)) \times$$

$$\times J_i'(H(x)) J_j(H(y)) dF^{(i)}(x) dF^{(j)}(y) -$$

$$- \iint_{x<y} F^{(j)}(x) g_i(H(x)) g_j(H(y)) \times$$

$$\times J_i'(H(x)) J_j(H(y)) dF^{(i)}(x) dF^{(j)}(y) +$$

$$+ \iint_{x<y} H(x)[1 - H(y)] g_i(H(x)) g_j(H(x)) \times$$

$$\times J_i'(H(x)) J_j'(H(y)) dF^{(i)}(x) dF^{(j)}(y) +$$

$$+ \iint_{x>y} H(y)[1 - H(x)] g_i(H(x)) g_j(H(y)) \times$$

$$\times J_i'(H(x)) J_j'(H(y)) dF^{(i)}(x) dF^{(j)}(y).$$

Notice that all the integrals in (9.9) are finite because $\operatorname{var} B_m^{(j)} < \infty$ for $j = 1, \ldots, c$.

Here, H, without loss of generality, can be assumed to be uniform on $[0, 1]$, after making $F^{(1)}, \ldots, F^{(c)}$ continuous. Further $\varphi_j = \dfrac{dF^{(j)}}{dH}$, is a bounded measurable function, $0 \leqslant \varphi_j \leqslant \dfrac{1}{\lambda_0}$ $(j = 1, 2, \ldots, c)$.

Then, we have the following proposition concerning the joint asymptotic normality of $B_m^{(j)}$ $(j = 1, 2, \ldots, c)$.

Proposition 9.1. *Let* P_m *be the joint distribution of* $B_m^{(j)}$ $(j = 1, 2, \ldots, c)$ *and* Q_m *be the c-variate normal distribution which has expectation zero and variance-covariance matrix* $\Gamma_m = ((\sigma_{m,i,j}))$ *where the* $\sigma_{m,i,j}$ *are given by (9.8) and (9.9). Then, for every* $\epsilon > 0$ *there exists* $m(\epsilon)$ *such that* $m \geqslant m(\epsilon)$ *implies*

$$|P_m(x) - Q_m(x)| < \epsilon$$

for every $x = (x_1, \ldots, x_c)$, *every* $(g_j, J_j) \in \mathscr{S}$, *and all* F_j $(j = 1, 2, \ldots, c)$. *In addition, there is* $m(\epsilon, a)$ *such that* $m > m(\epsilon, a)$ *and*

- 341 -

$\sigma^2_{m,j} \geqslant a$ $(j = 1, 2, \ldots, c)$ *implies that*

$$\sup_x |P_m(x) - Q_m(x)| < \epsilon$$

for every $(g_j, J_j) \in \mathscr{S}$, *and for all* F_j $(j = 1, 2, \ldots, c)$ *where*

$$P_m(x) = P_m\{(-\infty, x_1], \ldots, (-\infty, x_c]\}$$

and

$$Q_m(x) = Q_m\{(-\infty, x_1], \ldots, (-\infty, x_c]\}.$$

Proof. The first statement follows from the usual central limit theorem. The second statement follows from the first by considering the random vector $\dfrac{B^{(j)}_m}{\sigma_{m,j}}$ $(j = 1, 2, \ldots, c)$. This completes the proof of Proposition 9.1.

Now, we are ready to state a theorem giving the sufficient conditions for the joint asymptotic normality of $T_{m,j}$ $(j = 1, 2, \ldots, c)$ defined in (9.6).

Theorem 9.1. *Let* $(g_j, J_j) \in \mathscr{S}$, $(j = 1, 2, \ldots, c)$ *and let*

$$
(9.10) \quad
\begin{aligned}
T^*_{m,j} = \sqrt{m} \Big\{ &\int g_j(H) J_j\Big(\frac{m}{m+1} H_m\Big) dF^{(j)}_{n_j}(x) - \\
&- \int g_j(H) J_j(H) dF^{(j)}(x) \Big\}
\end{aligned}
$$

for $j = 1, 2, \ldots, c$. *Let* P_m *be the joint distribution of* $T^*_{m,j}$ $(j = 1, 2, \ldots, c)$ *and let* Q_m *be the* c*-variate normal distribution having expectation zero and variance-covariance matrix* $\Gamma_m = ((\sigma_{m,i,j}))$ *where* $\sigma^2_{m,j} = \sigma_{m,i,j}$ *and* $\sigma_{m,i,j}$ *are given by (9.8) and (9.9). Then, for every* $\epsilon > 0$ *there is an* $m(\epsilon)$ *such that* $m \geqslant m(\epsilon)$ *implies*

$$|P_m(x) - Q_m(x)| < \epsilon$$

for all x, *every* $(g_j, J_j) \in \mathscr{S}$ $(j = 1, 2, \ldots, c)$ *and every* 2c*-ple* $(F^{(1)}, \ldots, F^{(c)}, \lambda_1, \ldots, \lambda_c)$.

On sets such that $\sigma^2_{m,j}$ stays bounded away from zero, then

$|P_m(x) - Q_m(x)| < \epsilon$ *may be replaced by the Kolmogorov vertical distance.*

Proof. First one can expand $T^*_{m,j}$ as

(9.11) $T^*_{m,j} = B^{(j)}_m + C^{(j)}_{1,m} + C^{(j)}_{2,m}$

where

(9.12)
$$B^{(j)}_m = \sqrt{m} \int g_j(H)J_j(H)d(F^{(j)}_{n_j}(x) - F^{(j)}(x)) +$$
$$+ \sqrt{m} \int g_j(H)J'_j(H)(H_m - H)dF^{(j)}(x)$$

(9.13)
$$C^{(j)}_{1,m} = \sqrt{m} \int g_j(H) \times$$
$$\times \left[J_j\left(\frac{m}{m+1} H_m \right) - J_j(H) \right] d(F^{(j)}_{n_j}(x) - F^{(j)}(x))$$

and

(9.14)
$$C^{(j)}_{2,m} = \sqrt{m} \int g_j(H) \times$$
$$\times \left[J_j\left(\frac{m}{m+1} H_m \right) - J_j(H) - (H_m - H)J'_j(H) \right] dF^{(j)}(x).$$

$B^{(j)}_m$ is precisely the random term considered in (9.7). Hence, it suffices to show that $C^{(j)}_{1,m}$ and $C^{(j)}_{2,m}$ tend to zero in probability for all $(g_j, J_j) \in \mathcal{S}$. Toward this one can proceed as in Thorem 7.1. First, as in Proposition 6.1 and Corollary 6.1.1 one can show that the integral

(9.15)
$$\Delta_m(J_j, g_j, \tau) = \sqrt{m} \left\{ \int_A g_j(H)J_j\left(\frac{m}{m+1} H_m \right) dF^{(j)}_{n_j} - \right.$$
$$\left. - \int_A g_j(H)J_j(H)dF^{(j)} \right\}$$

with $A = (0, \tau] \cup [1 - \tau, 1)$ can be made arbitrarily small by choosing τ sufficiently small. Now, using arguments similar to those employed in Proposition 5.1 and Corollary 5.1.1, one can show that $C^{(j)}_{1,m}$ can be neglected. Further, since $F^{(j)}(x) \leqslant \dfrac{H(x)}{\lambda_0}$, it readily follows from Proposition 5.2 and Corollary 5.2.1 that $C^{(j)}_{2,m}$ can also be neglected.

The desired result readily follows from Proposition 9.1.

To proceed along with above steps, the necessary tools are: that, without loss of generality H can be assumed to be uniform on $[0, 1]$ and the appropriate versions of Lemmas 4.1, 4.2 and 4.3. Also since, $\varphi_j = \dfrac{dF^{(j)}}{dH}$ are bounded by $\dfrac{1}{\lambda_0}$, none of the deviations $\sqrt{m}\,[F_{n_j}^{(j)} - F^{(j)}]$ oscillates much on intervals, the probability content of which with respect H are small.

Remark 9.1. Remark 7.1.2 concerning the location and scale families of distributions is still applicable to Theorem 9.1. The class of pairs of functions (g, J_N) considered in Theorem 7.2 and Corollary 7.2.1 are also applicable to Theorem 9.1 since most of the arguments used in proving Theorem 9.1 do not depend on the observations but only on properties of order statistics from the uniform distribution.

Remark 9.2. If the subsample sizes are random, let there exist a nonstochastic sequence of positive integers N^* such that $\dfrac{m}{N^*}$ tends to unity in probability and $\dfrac{n_j}{N^*} \to p_j$ $(j = 1, \ldots, c)$ in probability as $N^* \to \infty$. We also assume that there exists a $p_0 \leqslant \dfrac{1}{c}$ such that $0 < p_0 \leqslant p_j \leqslant 1 - p_0 < 1$ for $j = 1, \ldots, c$. If

$$H^*(x) = \sum_{j=1}^{c} p_j F^{(j)}(x)$$

then, from Theorem 1 of P y k e and S h o r a c k [15] we have that

$$\sup_x \sqrt{N}\, h_2(F^{(j)}(x))[F_{n_j}^{(j)}(x) - F^{(j)}(x)]$$

is bounded in probability $(j = 1, \ldots, c)$. Using this result, we can establish the asymptotic joint normality of $(T_1, \ldots, T_c)$. The variances and covariances given by equations (9.8) and (9.9) need to be modified by replacing H by H^*, λ_j by p_j $(j = 1, \ldots, c)$.

10. APPLICATIONS

In this section we give some statistics the large-sample properties of which follow from the theorems developed in Sections 7, 8 and 9. Jung's [12] asymptotically optimal estimate of the scale parameter σ of a normal distribution is given by

$$(10.1) \qquad \hat{\sigma} = \frac{1}{N} \sum_{i=1}^{N} \Phi^{-1}\left(\frac{i}{N+1}\right) X_{i,N}.$$

In our notation for this statistic $g^*(x) = x$, $g(u) = \Phi^{-1}(u)$ and $J(u) = \Phi^{-1}(u)$ where Φ denotes the standard normal distribution function. The asymptotic normality of $\hat{\sigma}$ follows from Theorem 7.1, since $g(u) \sim$
$\sim \sqrt{-2 \ln u(1-u)}$ and $J'(u) \sim \dfrac{1}{u(1-u)\sqrt{-2 \ln u(1-u)}}$ and consequently $g(u)J'(u) \sim \dfrac{1}{u(1-u)}$ as $u \to 0$ or 1.

As a modification for small samples, let us consider σ as an estimate of $\hat{\sigma}$ which is given by

$$(10.2) \qquad \hat{\hat{\sigma}} = \frac{1}{N} \sum_{i=1}^{N} \mu_{i,N} X_{i,N}$$

where $\mu_{i,N}$ are expected values of standard normal order statistics in a random size N. The asymptotic normality of $\hat{\hat{\sigma}}$ follows from Theorem 7.2. The statistics (10.1) and (10.2) can be extended to c-sample situations and their large-sample properties studied. Also, the asymptotic normality of the Bell — Doksum [2] statistics arising from two or more samples easily follows from Theorem 9.1.

For another application set $F^{(1)} = \ldots = F^{(c)} = F$, $n_1 = \ldots = n_c = n$ and $c = 2$. Then the bivariate asymptotic normality of $T^*_{m,1}$ and $T^*_{m,2}$ follows from Theorem 9.1. One can also obtain the asymptotic expression for the variance-covariance matrix of the bivariate statistic. Such pairs of statistics and the ratio of $T^*_{m,1}$ to $T^*_{m,2}$ have been considered by Sarkadi [20] in goodness of fit problems.

Concluding remarks. Bickel [3], Govindarajulu [7], Moore [14], Stigler [23], [24], Shorack [22], Ruymgaart and Van

Zuijlen [19] and Mason [13] have also made significant contributions to the asymptotic normality of linear combination of functions of order statistics. Some of them do not seem to require the absolute continuity of the *J*-function. However, they work with smooth score functions and impose some other regularity conditions. Stigler [24], Ruymgaart and Van Zuijlen [19] and Mason [13] consider the case when the *X*'s are independent but are not necessarily identically distributed. Wellner [28] and Van Zwet [27] have obtained strong laws for linear functions of order statistics.

Acknowledgements. I thank Professor Herman Chernoff for arousing my interest in developing asymptotic theory for linear combinations of order statistics. My thanks are also due to Professor Lucien LeCam for some of his inspiring ideas shared by me in GLR [9] and extensively used in this investigation. I also thank Professor K. Sarkadi for a careful reading of the manuscript.

REFERENCES

[1] F.J. Anscombe, Large-sample theory of sequential estimation, *Proc. Cambridge Philos. Soc.*, 48 (1952), 600-607.

[2] C.B. Bell − K.A. Doksum, Some new distribution-free statistics, *Ann. Math. Statist.*, 36 (1965), 203-214.

[3] P. Bickel, Some contributions to the theory of order statistics, *Proc. Fifth Berkeley Symposium on Mathematical Statistics and Probability*. (University of California Press at Berkeley and Los Angeles), 1 (1967), 575-591.

[4] G. Blom, *Statistical estimates and transformed beta variables*, John Wiley and Sons − Almquist and Wicksell, 1958.

[5] H. Chernoff − J.T. Gastwirth − M.V. Johns Jr., Asymptotic distribution of linear combinations of functions of order statistics with applications to estimation, *Ann. Math. Statist.*, 38 (1967), 52-72.

[5a] F. Eicker − M.L. Puri, On linear combinations of order statistics, *Essays in Probability and Statistics,* Ogawa Volume, (Ed. Ikeda et. al), Shinko Tsusho Co. Ltd. Tokyo, 433-450.

[5b] M. Ghosh, On the representation of linear functions of order statistics, *Sankhyā A,* 34 (1972), 349-356.

[6] Z. Govindarajulu, Asymptotic normality of linear functions of order statistics in one and multi-samples, Technical Report No. 216 U.S. Air Force Office Grant No. AF-AFOSR-741-65, (University of California, Berkeley, California) 1965, 1-26.

[7] Z. Govindarajulu, Asymptotic normality of linear combinations of functions of order statistics II, *Proc. Nat. Acad. Sci.,* 59 (1968), 713-719.

[8] Z. Govindarajulu, Weak laws of large numbers for dependent summands, *Indian Journal of Pure and Applied Mathematics,* 6 (9) (1975), 1013-1022.

[9] Z. Govindarajulu − L. LeCam − M. Raghavachari, Generalizations of theorems of Chernoff and Savage on the asymptotic normality of tests statistics, *Proceedings of the Fifth Berkeley Symposium on Mathematical Statistics and Probability,* (University of California Press at Berkeley and Los Angeles), 1 (1967), 609-638.

[10] Z. Govindarajulu − D.M. Mason, A strong representation for linear combinations of order statistics with application to fixed-width confidence intervals for location and scale parameters, Technical Report No. 154, Department of Statistics, University of Kentucky, 1980, 1-24.

[11] J. Hajek, Brownian Bridge, Lecture delivered at the IMS Summer Institute held at Michigan State University, (unpublished).

[12] J. Jung, On linear estimates defined by a continuous weight function, *Arkiv. For Matematik,* 3 (15) (1955), 199-209.

[13]   D.M. Mason, Asymptotic normality of linear combinations of order statistics with a smooth score function, Department of Statistics, University of Kentucky Technical Report No. 149 (1979), 1-18.

[14]   D.S. Moore, An elementary proof of asymptotic normality of linear functions of order statistics, *Ann. Math. Statist.*, 39 (1968), 263-265.

[15]   R. Pyke – G.R. Shorack, Weak convergence and a Chernoff – Savage theorem for random sample sizes, *Ann. Math. Statist.*, 39 (1968), 1675-1685.

[16]   P. Révész, *The laws of large numbers,* Academic Press, New York, 1968, 149-150.

[17]   A. Rényi, On the theory of order statistics, *Acta Math. Acad. Sci. Hungar.*, 4 (1953), 191-232.

[18]   A. Rényi, On the asymptotic distribution of the sum of a random number of independent random variables, *Acta. Mathematica,* 8 (1957), 193-199.

[19]   F.H. Ruymgaart – M.C.A. Van Zuijlen, Asymptotic normality of linear combinations of order statistics in the non i.i.d. case, *Proc. Koninklijke Nederlandse Akademie van Wettenschappen Ser. A.*, Mathematical Sciences, 80 (5) (1977), 432-447.

[20]   K. Sarkadi, The asymptotic distribution of certain goodness of fit test statistics, *The First Pannon Symposium on Math. Statist.*, (Ed. W. Wertz), Lecture Notes in Statistics, Vol. 8., Springer-Verlag, New York, 1981, 245-253.

[21]   P.K. Sen, On Wiener process embedding for linear combinations of order statistics, *Sankhyā Ser. A,* 39 (1977), 138-143.

[22]   G.R. Shorack, Functions of order statistics, *Ann. Math. Statist.*, 43 (1972), 412-427.

[23]   S.M. Stigler, Linear functions of order statistics, *Ann. Math. Statist.*, 40 (1969), 770-788.

[24]   S.M. Stigler, Linear functions of order statistics with smooth weight functions, *Ann. Statist.*, 2 (1974), 676-693.

[25]   J.W. Tukey, The future of data analysis, *Ann. Math. Statist.*, 33 (1962), 1-67.

[26]   J.W. Tukey – D.H. McLaughlin, Less vulnerable confidence and significance procedures for location based on a single sample: Trimming/Winsorization, *Sankhyā Ser. A*, 35 (1963), 331-352.

[27]   W.R. Van Zwet, A strong law for linear functions of order statistics, *Ann. Prob.*, 8 (1980), 986-990.

[28]   J.A. Wellner, A Glivenko – Cantelli theorem and strong laws of large numbers for functions of order statistics, *Ann. Statist.*, 5 (1977), 473-480.

Z. Govindarajulu

Dept. of Stat., Univ. of Kentucky, Lexington, Kentucky 40506, USA.

TRANSFORMATION OF VARIABLES VERSUS NONPARAMETRICS

H. GRIMM

The problem of transformation of data before analysing them with well known methods of analysis of variance and of their use instead of nonparametric methods is evergreen in the controversy between parametrics and nonparametrics. The question for the practical statistician is, whether he should give up the familiar and simple models, such as analysis of variance of regression or the high-reaching covariance matrices in multivariate analysis.

I remember the discussion inaugurated by J u v a n c z [23] at the Symposium on Quantitative Methods in Pharmacology, held in Leyden 1960. Since my first survey on transformations of variables (G r i m m [18]), some new aspects and ideas appeared and some further reviews were given by L i e n e r t [24], B o x and C o x [11], T h ö n i [29], H o y l e [22], B i c k e l and D o k s u m [8] and C a r r o l l [12]. A survey of evaluation of data from well-designed experiments is given in Figure 1.

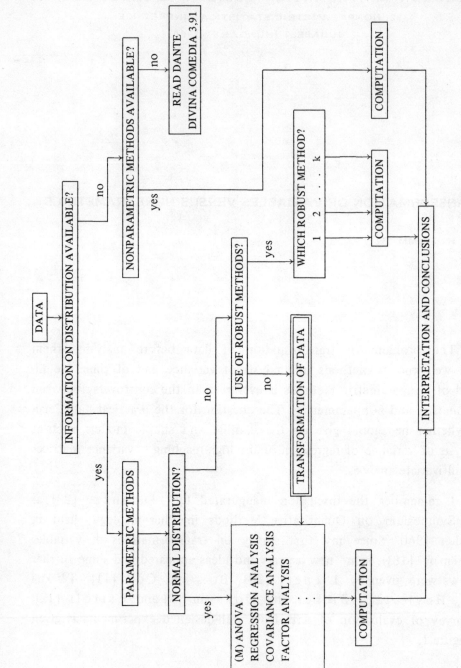

Figure 1. Diagram for evaluation of data

* Abandon hope all ye who enter here.

In a really new research project we start from ignorance. So we do not know anything about distributions and parameters and we begin with a nonparametric method. After some experience we have more information about these and we are able to apply more powerful (parametric) methods. In the final report of our research we mention the methods of the last phase, so one has the impression that parametric methods have been applied in most cases and we forget that nonparametric methods have been very useful in initial phases of research.

Often the distributions are nonnormal. Then we use some transformations. This is the topic of our following considerations.

The main goals for transforming data are to achieve the fulfilment of the following assumptions for a linear model (if the original data fail to fulfil them):

(1) additivity, i.e. the main effects combine linearly to explain the observations,

(2) homogeneity (= constant variance), i.e. the observations are assumed to have a constant variance about their (varying) means, so that the variance is independent of the expected value of the observations and of the sample size,

(3) normality, i.e. the observations are assumed to have a normal distribution.

D r a p e r and H u n t e r [14] indicated methods of finding transformations which simultaneously satisfy (more or less well) these three assumptions. This can usually be done as long as one does not press to hard for any one of them. But it may even happen that through a transformation which leads to one of the desiderata some other desirable characteristic gets lost. A lot of methods have been derived for testing additivity, homogeneity and normality. Very often simple graphical methods indicate an appropriate transformation (G r i m m [17], O r d [27]. Tests for additivity were developed by T u k e y [30], M o o r e and T u k e y [25], A n s c o m b e [3], [4], E l s t o n [15], A n s c o m b e and T u k e y [5]. The examination of residuals plays an important role. For tests for homogeneity of variance, we refer to Bickel [7] and Carroll [12]. Recent contributions to the testing for normality are to be found in Atkinson [6], Carroll [12], Pearson, d'Agostino and Brown [28], Vasicek [33].

The power transformation family studied by Box and Cox [11] has often been used. They assumed that for some λ

$$\vec{Y}^{(\lambda)} = \vec{X}\vec{\beta} + \vec{\epsilon},$$

where $\vec{X}$ is an $n \times p$ full rank matrix, $\vec{\beta}$ is a $p \times 1$ vector of constants, the errors $\vec{\epsilon}$ are normal with covariance matrix $\sigma^2 \vec{I}$, and individual elements $Y_i^{(\lambda)}$ are defined by

$$Y_i^{(\lambda)} = \begin{cases} \dfrac{Y_i^\lambda - 1}{\lambda} & \text{for} \quad \lambda \neq 0 \\[2mm] \log Y_i & \text{for} \quad \lambda = 0. \end{cases}$$

They propose maximum likelihood estimates for λ and $\vec{\beta}$ when F is the normal distribution. There are numerous alternative methods as well as proposals for testing hypotheses of the form $H_0 : \lambda = \lambda_0$ (Hinkley [21], Andrews [1], Carroll [12].) Carroll [12] studies the testing problem via Monte-Carlo, by allowing F to be nonnormal; he approximated a problem with outliers and found that the chance of rejecting the null hypothesis erroneously can be very high, indeed. Bickel and Doksum [8] show that $\hat{\beta}(\hat{\lambda})$ and $\hat{\lambda}$ are unstable and highly correlated, a problem similar in nature to that of multicollinearity in regression. This problem caused some controversy, e.g. the following: Should one make unconditional inference about the regression parameter in the correct but unknown scale or a conditional inference for an appropriately defined "regression parameter" in an estimated scale?

Tukey [32] considered mixtures of the standard normal distribution Φ with small fractions of contaminating normals:

$$F(x) = (1 - \epsilon)\Phi(x) + \epsilon\Phi\left(\frac{x}{k}\right) \quad (0 \leqslant \epsilon \leqslant 0,1).$$

As the main aim of robust estimation, we can consider building in safeguards against unexpectedly large amounts of gross errors, putting a bound on the influence of hidden contamination and questionable outliers, isolating clear outliers for separate treatment (if desired), and still being nearly optimal at the strict parametric model.

Unfortunately the definition of robustness is not unique in literature and was rather vaguely from the start by B o x and A n d e r s o n [10]. But most scientists of robustness tend to parametrics.

A large part of classical statistics is nonrobust (in varying degree). Nonrobust (= sensitive to small changes of the model) are e.g. the arithmetic mean and the method of least squares, standard deviation, mean deviation and range, covariance and correlation. Robust are: the same methods with sensible "looking at the data" setting aside outliers etc.

Robust (under some restrictions) are also estimators combined with reasonable formal rejection procedures (A n d r e w s et al. [2]). The median is in some sense the most robust estimator of location. C a r r o l l [12] and B i c k e l and D o k s u m [8] suggest that transformation methods based on normal theory of likelihhod have suspecious statistical properties. In addition, there are situations in which data transformations to achieve symmetry are counterproductive or even not possible. In regression it might be conjectured that asymmetry has different effects on the intercept and slopes or contrasts. In fact the asymptotic properties of certain estimates of parameters in the linear model are not grossly influenced by asymmetry, and when there is interest in these parameters, it makes little sense to invest great effort in transformations.

In linear regression, a similar conclusion holds true for the intercept. Theoretical and Monte-Carlo results show that slope parameters are influenced only neglegibly by skewness. Jackknifing for the intercept does not work as well here, the variance estimates now tending to be slightly too large.

A border area of robustness which contains no (at least no simple) parametric model and yet the possibility of robustifying is smoothing. After polynomials were deemed unreasonable, spline smoothing (with piecewise polynomials) is now very popular, and yet this may also only be an intermediate step towards truely "smooth" curves. But in all cases including freehand smoothing, one would like to ignore single outliers or short stretches of outliers which do not fit into a nice smooth curve starting from "running medians". Tukey has developed intricate resistant smoothing schemes whose properties are not easy to understand; a recent Princeton thesis (Velleman [34]) contains a survey of such methods.

Various problems arise in presenting results in the original variables when the analysis has been performed on the transformed variable. E.g. it is interesting to note that the results of Neyman and Scott [26] indicate a theoretical difficulty with the reciprocal transformation $z = \frac{1}{x}$. In this case usually $\Theta = E(x) = E\left(\frac{1}{z}\right)$ does not exist, so it cannot be estimated. However Box [9] has suggested defining of a pseudo expectation Θ' of x as

$$\hat{\Theta}' = \widehat{PE}(x) = \lim_{\epsilon \to 0} \int_\epsilon^\infty \frac{1}{z} E^{-1}\left(\frac{1}{\sigma\sqrt{2\pi}} \exp\left\{-\frac{(z-u)^2}{2\sigma^2}\right\}\right) dz.$$

In general the formula of Neyman and Scott [26] is complicated, but they point out that the commonly used transformations (as $\sqrt{\ }$, lg, angular and hyperbolic) are linked by a simple differential equation concerning their inverse functions.

If we consider the situation dialectically we may set the contrasts in the following way:

$$parametrics \longleftrightarrow nonparametrics$$
$$optimist \longleftrightarrow pessimist.$$

Pessimists were lately called data analysts (Bickel [7]). They use models only provisionally and are always ready to change their view of the structure of the data in light of the values they see.

Fortunately purists of either persuasion are rare. Scientific research is an interative and complex process. Extreme positions in theory are necessary for promotion of knowledge; for practical application we have to find reasonable compromises.

Figure 1 is rather general. It might be laid out in more detail for different aimes as estimation, testing of hypotheses and prediction. Another difficult problem is data transformation in the presence of outliers.

REFERENCES

[1] D.F. Andrews, A note on the selection of data transformations, *Biometrika*, 58 (1971), 249-254.

[2] D.F. Andrews – P.J. Bickel – F.R. Hampel – P.J. Huber – W.H. Rogers – J.W. Tukey, *Robust estimation of location: survey and advances,* Princeton Univ. Press, Princeton, 1972.

[3] F.J. Anscombe, Rejection of outliers, *Technometrics*, 2 (1962), 123-166.

[4] F.J. Anscombe, Examination of residuals, *Proc. Fourth Berkeley Symp. Math. Stat. Prob.*, 1 (1961), 1-36.

[5] F.J. Anscombe – J.W. Tukey, Analysis of residuals, *Technometrics*, 5 (1963), 141-160.

[6] A.C. Atkinson, Testing transformations to normality, *JRSS*, B 35 (1973), 473-479.

[7] P.J. Bickel, Another look at robustness: a review of reviews and some new developments, *Scand. J. Statist.*, 3 (1976), 145-168.

[8] P.J. Bickel – K.A. Doksum, An analysis to transformations revisited, Dept. of Statist. Univ. of California, Berkeley, 1978, unpublished).

[9] M.J. Box, Bias in nonlinear estimation, *JRSS*, B 33 (1971), 171-201.

[10] G.E.P. B o x – S.E.L. A n d e r s o n , Permutation theory in the derivation of robust criteria and the study of departures from assumption, *JRSS, B* 17 (1955), 1-26.

[11] G.E.P. B o x – D.R. C o x , An analysis of transformations, *J. of Roy. Stat. Soc., B* 26 (1964), 211-234, 244-252.

[12] R.J. C a r r o l l , A robust method for testing transformations to achieve approximate normality, 1978, (to appear in *JRSS, B*).

[13] R.J. C a r r o l l – D. R u p p e r t , On prediction and the power transformation family, Univ. of North Carolina, 1980 (unpublished).

[14] N.R. D r a p e r – W.G. H u n t e r , Transformations: some examples revisited, *Technometrics,* 11 (1969), 23-40.

[15] R.E. E l s t o n , On additivity in the analysis of variance, *Biometrics,* 17 (1961), 209-219.

[16] H. G r i m m , Transformation von Zufallsvariablen, *Biometrische Zeitschrift,* 2 (1960), 164-183.

[17] H. G r i m m , Graphical methods for the determination of type and parameters of some discrete distributions, 1970 (= pp. 194-206 in G.P. Patil (edit.): *Random coupts in models and structures,* The Pennsylvania State Univ. Press, University Park of London, 1970).

[18] F.R. H a m p e l , Robust estimation: a condensed partial survey, *Z. Wahrscheinlichkeitstheorie verw. Geb.,* 27 (1973), 87-104.

[19] F.R. H a m p e l , The influence curve and its role in robust estimation, *JASA,* 69 (1974), 383-393.

[20] F.R. H a m p e l , Modern trends in the theory of robustness, *Math. Operationsforschung Ser. Statist.,* 9 (1978), 425-442.

[21] D.V. H i n k l e y , On power transformations to symmetry, *Biometrika,* 62 (1975), 101-111.

[22] M.H. Hoyle, Transformations – an introduction and a bibliography, *Intern. Statist. Rev.*, 41 (1973), 203-223.

[23] I. Juvancz, Contraindications of non-parametric methods in medical experimentation, 1961 (= pp. 159-171 in H.D.E. Jonge (edit.): *Quantitative methods in Pharmacology*, North-Holland Publ. Comp. Amsterdam).

[24] G. Lienert, Über die Anwendung der Variablen-Transformationen in der Psychologie, *Biom. Zeitschr.*, 4 (1962), 145-181.

[25] P.G. Moore – J.W. Tukey, Answer to query 112, *Biometrics*, 10 (1954), 562-568.

[26] J. Neyman – E.L. Scott, Correlation for bias introduced by a transform of variables, *AMST*, 31 (1960), 643-655.

[27] J.K. Ord, *Families of frequency distributions*, Griffin, London, 1972.

[28] E.S. Pearson – R.B. d'Agostino – K.O. Bowman, Tests for departure from normality. Comparison of powers. *Biometrika*, 64 (1977), 231-246.

[29] H. Thöni, Transformations of variables used in the analysis of experimental and observational data: a review. Techn. Rep. No. 7, Statist. Labor. Iowa State Univ., Ames, Iowa, 1967.

[30] J.W. Tukey, One degree of freedom for non-additivity, *Biometrics*, 5 (1949), 232-242.

[31] J.W. Tukey, Conclusions vs. decisions, *Technometrics*, 2 (1960), 1-11.

[32] J.W. Tukey, *Exploratory data analysis*, Addison-Wesley Publishing Comp., 1977.

[33] O. Vasicek, A test for normality based on sample entropy, *Journ. Royal Statist. Soc. B*, 38 (1976), 54-59.

[34]   P.F. Velleman,   Robust nonlinear data smoothness: theory, definitions and applications,   Unpubl. dissertation, Princeton Univ., 1975.

H. Grimm

DDR-6900 Jena, Grete-Unrein–Str. 2.

COLLOQUIA MATHEMATICA SOCIETATIS JÁNOS BOLYAI
32. NONPARAMETRIC STATISTICAL INFERENCE,
BUDAPEST (HUNGARY), 1980.

NONPARAMETRIC PROCEDURES IN MULTIPLE DECISIONS*
(RANKING AND SELECTION PROCEDURES)

S.S. GUPTA — G.C. MCDONALD

1. INTRODUCTION TO SELECTION AND RANKING PROCEDURES

A common problem faced by an experimenter is one of comparing several categories or populations. There may be, for example, different varieties of a grain, different competing manufacturing processes for an industrial product, or different drugs (treatments) for a specific disease. In other words, we have $k (\geqslant 2)$ populations and each population is characterized by the value of a parameter of interest θ, which may be, in the example of drugs, an appropriate measure of the effectiveness of a drug. The classical approach to this problem is to test the homogeneity (null) hypothesis $H_0 : \theta_1 = \ldots = \theta_k$, where $\theta_1, \ldots, \theta_k$ are the values of the parameter for these populations. In the case of normal populations with means $\theta_1, \ldots, \theta_k$ and a common variance σ^2, the test can be carried out using the F-ratio of the analysis of variance.

*The writing of this paper was supported in part by the Office of Naval Research Contract N00014-75-C-0455 at Purdue University.

The above classical approach is inadequate and does not answer a frequently encountered experimenter's question, namely, how to identify the best category? In fact, the method of least significant differences based on *t*-tests has been used in the past to detect differences between the average yields of different varieties and thereby choose the 'best' variety. But this method (and others related to it) is indirect and does not easily provide an overall probability of a correct selection. Also the multiple comparison techniques developed largely by T u k e y and S c h e f f é arose from the desire to draw inference about the populations when the homogeneity hypothesis is rejected.

Selection and ranking procedures

The formulation of a *k*-sample problem as a multiple decision problem enables the experimenter to answer questions regarding the best category. The formulation of multiple decision procedures in the framework of selection and ranking procedures has been accomplished generally using either the indifference zone approach or the (random sized) subset selection approach. The former approach was introduced by B e c h h o f e r [2]. Substantial contribution to the early and subsequent developments in the subset selection theory has been made by G u p t a starting from his work in 1956 ([6]). For more details about the numerous contributions and the related bibliography, reference should be made to a recently published book by G u p t a and P a n c h a p a k e s a n [15]. This monograph discusses both approaches.

A brief description of the two approaches

B e c h h o f e r [2] considered the problem of ranking *k* normal means. In order to explain the basic formulation, consider the problem of selecting the population with the largest mean from *k* normal populations with unknown means μ_i $(i = 1, \ldots, k)$, and a common known variance σ^2. Let $\bar{x}_i$ $(i = 1, \ldots, k)$, denote the means of independent samples of size *n* from these populations. The 'natural' procedure (which can be shown to have optimum properties) will be to select the population that yields the largest $\bar{x}_i$. The experimenter would, of course, need a guarantee that this procedure will pick the population with the largest μ_i with a probability not less than a specified level P^*. For the problem to be meaningful P^* lies between $\frac{1}{k}$ and 1. Since we do not know the true configuration of the μ_i, we look for the *least favorable configuration* (LFC) for which the probability of a *correct selection*, P(CS), will be at least P^*. Since the LFC is given by $\mu_1 = \ldots = \mu_k$, the probability guarantee cannot be met whatever be the sample size n.

A natural modification is to insist on the minimum probability guarantee whenever the best population is sufficiently superior to the next best. In other words, the experimenter specifies a positive constant Δ^* and requires that the P(CS) is at least P^* whenever $\mu_{[k]} - \mu_{[k-1]} \geq \Delta^*$, where $\mu_{[1]} \leq \ldots \leq \mu_{[k]}$ denote the ordered means. Now the minimization of P(CS) is over the part Ω_{Δ^*} of the parameter space in which $\mu_{[k]} - \mu_{[k-1]} \geq \Delta^*$. The complement of Ω_{Δ^*} is called the indifference zone for the obvious reason. The LFC in Ω_{Δ^*} is given by $\mu_{[1]} = \ldots = \mu_{[k-1]} = \mu_{[k]} - \Delta^*$. The problem then reduces to determining the minimum sample size required in order to have P(CS) $\geq P^*$ for the LFC.

B e c h h o f e r's formulation can be generalized from that described above. His general ranking problem includes, for example, selection of the t best populations.

In the subset selection approach, the goal is to select a nonempty subset of the populations so as to include the best population. Here the size of the selected subset is random and is determined by the observations themselves. In the case of normal populations with unknown means $\mu_1, \ldots, \mu_k$, and a common variance σ^2, the rule proposed by G u p t a [6] selects the population that yields $\bar{x}_i$ if and only if

$$\bar{x}_i \geq \max_{1 \leq j \leq k} \bar{x}_j - \frac{d\sigma}{\sqrt{n}},$$

where $d = d(k, P^*) > 0$ is determined so that the P(CS) is at least P^*. Here a correct selection is selection of any subset that includes the population with the largest μ_i. Thus, the LFC is with regard to the whole parameter space Ω. Under this formulation, for given k and P^* we determine d. The rule explicitly involves n. In general, the rule will involve a constant which depends on k, P^* and n. The performance of a subset selection procedure is studied by evaluating the expected subset size and its supremum over the parameter space Ω.

Nonparametric techniques in multiple decision theory

In the present paper, we describe selection and ranking (ordering) procedures which are nonparametric or distribution-free. Such procedures have the desirable property that their applicabbility is valid under relatively mild assumptions regarding the underlying population(s) from which the data are obtained. Although the importance of nonparametric methods as a significant branch of modern statistics is recognized by statisticians, modern nonparametric techniques are usually restricted to hypothesis testing, point estimators, confidence intervals, and multiple comparison procedures. Other recent advances in nonparametric tests can be found in Hollander and Wolfe [17] and Lehmann [20]. The development of nonparametric methods for multiple decision procedures is important in statistical research. The present paper deals with selection procedures with special emphasis on the subset selection approach related to the largest unknown parameter. Analogous procedures (with proper modifications) are available for the selection in terms of the smallest parameter.

In Section 2, we discuss procedures based on the ranks in the combined sample. Section 3 deals with bounds on the probability of a correct selection associated with the first two procedures $R_1(G)$ and $R_2(G)$ of Section 2. In Section 4, the exact and asymptotic distribution of the (appropriate) statistic based on rank sums is discussed. In Section 5, we provide comparisons between R_1 and R_3 and certain parametric procedures in terms of their asymptotic relative efficiencies. Selection procedures based on pairwise ranks are discussed briefly in Section 6. Section 7 deals with selection procedures based on vector ranks. In Section 8, procedures based on Hodges — Lehmann estimators are discussed.

2. PROCEDURES BASED ON COMBINED RANKS

Let $\pi_1, \ldots, \pi_k$ be $k\ (\geqslant 2)$ independent populations. The associated random variables $X_{ij}\ (j = 1, \ldots, n_i, i = 1, \ldots, k)$, are assumed independent and to have a continuous distribution $F_{\theta_i}(x)$, where θ_i belongs to some interval Θ on the real line. Suppose $F_\theta(x)$ is a *stochastically increasing* (SI) family of distributions, i.e. if θ_1 is less than θ_2, then $F_{\theta_1}(x)$ and $F_{\theta_2}(x)$ are distinct and $F_{\theta_2}(x) \leqslant F_{\theta_1}(x)$ for all x.

Examples of such families of distributions are:

(1) any location parameter family, i.e. $F_\theta(x) = F(x - \theta)$;

(2) any scale parameter family, i.e. $F_\theta(x) = F\left(\frac{x}{\theta}\right),\ x > 0,\ \theta > 0$;

(3) any family of distribution functions whose densities possess the monotone likelihood ratio (or TP_2) property.

Let R_{ij} denote the rank of the observation x_{ij} in the combined sample; i.e. if there are exactly r observations less than x_{ij} then $R_{ij} = = r + 1$. These ranks are well-defined with probability one, since the random variables are assumed to have a continuous distribution. Let $Z(1) \leqslant Z(2) \leqslant \ldots \leqslant Z(N)$ denote an ordered sample of size $N = \sum_{i=1}^{k} n_i$ from any continuous distribution G, such that

$$- \infty < a(r) \equiv \mathsf{E}\,(Z(r) \mid G) < \infty \qquad (r = 1, \ldots, N).$$

With each of the random variables X_{ij} associate the number $a(R_{ij})$ and define

$$(2.1) \qquad H_i = \frac{1}{n_i} \sum_{j=1}^{n_i} a(R_{ij}) \qquad (i = 1, \ldots, k).$$

Using the quantities H_i, Gupta and McDonald [11] have defined procedures for selecting a subset of the k populations. Letting $\theta_{[i]}$ denote the i-th smallest unknown parameter, we have

$$(2.2) \qquad F_{\theta_{[1]}}(x) \geqslant F_{\theta_{[2]}}(x) \geqslant \ldots \geqslant F_{\theta_{[k]}}(x),\ \forall x.$$

The population whose associated random variables have the distribution $F_{\theta_{[k]}}(x)$ will be called the best population. In case several populations possess the largest parameter value $\theta_{[k]}$, one of them is tagged at random and called the best. A 'Correct Selection' (CS) is said to occur if and only if the best population is included in the selected subset. In the usual subset selection problem one wishes to select a subset such that the probability is at least equal to a preassigned constant P^* $\left(\frac{1}{k} < P^* < 1\right)$ that the selected subset includes the best population. Mathematically, for a given selection rule R,

(2.3) $\inf_{\Omega} P(CS \mid R) \geqslant P^*$,

where

(2.4) $\Omega = \{\vec{\theta} = (\theta_1, \ldots, \theta_k): \theta_i \in \Theta, \ i = 1, 2, \ldots, k\}$.

The following three classes of selection procedures, which choose a subset of the k given populations, and which depend on the given distribution G, have been considered:

(2.5)
$$R_1(G): \text{ Select } \pi_i \text{ if and only if } H_i \geqslant \max_{1 \leqslant j \leqslant k} H_j - d$$
$$(i = 1, \ldots, k, \ d \geqslant 0),$$

(2.6)
$$R_2(G): \text{ Select } \pi_i \text{ if and only if } H_i \geqslant \frac{1}{c} \max_{1 \leqslant j \leqslant k} H_j$$
$$(i = 1, \ldots, k, \ c \geqslant 1),$$

(2.7)
$$R_3(G): \text{ Select } \pi_i \text{ if and only if } H_i \geqslant D$$
$$(i = 1, \ldots, k, \ -\infty < D < \infty).$$

It should be noted that rules $R_1(G), R_2(G)$ and $R_3(G)$ are equivalent if $k = 2$. The procedures $R_1(G)$ (and their randomized analogs) have been suggested by B a r t l e t t and G o v i n d a r a j u l u [1] for continuous distributions differing by a location parameter. The procedure $R_2(G)$ will be studied in this paper only for the case where $H_i \geqslant 0$ for all i. The constants d and c are usually chosen to be as small as possible, D as large as possible while satisfying the probability requirement (2.3). The number of populations included in the selected subset is a random variable which takes values 1 to k inclusive for rules $R_1(G)$ and $R_2(G)$. The subset chosen by rule $R_3(G)$, however, could possibly be empty. This aspect will be addressed further at the end of Section 3.

It has been shown by Gupta and McDonald [11] that the infimum of $P(CS \mid R_i(G))$ $(i = 1, 2, 3)$ over Ω is attained for $\theta \in \Omega_k = \{\vec{\theta}: \theta_{[k-1]} = \theta_{[k]}\}$. This shows that for $k = 2$ the infimum occurs at an equiparameter configuration.

For $k \geqslant 3$ the least favorable configuration (LFC) is not given by the equiparameter configuration for $R_1(G)$ and $R_2(G)$ as can be seen from the counterexamples of Rizvi and Woodworth [35]. The counterexample distribution is a mixture of two distinct uniform random variables and is established for P^* near 1.

For the procedure $R_3(G)$ we can say more about the infimum of the probability of a correct selection. The LFC is given by the equiparameter configuration and so

$$\inf_{\Omega} P(CS \mid R_3(G)) = \inf_{\Omega_0} P(CS \mid R_3(G)),$$

where $\Omega_0 = \{\vec{\theta} \in \Omega: \theta_{[1]} = \ldots = \theta_{[k]}\}$.

These selection rules are called distribution-free (or nonparametric) if the constants required for implementation are computed from $P(CS \mid R_i(G)) = P^*$ for $\theta \in \Omega_0$. In this case the probability does not depend on the common parameter value and on the underlying distribution functions. The probability computation is based on a random assignment of rank scores.

3. BOUNDS ON $P(CS \mid R_i(G))$ $(i = 1, 2)$

Since the exact LFC for the selection rules $R_1(G)$ and $R_2(G)$ is unknown for $k > 2$, it is useful to have bounds for the probabilities of correct selection. We will assume $n_i = n$ $(i = 1, \ldots, k)$. First consider rule $R_1(G)$. Since

$$(3.1) \qquad \frac{1}{k-1} \sum_{j=1}^{k-1} H_{(j)} \leqslant \max_{1 \leqslant j \leqslant k-1} H_{(j)} \leqslant \frac{1}{n} \sum_{r=N-n+1}^{N} a(r),$$

and

$$\sum_{j=1}^{k} H_{(j)} = \frac{A}{n},$$

where $\displaystyle\sum_{r=1}^{N} a(r) \equiv A$, it follows that

$$(3.2) \qquad \inf_{\Omega} P(H_{(k)} \geqslant v) \leqslant \inf_{\Omega} P(CS \,|\, R_1(G)) \leqslant \inf_{\Omega} P(H_{(k)} \geqslant u).$$

The quantities u and v are defined by

$$(3.3) \qquad u \equiv u(d, k, n) = \frac{[A - nd(k-1)]}{nk},$$

and

$$(3.4) \qquad v = v(d, k, n) = \frac{1}{n} \sum_{r=N-n+1}^{N} a(r) - d.$$

For the rule $R_2(G)$, we get a similar expression:

$$(3.5) \qquad \inf_{\Omega} P(H_{(k)} \geqslant v') \leqslant \inf_{\Omega} P(CS \,|\, R_2(G)) \leqslant \inf_{\Omega} P(H_{(k)} \geqslant u'),$$

where

$$(3.6) \qquad u' \equiv u'(d, k, n) = \frac{1}{n} \frac{A}{[1 + c(k-1)]}$$

and

$$(3.7) \qquad v' \equiv v'(d, k, n) = \frac{1}{nc} \sum_{r=N-n+1}^{N} a(r).$$

The important point to note from the inequalities (3.2) and (3.5) is that the infima over Ω of expressions of the form $P(H_{(k)} \geqslant K)$ are attained when $\theta_{[1]} = \ldots = \theta_{[k]}$.

For the particular case when $a(r) = r$, $nH_i \equiv T_i$, the rank sum statistic associated with π_i. Denoting $R_i(G)$ by R_i in this case, the infimum of $P(CS \,|\, R_i)$ can be related to the Mann – Whitney statistic. If U is the

Mann — Whitney statistic associated with samples of size n and $(k-1)n$ taken from identically distributed populations, then

(3.8) $\inf_{\Omega} P(CS \mid R_1) \geqslant P(U \geqslant nd).$

A similar expression can be derived for R_2. The Mann — Whitney U-statistic has been tabulated by M i l t o n [30] among others.

Since $\sum_{j=1}^{k} H_{(j)} = \dfrac{A}{n}$, we see that

(3.9) $\max_{1 \leqslant j \leqslant k} H_j \geqslant \dfrac{A}{nk}.$

Hence, a sufficient, but not necessary, condition for the selection rule $R_3(G)$ to select a nonempty subset is that P^* be sufficiently large so that

(3.10) $D \leqslant \dfrac{A}{N}.$

For large n, this sufficiency condition for rule $R_3(G)$ is satisfied if $P^* > \dfrac{1}{2}$. For rule R_3, i.e. when $a(r) = r$, the condition is $D \leqslant \dfrac{N+1}{2}$. As an example, with $k = 3$, $n = 5$ the sufficient condition $D \leqslant 8$ is satisfied for $P^* \geqslant 0.523$ and for such values a nonempty subset will be selected.

The evaluation of the constants $D = D(k, n, P^*)$ for the rule R_3 can be effected as follows:

(3.11) $P^* \leqslant P(T_i \geqslant Dn) = P\left(U \leqslant n^2\left(k - \dfrac{1}{2}\right) - n\left(D - \dfrac{1}{2}\right)\right),$

where now we consider all populations identically distributed. Hence, Dn is the largest integer satisfying the inequality (3.10).

4. THE EXACT AND ASYMPTOTIC DISTRIBUTION OF $\max\limits_{1 \leqslant j \leqslant k} T_j - T_i$ FOR IDENTICALLY DISTRIBUTED POPULATIONS

In this section the random variables X_{ij} $(j = 1, \ldots, n_i; \ i = 1, \ldots, k)$ are assumed independent identically distributed with a continuous distribution $F(x)$. In this case the H_i are exchangeable random variables if $n_i = n$ $(i = 1, \ldots, k)$. It should be noted that in a slippage-type configuration, the constants required to implement rules $R_i(G)$ $(i = 1, 2, 3)$, are determined from the basic probability requirement $P(CS \mid R_i(G)) \geqslant P^*$ calculated with identically distributed populations. In the case $a(R_{ij}) = R_{ij}$ the procedures $R_i(G)$ reduce to the rank sum procedures R_i $(i = 1, 2, 3)$. The distribution of the statistic $\max\limits_{1 \leqslant j \leqslant k} T_j - T_1$, both exact and asymptotic, is somewhat easier to obtain than the corresponding distribution of the statistic $\max\limits_{1 \leqslant j \leqslant k} \dfrac{T_j}{T_1}$. For some results concerning the latter statistic, see McDonald [22]. Our concern here will be the former which is tantamount to considering rule R_1. Corresponding to rule R_3 is the statistic T_1, the distribution of which has been well-treated elsewhere in the Mann — Whitney format.

Gupta and McDonald [11] have tabulated the quantity $P(T_1 \geqslant \max\limits_{1 \leqslant j \leqslant 3} T_j - m)$ for $n = 2(1)5$ and $m = 0, 1, \ldots, 2n^2$ (which covers the entire distribution). Asymptotically (as $n \to \infty$), they show

$$P(T_k \geqslant \max\limits_{1 \leqslant j \leqslant k} T_j - m) \to \int_{-\infty}^{\infty} \left(\Phi\left(x + \frac{m}{z}\right) \right)^{k-1} \varphi(x)\, dx$$

(4.1)

$$(m \geqslant 0),$$

where $\Phi(\cdot)$ and $\varphi(\cdot)$ are the cumulative distribution function and density of a standard normal random variable, respectively, and

(4.2) $\quad z = z(n, k) = n\sqrt{\dfrac{k(nk + 1)}{12}}.$

Integrals of the type

(4.3) $$\int_{-\infty}^{\infty} (\Phi(x + h\sqrt{2}))^{k-1} \varphi(x)\, dx = P^*$$

have been considered in several publications. The h quantity appearing in this expression has been tabulated (to 3 dp) by G u p t a [8] in Table I for $k = 2(1)51$ and $P^* = 0.75, 0.90, 0.95, 0.975$ and 0.99. Similar values are provided (to 4 dp) in Table 1 of G u p t a, N a g e l and P a n c h a p a k e s a n [14] for the same P^* and $k = 2(1)11(2)51$. Additional tabulation of h is provided by M i l t o n [29]. In Table IB of Milton's report, the h quantity is tabulated (to 6 dp) for $k = 3(1)10(5)25$ and $P^* = 0.3(0.05)0.95, 0.975, 0.99, 0.995, 0.999, 0.9995$ and 0.9999. In Table II of the same publication P^* values are given (to 8 dp) for $h = 0(0.05)5.15$ for all the previously mentioned values of k. Thus, this asymptotic value can be obtained from a variety of sources and can be applied directly to very large data sets – up to 51 populations and any (large) sample size.

Matching the right hand side of (4.1) with (4.3) yields an asymptotic approximation to $m = nd$ given by

(4.4) $$\tilde{m} = hn \sqrt{\frac{k(nk + 1)}{6}},$$

h being the appropriate solution to (4.3) corresponding to the given P^* and k. In the selection rule the smallest integer not less than $\tilde{m}$ should be taken.

Approximations to the constant m for use with R_1

We saw that $\inf P(CS \mid R_1)$ over

$$\Omega' = \{\theta: \theta_{[1]} = \ldots = \theta_{[k-1]} \leqslant \theta_{[k]}\}$$

is attained when $\theta_1 = \ldots = \theta_k$. Suppose we want to evaluate d for which this infimum is at least P^*. Using the rank sum statistics, this means that we want the smallest integer $m = nd$ such that

(4.5) $$P(T_k \geqslant T_{[k]} - m) \geqslant P^*$$

where the T_i are i.i.d. random variables. M c D o n a l d [23] has discussed two methods of approximating the solution. The first method uses the asymptotic $(n \to \infty)$ expression for the probability given by (4.1).

The second approximation is for large P^* (near 1). Suppose $Z_1, \ldots, Z_k$ are $N(0, 1)$ random variables with the correlation matrix Σ. Let

$$(4.6) \qquad P(Z_{[1]} \geqslant - \delta) = P^*.$$

D u d e w i c z [4] has shown that, for large P^* (near 1), an approximation to δ is given by

$$(4.7) \qquad \delta^2 \approx - 2(\log (1 - P^*))$$

in the sense that the ratio tends to 1 as $P^* \to 1$. Using his approximation and the joint asymptotic normal distribution of

$$\left\{ \frac{1}{6} n^2 k(nk + 1) \right\}^{-\frac{1}{2}} (T_k - T_i) \qquad (i = 1, \ldots, k - 1),$$

we obtain the approximation

$$(4.8) \qquad m^2 \approx - \left[n^2 k \frac{(nk + 1)}{3} \right] \log (1 - P^*).$$

One can also obtain this approximation from (4.1) by noting that $\frac{m}{z} \approx$ $\approx \sqrt{2} \Phi^{-1}(P^*)$ as $P^* \to 1$, a result of R i z v i and W o o d w o r t h [35] and using the well-known fact that

$$(4.9) \qquad \Phi^{-1}(P^*) \approx \sqrt{- 2 \log (1 - P^*)} \qquad \text{as} \qquad P^* \to 1.$$

The two approximations have been compared by M c D o n a l d [23] in the case of $P^* = 0.99$ for $k = 2(1)5$, $n = 5(5)25$.

Let $\hat{m}_1$ and $\hat{m}_2$ denote the approximate values of m given by (4.4) and (4.8), respectively. The numerical evaluations of $\hat{m}_1$ and $\hat{m}_2$ show that

(a) $\hat{m}_2 - \hat{m}_1$ increases in n for fixed k, and decreases in k for fixed n,

(b) $\dfrac{\hat{m}_1}{\hat{m}_2}$ increases in k for fixed n, and is constant for fixed k over various values of n,

(c) both approximations are conservative, $\hat{m}_2$ being more so than $\hat{m}_1$.

For $k = 2$, McDonald [23] has analytically shown that $\hat{m}_2 - \hat{m}_1$ is positive and increasing in n, and that $\dfrac{\hat{m}_2}{\hat{m}_1}$ is independent of n.

5. COMPARISONS BETWEEN R_1 AND R_3 AND WITH PARAMETRIC PROCEDURES

Recall that for $k = 2$ the rules $R_i(G)$ $(i = 1, 2, 3)$, are equivalent. For the special case of rank sum statistics based on equal sample sizes, Gupta and McDonald [11] have studied the asymptotic efficiency of R_1 relative to the means procedure of Gupta [6] for normal populations and the efficiency of R_2 relative to the procedure of Gupta [7] for gamma populations both in the case of $k = 2$ populations.

Let π_1 and π_2 be independent normal populations with means θ_0 and $\theta_0 + \theta$ $(\theta \geqslant 0)$ and common unit variance. Let R denote Gupta's means procedure. For both R_1 and R satisfying the P^*-condition, the *asymptotic efficiency of R_1 relative to R* is

$$\text{ARE}(R_1, R; \theta) = \lim_{\epsilon \to 0} \frac{n_R(\epsilon)}{n_{R_1}(\epsilon)},$$

where $n_R(\epsilon)$ and $n_{R_1}(\epsilon)$ are the sample sizes for which $\text{E}(S) - \text{P(CS)} = \epsilon$ for R and R_1, respectively, where $\text{E}(S)$ is the expected subset size. It is shown by Gupta and McDonald that

$$(5.1) \qquad \text{ARE}(R_1, R; \theta) = \left[\frac{2\Phi\left(\dfrac{\theta}{\sqrt{2}}\right) - 1}{2\theta B(\theta)} \right]^2,$$

where

$$B^2(\theta) = \int_{-\infty}^{\infty} \Phi^2(x + \theta)\varphi(x)\,dx - \Phi^2\left(\frac{\theta}{\sqrt{2}}\right).$$

As θ decreases to zero, we see that $\mathrm{ARE}\,(R_1, R; \theta) \to \dfrac{3}{\pi} = 0.9549$.

Some exact calculations for the probabilities of choosing π_1 and π_2 using rank sum procedures can be made using Table C-1 of M i l t o n [31] for $\theta = 0.2(0.2)1.0,\ 1.5, 2.0$ and 3.0. This table tabulates the distribution of the Wilcoxon two-sample statistic under the normal shift alternative specified by θ. As an example, for $k = 2$, $n = 6$ and $P^* = 0.910177$, the rank sum selection rules take the form: select π_i if and only if $T_i \geqslant 31$ $(i = 1, 2)$. If the underlying distributions are normal with means 0 and $\theta = 0.2$ with unit variances, then by summing the appropriate columns in Table C-1 we find $\mathrm{P}(T_1 \geqslant 31) = \mathrm{P}(\text{Choosing } \pi_1) \doteq 0.8465$ and $\mathrm{P}(T_2 \geqslant 31) = \mathrm{P}(\text{Choosing } \pi_2) \doteq 0.9518$.

Let R' denote Gupta's procedure for gamma populations. Let the scale parameters of π_1 and π_2 be θ_0 and $\theta_0\theta$, $\theta > 1$. In this case

$$(5.2) \qquad \mathrm{ARE}\,(R_2, R'; \theta) = \frac{1}{2}\left[\frac{(\theta - 1)}{(\theta + 1)B(\theta)\log\theta}\right]^2,$$

where now

$$B^2(\theta) = 1 - \frac{2}{1 + \theta} + \frac{1}{2\theta + 1} + \frac{\theta}{2 + \theta} - \frac{2\theta^2}{(1 + \theta)^2}.$$

As θ decreases to 1, we have $\mathrm{ARE}\,(R_2, R'; \theta) \to \dfrac{3}{4}$.

In another paper G u p t a and M c D o n a l d [12], have compared the procedures R_1, R_2 and R_3 based on rank sum statistics with a procedure R_m which they proposed for selection from gamma populations in terms of the guaranteed life. Let π_i have the associated density function

$$f(x - \theta_i) = \begin{cases} \dfrac{1}{\lambda\Gamma(r)}\left(\dfrac{x - \theta_i}{\lambda}\right)^{r-1} e^{-\frac{x - \theta_i}{\lambda}}, & x \geqslant \theta_i \\ 0 & \text{elsewhere,} \end{cases}$$

where $r(> 0)$ and $\lambda (> 0)$ are known common parameters. In life-testing problems, the parameter θ is called the guaranteed life time. We can assume with no loss of generality that $\lambda = 1$. Let $Y_i = X_{i[1]}$, be the smallest order statistics based on n independent observations from π_i. It is known that Y_i is complete and sufficient statistic for θ_i. The procedure R_m of Gupta and McDonald for selecting a subset containing the population with the largest guaranteed life is

(5.3) R_m : select π_i if and only if $Y_i \geqslant Y_{[k]} - b$,

where $b = b(k, n, P^*) > 0$ is chosen to satisfy the P^*-requirement. They have shown that

(5.4) $\inf_{\Omega} P(CS \mid R_m) = \int_0^\infty H^{k-1}(x + b)\, dH(x),$

where $H(x)$ is the cumulative distribution function of Y_i when $\theta_i = 0$.

In the special case of $r = 1$, the exponential case, (5.4) reduces to

(5.5) $\inf_{\Omega} P(CS \mid R_m) = \dfrac{1 - w^k}{k(1 - w)},$

where $w = 1 - e^{-nb}$. For this special case, Gupta and McDonald [12] have tabulated the values of b for $k = 5, 10$; $n = 2(1)25$; and $P^* = 0.75, 0.90, 0.95, 0.975, 0.99$.

Consider three exponential populations with location parameters $\theta_1 = 0$, $\theta_2 = \theta_3 = \theta \geqslant 0$. In this case, Gupta and McDonald [12] have compared the expected subset sizes for the procedures R_1, R_2, R_3 and R_m for $\theta^* = 0(0.1)1.5$ and $P^* = 0.6, \dfrac{14}{15}$. The computations indicate that

(1) R_1 and R_2 perform equally well for $P^* = 0.6$,

(2) R_2 and R_3 perform equally well for $P^* = \dfrac{14}{15}$,

(3) $E(S \mid R_2) = E(S \mid R_3) \leqslant E(S \mid R_1)$ for all θ, equality holding for $\theta = 0$,

(4) R_m performs better than all the distribution-free procedures for the smaller value of P^*,

(5) for the larger P^*, the distribution-free procedures are better than R_m for $\theta \leqslant 0.5$,

(6) for larger values of θ ($\theta \geqslant 1.1$) R_m is the best among the four rules.

Ofosu [32] has studied the procedure R_m and compares its performance with a procedure that excludes from the selected subset those populations for which Y_i is sufficiently below $\bar{Y}$, the average of the Y_i. Based on a comparison of the expected subset sizes, Ofosu concludes that R_m is superior to the rules based on averaged Y_i in almost all situations. For those rare situations where R_m is not superior, it is only slightly inferior.

Gupta and McDonald [11] compare the performance of selection rules R_1 and R_3 in some Monte Carlo studies. Normal and logistic distributions with variance unity were studied for different configurations of their means. For $k = 3$ and $n = 2, 3, 4$, these configurations were taken to be

$$(0 \cdot 1, 0, 0), (0 \cdot 2, 0, 0), (0 \cdot 5, 0, 0), (1 \cdot 0, 0, 0), (2 \cdot 0, 0, 0),$$
$$(0 \cdot 1, 0 \cdot 1, 0), (0 \cdot 2, 0 \cdot 2, 0), (0 \cdot 5, 0 \cdot 5, 0),$$
$$(1 \cdot 0, 1 \cdot 0, 0), (2 \cdot 0, 2 \cdot 0, 0).$$

The number of simulations were 500 or 1000. The logistic distribution was chosen because equally spaced scores such as ranks yield locally most powerful tests for the location parameter of this distribution. The constants d and D were chosen to yield approximately the same P^* in the case of identical distributions. Then the ratio of $kP(CS \mid R)$ and $E(S \mid R)$ was computed for both rules R_1 and R_3. The bigger ratio for a rule indicates it to be better than the other. For example, for $k = 3$, $n = 2$, then $D = 2$ and $d = 3$ give the probability $\frac{14}{15}$ for the identical case. Using the configuration $(0 \cdot 1, 0, 0)$ for the normal means, the two ratios are $1 \cdot 012$ for R_1 and $1 \cdot 005$ for R_3 so that R_1 seems slightly better than R_3.

Using the configuration $(0 \cdot 5, 0, 0)$, R_3 was slightly better than R_1; the ratios being $1 \cdot 045$ for R_1 and $1 \cdot 049$ for R_3.

These Monte Carlo studies showed no significant uniform superiority of either of these procedures. However, R_3 seemed to perform slightly better than R_1 in the cases where the two highest parameters are equal. No difference in the performance of R_1 and R_3 was noticeable when the distribution changed from logistic to normal. In all cases the frequency of correct selections for R_1 was higher than the theoretical value exactly calculated for the identical distributions. Thus, there was no indication that the infimum of the probability of a correct selection does not take place when all populations are identically distributed as normal or logistic distributions under shift in location.

6. SELECTION PROCEDURES BASED ON PAIRWISE RANKS

As noted earlier the least favorable configuration over Ω for the selection rule $R_1(G)$ is not known and a counterexample exists showing that the infimum of the probability of a correct selection does not occur when all populations are identically distributed (Rules of the form $R_3(G)$ do not share this difficulty). H s u [18] overcomes this difficulty by constructing a rule based on pairwise rather than joint ranking of the samples.

Let $R_{jl}^{(i)}$ denote the rank of X_{jl} within $X_{i1}, \ldots, X_{in}, X_{j1}, \ldots, X_{jn}$, and let $R_j^{(i)} = \sum_{l=1}^{n} R_{jl}^{(i)}$ be the rank sum statistic for π_j compared to π_i. Let $\{D_l^{(ji)}, l = 1, \ldots, n^2\}$ denote the collection of n^2 ordered differences $X_{iu} - X_{jv}$, $u, v = 1, \ldots, n$ and set $D_{\text{med}}^{(ji)} = \text{median}\{D_l^{(ji)}\}$, i.e., $D_{\text{med}}^{(ji)}$ is the usual Hodges $-$ Lehmann (H–L) estimator of $\theta_i - \theta_j$. For $i = 1, \ldots, k$, let $M_i = \frac{1}{k} \sum_{j=1}^{k} D_{\text{med}}^{(ji)}$, where $D_{\text{med}}^{(ii)} = 0$. The procedure proposed by Hsu for selection of the population with the largest θ_i, denoted by R_R, is as follows:

R_R: Select π_i if and only if $M_i = \max_{1 \leqslant j \leqslant k} M_j$ or

(6.1)
$$\max_{j \neq i} R_j^{(i)} < r_n(P^*),$$

where $r_n(P^*)$ is the smallest integer such that $\mathsf{P}_0(\max_{j \neq i} R_j^{(i)} < r_n(P^*)) \geqslant$
$\geqslant P^*$ and $\mathsf{P}_0(\cdot)$ indicates the probability is computed assuming all populations are identically distributed.

The procedure R_R does not depend on the pairwise ranks alone. However, the contribution from the "$M_i = \max_{1 \leqslant j \leqslant k} M_j$" portion is small when n is large, and is included to insure that a nonempty subset is selected. The constants $r_n(P^*)$ can be obtained from Steel [36] for $P^* = 0.95, 0.99$, $k = 3(1)10$, $n = 4(1)20$; from Miller [28] (Table VIII) for $P^* = 0.95, 0.99$, $k = 3(1)11$, $n = 6(1)20(5)50, 100$.

Hsu also investigates the Pitman efficiency of the R_R procedure compared to a means procedure (with common unknown variance) and shows it to be the same as the Pitman efficiency of the Mann – Whitney test relative to the usual t-test.

Letting $D_{(1)}^{(ji)} \leqslant D_{(2)}^{(ji)} \leqslant \ldots \leqslant D_{(n^2)}^{(ji)}$ denote the n^2 ordered values of $D_l^{(ji)}$, $m = r_n(P^*) - \dfrac{n(n+1)}{2}$ and $D_{(m)}^{(\cdot i)} = \dfrac{1}{k} \sum_{j \neq i} D_{(m)}^{(ji)}$, an alternative procedure was also suggested by Hsu and is given by

$$R_R': \text{Select } \pi_i \text{ if and only if } \min_{j \neq i} (D_{(m)}^{(\cdot i)} - M_j) > 0 \text{ or}$$

(6.2)

$$\max_{j \neq i} R_j^{(i)} < r_n(P^*).$$

The subset selected by R_R' always contains the subset selected by R_R; however, the two rules are equiefficient in terms of their Pitman efficiencies under the location model.

7. SELECTION PROCEDURES BASED ON VECTOR RANKS

In the preceding procedures R_1, R_2 and R_3 the statistics H_i are defined using the ranks of the observations in the pooled sample. In cases with equal sample sizes, vector-at-a-time sampling can be used effectively to remove block effects, such as in a two-way layout, and to reduce data storage requirements. These procedures cover, for example, models of the form

(7.1) $X_{ij} = \mu + \theta_i + \beta_j + \epsilon_{ij}$ $(i = 1, \ldots, k, \; j = 1, \ldots, n)$,

where θ refers to a population effect, β to a block effect, and ϵ to an error term with any (not necessarily normal) continuous distribution.

Let $(X_{1j}, X_{2j}, \ldots, X_{kj})$ be the j-th vector and R_{ij} be the rank of X_{ij} among the k observations of the vector. Let $Z(1) \leqslant Z(2) \leqslant \ldots \leqslant Z(k)$ denote an ordered sample of size k from a continuous distribution G. Define $a(r)$ as in Section 2, i.e.

$$a(r) \equiv E[Z(r) \mid G] \quad (r = 1, \ldots, k),$$

and set

(7.2) $J_i = \dfrac{1}{n} \displaystyle\sum_{j=1}^{n} a(R_{ij})$ $(i = 1, \ldots, k)$.

M c D o n a l d [24] investigated the classes of procedures $R_1'(h; G)$ and $R_2'(g; G)$ which are defined using the two classes of functions $\{h(x)\}$ and $\{g(x)\}$, where h and g are nondecreasing real valued functions defined on the interval $I = [b(1), b(k)]$ and h satisfies the additional property that $h(x) \geqslant x$ for all $x \in I$. The two classes of procedures are

(7.3) $R_1'(h; G)$: Select π_i if and only if $h(J_i) \geqslant J_{[k]}$,

and

(7.4) $R_2'(g; G)$: Select π_i if and only if $g(J_i) \geqslant 0$.

Particular members of these classes that are of special interest are $R_1'(G)$ with $h(x) = x + b$, $b > 0$ and $R_2'(G)$ with $g(x) = x - d$, d real. Of course, $R_2'(g; G)$ can select an empty set; however, the rule $R_2(G)$ will necessarily choose a nonvoid subset if $P^* > 0.5$ and n is large. The treatment of $R_1'(G)$ and $R_2'(G)$ parallels that of $R_1(G)$ and $R_2(G)$ described earlier. The infimum of $P(CS)$ over Ω is attained at a point in Ω_k in the case of $R_1'(h; G)$. However as in the case of $R_1(G)$, it is not generally true that the infimum is attained at an equiparameter configuration. But the statement is true in the case of $R_2'(g; G)$.

When $b(r) = r$, $nH_i = T_i$, the rank sum statistic associated with

π_i. McDonald [25] has discussed the related distribution of $U = = \max_{1 \leqslant j \leqslant k} T_j - T_1$, where the distributions F_i are identical. He has tabulated $P(U \leqslant b)$ for $k = 2$, $n = 2(1)20$; $k = 3$, $n = 2(1)8$; $k = 4$, $n = 2(1)5$; and $k = 5$, $n = 2, 3$. For $P(U \leqslant b) = P^* = 0.75, 0.90, 0.95$, 0.975 and 0.99, he has tabulated the asymptotic value of b for $k = 2$, $n = 10(5)20$; $k = 3$, $n = 6(1)8$; $k = 4$, $n = 3(1)5$; and $k = 5$, $n = 3$.

The investigations of McDonald [25] with respect to slippage configuration based on simulations show that R_1' and R_2' (which are $R_1'(G)$ and $R_2'(G)$, respectively, in the special case with $b(r) = r$) are roughly equivalent when the underlying distribution has a long tail and the slippage is small, and that R_1 is better otherwise. These rules have been used by McDonald [27] in an analysis of state traffic fatality rates recorded by year.

Lorenzen and McDonald [21] further investigate the probability of a correct selection using rule R_1' by Monte Carlo simulations covering a wide range of distributions and parameter configurations (both location and scale). In all cases investigated the LFC, i.e., the configuration minimizing $P(CS)$, appeared to be the equiparameter configuration. This suggests that the practical inference corresponding to the selection procedure need not be restricted to the slippage configurations.

In another paper, McDonald [26] considered the case of three exponential distributions with parameters (guaranteed lives) $0 = = \theta_1 \leqslant \theta_2 = \theta_3 = \theta$ with samples of size two. For the rules R_1, R_2 and R_3 (using rank sum statistics T_i) the infimum of $P(CS)$ takes place when $\theta_1 = \theta_2 = \theta_3$. However, it is shown that the expected subset size is not bounded above by kP^*, a property enjoyed by many parametric procedures [see Gupta [9]] for the location parameter case under monotone likelihood ratio conditions.

Within the context of a block design (2-way classification) Lee [19] considers another type of selection rule based on the statistics $Y_i = \sum_{j=1}^{n} Y_{ij}$ $(i = 1, \ldots, k)$, where

$$(7.5) \qquad Y_{ij} = \begin{cases} 1 & \text{if} \quad X_{ij} = \max_{1 \leqslant l \leqslant k} X_{lj} \\ 0 & \text{otherwise.} \end{cases}$$

The selection rule is stated as

$$(7.6) \qquad R_{MS}: \text{Select } \pi_i \text{ if and only if } Y_i \geqslant \max_{1 \leqslant j \leqslant k} Y_j - d_{MS},$$

where d_{MS} is the smallest nonnegative integer required to insure the probability of a correct selection is no less than a prescribed P^*. The procedure is a *multinomial selection* rule (hence the subscript MS) designed to choose a subset to contain the population having the highest probability of yielding the largest observation. An analogous rule for choosing a subset to contain the population having the highest probability of yielding the smallest observation is also defined.

The constants d_{MS} required to implement the procedure R_{MS} have been determined by Lee [19] using Monte Carlo simulation, assuming the underlying distributions are identical, for $k = 49$ and $n = 17$. These values were then used to select subsets of states on the basis of traffic fatality rates recorded over a period of 17 years. Gupta and Nagel [13] investigated the least favorable configuration in a corresponding multinomial formulation and concluded, based on some numerical case studies, that the identically distributed case appears least favorable. Panchapakesan [33] proved that the identically distributed configuration is asymptotically least favorable.

8. SELECTION PROCEDURES BASED ON HODGES – LEHMANN ESTIMATORS

Let X_{ij} $(j = 1, \ldots, n, i = 1, 2, \ldots, k)$, $k \geqslant 2$, be independent random observations from k populations with continuous cumulative distribution functions $F(x - \theta_i)$ $(i = 1, 2, \ldots, k)$, with common variance $\sigma^2 = 1$. The following problems have been considered by Bechhofer [2] under the normality assumption:

(i) Select a "good" population, the i-th population being regarded as good if $\theta_i \geqslant \theta_{[k]} - \Delta$, for some preassigned $\Delta > 0$ $(i = 1, 2, \ldots, k - 1)$;

(ii) select the best t populations, i.e., the populations with location parameters $\theta_{[k-t+1]}, \ldots, \theta_{[k]}$ without regard to order;

(iii) select the best t populations with regard to order.

His approach, now known as the "indifference zone" approach selects the "best" populations with a guaranteed minimum probability P^* (preassigned) of correct selection when $(\theta_1, \ldots, \theta_k)$ lies in a subset, say Ω' of the parameter space. The region Ω' is called the preference zone and $R^k - \Omega'$ is the indifference zone. Some of the procedures discussed earlier use rank statistics for selection purposes.

However, when formulated for the problems discussed in this section, the slippage configuration of parameters defined by the indifference zone is not necessarily the LFC. The slippage configuration as pointed out by P u r i and P u r i [34] is least favorable when the parameters satisfy the relation $\theta_{[i]} - \theta_{[j]} = o(n^{-\frac{1}{2}})$ for all $1 \leqslant i, j \leqslant k, i \neq j$.

G h o s h [5] has proposed alternate procedures, based on one-sample Hodges – Lehmann estimators of θ_i's under the additional assumption that F is symmetric about the origin. Ghosh's procedures give in all these cases least favorable configurations for finite n without needing any restriction on the parameters.

G u p t a and H u a n g [10] have proposed some procedures to select a subset of the given k populations which is guaranteed to exclude all bad populations with probability not less than some preassigned P^*.

Let

$$R_{ij} = \frac{1}{2} + \sum_{l=1}^{n} u(|X_{ij}| - |X_{il}|)$$

$$(j = 1, 2, \ldots, n, \ i = 1, 2, \ldots, k),$$

where

$$u(t) = \begin{cases} 1 & \text{if} \quad t > 0, \\ \frac{1}{2} & \text{if} \quad t = 0, \\ 0 & \text{if} \quad t < 0. \end{cases}$$

Thus R_{ij} is the rank of $|X_{ij}|$ among $|X_{i1}|, \ldots, |X_{in}|$ $(1 \leqslant i \leqslant k,$ $1 \leqslant j \leqslant n)$. Let $\vec{X}'_i = (X_{i1}, \ldots, X_{in})$. Consider the one-sample signed rank statistics

$$(8.1) \qquad h(\vec{X}_i) = \sum_{j=1}^{n} \operatorname{sgn}(X_{ij}) E J(U_{nR_{ij}}) \qquad (i = 1, 2, \ldots, k),$$

where $\operatorname{sgn}(t) = 1, 0$ or -1 according as $t > 0$, $t = 0$ or $t < 0$; $U_{n1} \leqslant U_{n2} \leqslant \ldots \leqslant U_{nn}$ are the n ordered random variables from a rectangular $(0, 1)$ distribution, and $J(u) = \Psi^{-1}\left(\frac{1 + u}{2}\right)$, where $\Psi(x)$ is the distribution function of a random variable satisfying $\Psi(x) + \Psi(-x) = = 1$ for all real x.

The one-sample H–L estimators are given by

$$(8.2) \qquad \hat{\theta}_i(\vec{X}_i) = \frac{1}{2}\{\hat{\theta}_{i1}(\vec{X}_i) + \hat{\theta}_{i2}(\vec{X}_i)\} \qquad (i = 1, 2, \ldots, k),$$

where

$$\hat{\theta}_{i1}(\vec{X}_i) = \sup\{a: h(\vec{X}_i - a\vec{1}_n) > 0\},$$

$$\hat{\theta}_{i2}(\vec{X}_i) = \inf\{a: h(\vec{X}_i - a\vec{1}_n) < 0\},$$

and $\vec{1}_n$ is an n-tuple with all elements 1. All these statistics and estimators depend on n. The following property of location invariance (see Hodges and Lehmann [16]) is satisfied by these estimators:

$$(8.3) \qquad \hat{\theta}_i(\vec{X}_i + c\vec{1}_n) = \hat{\theta}_i(\vec{X}_i) + c \qquad (i = 1, 2, \ldots, k),$$

c being any constant. In the particular case when $J(u) = u$ or $\chi_1^{-1}(u)$ (the inverse of a chi-distribution with one degree of freedom) the statistics become the Wilcoxon signed rank or normal score statistics. In the former case

$$\hat{\theta}_i(\vec{X}_i) = \operatorname*{med}_{1 \leqslant j \leqslant j' \leqslant n} \frac{X_{ij} + X_{ij'}}{2} \qquad (i = 1, 2, \ldots, k).$$

Let $\hat{\theta}_{[1]} \leqslant \hat{\theta}_{[2]} \leqslant \ldots \leqslant \hat{\theta}_{[k]}$ denote the ordered estimators and let $\hat{\theta}_{(i)}$ be the unknown estimator associated with $\theta_{[i]}$ $(1 \leqslant i \leqslant k)$.

An elimination type procedure to select a subset excluding all
"Strictly Non t Best" populations

Let $d(\theta_i, \theta_j)$ be a suitable distance measure between θ_i and θ_j; the population π_i is "strictly non t best" if $d(\theta_{[k-t+1]}, \theta_i) = \theta_{[k-t+1]} - \theta_i > \Delta$, where Δ is a given positive constant. Let m denote the unknown number of "strictly non t best" populations in the given collection of k populations. Clearly, we have $0 \leqslant m \leqslant k - t$. Let

$$\Omega_m = \{\vec{\theta}: \theta_{[1]} \leqslant \ldots \leqslant \theta_{[m]} < \theta_{[k-t+1]} - \Delta \leqslant \theta_{[m+1]} \leqslant \ldots$$

$$\ldots \leqslant \theta_{[k-t+1]} \leqslant \ldots \leqslant \theta_{[k]}\}.$$

Then

$$\Omega = \bigcup_{m=0}^{k-t} \Omega_m .$$

Let CD stand for a correct decision, which is defined to be the selection of a subset which excludes all the "strictly non t best" populations. Gupta and Huang [10] define the rule R as follows:

(8.4)
$$R: \text{Reject } \pi_i \text{ if and only if } \hat{\theta}_i < \hat{\theta}_{[k-t+1]} - \Delta + d_1$$

$$(0 < d_1 < \Delta).$$

The constant d_1 is chosen to be the smallest number such that

$$\inf_{\vec{\theta} \in \Omega} P_{\vec{\theta}}(\text{CD} \mid R) \geqslant P^*.$$

Gupta and Huang [10] have shown that $P_{\vec{\theta}}(\text{CD} \mid R)$ is a nonincreasing function of $\theta_{[i]}$ $(i = 1, \ldots, m)$ and a nondecreasing function of $\theta_{[i]}$ $(i = m + 1, \ldots, k)$. Hence

$$\inf_{\vec{\theta} \in \Omega} P_{\vec{\theta}}(\text{CD} \mid R) = \inf_{0 \leqslant m \leqslant k-t} \inf_{\vec{\theta} \in \Omega_m} P_{\vec{\theta}}(\text{CD} \mid R).$$

It is known that if θ_i's are true values of the parameters, then under some regularity assumptions $\sqrt{n}(\hat{\theta}_i(\vec{X}_i) - \theta_i) \frac{B(F)}{A}$ tends asymptotically (as $n \to \infty$) to Y_i with $N(0, 1)$ where

$$A^2 = \frac{1}{4} \int_0^1 J^2(u)\, du, \quad B(F) = \int_0^\infty \frac{d}{dx} J(2F(x) - 1)\, dF(x).$$

These statistics Y_i's are mutually independent. This leads to a lower bound on the infimum of the probability of a correct decision for large n as follows:

(8.5)
$$\inf_{\vec{\theta} \in \Omega} \mathbf{P}_{\vec{\theta}}(CD \mid R) \geqslant \frac{t!}{r!(t-r-1)!} \int_{-\infty}^\infty \Phi^{k-t}(x + d\sqrt{n})\Phi^r(x) \times$$
$$\times [1 - \Phi(x)]^{t-r-1} \varphi(x)\, dx,$$

where $r = \min(t, k - t - 1)$, $d = \sqrt{0.864}\, d_1$ (or d_1) for the Wilcoxon (or normal score) case. For the case $F(x) = \Phi(x)$, then using normal scores the inequality (8.5) is an equality and the result agrees with that obtained by C a r r o l l, G u p t a and H u a n g [3].

It has also been shown that

$$\lim_{n \to \infty} \inf_{\vec{\theta} \in \Omega} \mathbf{P}_{\vec{\theta}}(CD \mid R(n)) =$$

$$= \lim_{n \to \infty} \frac{t!}{r!(t-r-1)!} \int_{-\infty}^\infty \Phi^{k-t}\left(x + \frac{B(F)}{A} d_1 \sqrt{n}\right) \times$$

$$\times \Phi^r(x)[1 - \Phi(x)]^{t-r-1} \varphi(x)\, dx = 1, \quad \text{since} \quad \frac{B(F)}{A} d_1 > 0,$$

so that the sequence of rules $\{R(n)\}$ is consistent with respect to Ω. Since the cumulative distribution function of each $\theta_i(\vec{X}_i)$ is stochastically non-decreasing in θ_i, it follows that for every $\vec{\theta} \in \Omega$ and $1 \leqslant i < j \leqslant k$.

$$\mathbf{P}_{\vec{\theta}}\{R(n) \text{ rejects } \pi_{(i)}\} \geqslant \mathbf{P}_{\vec{\theta}}\{R(n) \text{ rejects } \pi_{(j)}\},$$

and thus $R(n)$ is a so-called monotone procedure.

REFERENCES

[1] N.S. B a r t l e t t – Z. G o v i n d a r a j u l u, Some distribution-free statistics and their application to the selection problem, *Ann. Inst. Statist. Math.*, 20 (1968), 79-97.

[2] R.E. Bechhofer, A single sample decision procedure for ranking means of normal populations with known variances, *Ann. Math. Statist.*, 25 (1954), 16-39.

[3] R.J. Carroll — S.S. Gupta — D.Y. Huang, Selection procedures for the t best populations and some related problems, *Comm. Statist.*, 4 (1975), 987-1008.

[4] E.J. Dudewicz, An approximation to the sample size in selection problems, *Ann. Math. Statist.*, 40 (1969), 492-497.

[5] M. Ghosh, Nonparametric selection procedures for symmetric location parameter populations, *Ann. Statist.*, 1 (1973), 773-779.

[6] S.S. Gupta, On a decision rule for a problem in ranking means, Tech. Report No. 150, Institute of Statistics, University of North Carolina, 1956.

[7] S.S. Gupta, On a selection and ranking procedure for gamma populations, *Ann. Inst. Statist. Math.*, 14 (1963), 199-216.

[8] S.S. Gupta, Probability integrals of multivariate normal and multivariate t, *Ann. Math. Statist.*, 34 (1963), 792-828.

[9] S.S. Gupta, On some multiple decision (selection and ranking) rules, *Technometrics*, 7 (1965), 225-245.

[10] S.S. Gupta — D.Y. Huang, Nonparametric subset selection procedures for the t best populations, *Bull. Inst. Math. Academia Sinica*, 2 (1974), 377-386.

[11] S.S. Gupta — G.C. McDonald, On some classes of selection procedures based on ranks, *Nonparametric Techniques in Statistical Inference*, Cambridge University Press, Cambridge, England, (1970), 491-514.

[12] S.S. Gupta — G.C. McDonald, Some selection procedures with applications to reliability problems, *Operations Research and Reliability*, Proceedings of the NATO Conference, Italy, 1969, Gordon and Breach, New York, (1972), 421-439.

[13]   S.S. Gupta – K. Nagel, On selecting and ranking procedures and order statistics from the multinomial distribution,
Sankhyā, B, 29 (1967), 1-34.

[14]   S.S. Gupta – K. Nagel – S. Panchapakesan, On the
order statistics from equally correlated normal random variables,
Biometrika, 60 (1973), 403-413.

[15]   S.S. Gupta – S. Panchapakesan, *Multiple Decision
Procedures: Theory and Methodology of Selecting and Ranking
Populations,* John Wiley, New York, 1979.

[16]   J.L. Hodges, Jr. – E.L. Lehmann, Estimates of location based on rank tests, *Ann. Math. Statist.,* 34 (1963), 598-
611.

[17]   M. Hollander – D.A. Wolfe, *Nonparametric Statistical
Methods,* John Wiley, New York, 1973.

[18]   J.C. Hsu, Robust and nonparametric subset selection procedures, *Commun. Statist.-Theor. Meth.,* A9 (1980), 1439-1459.

[19]   Y.J. Lee, Nonparametric selections in blocked data: application to motor vehicle fatality rate data, *Technometrics* 22 (1980),
535-542.

[20]   E.L. Lehmann, *Nonparametrics: Statistical Methods Based on
Ranks,* McGraw-Hill, New York, 1975.

[21]   T.J. Lorenzen – G.C. McDonald, A nonparametric
analysis of urban and rural traffic fatality rates, General Motors
Research Publication GMR-3341, General Motors Research Laboratories, Warren, Michigan 48090, 1980.

[22]   G.C. McDonald, On some distribution-free ranking and
selection procedures, Department of Statistics, Purdue University,
Mimeograph Series No. 174, 1969.

[23] G.C. McDonald, On approximating constants required to implement a selection procedure based on ranks, *Statistical Decision Theory and Related Topics,* Academic Press, New York, (1971), 299-312.

[24] G.C. McDonald, Some multiple comparison selection procedures based on ranks, *Sankhyā, A,* 34 (1972), 53-64.

[25] G.C. McDonald, The distribution of some rank statistics with applications in block design selection problems, *Sankhyā, A,* 35 (1973), 187-204.

[26] G.C. McDonald, Characteristics of three selection rules based on ranks in the 3 × 2 exponential case, *Sankhyā, B,* 36 (1975), 261-266.

[27] G.C. McDonald, Nonparametric selection procedures applied to state traffic fatality rates, *Technometrics,* 21 (1979), 515-523.

[28] R.G. Miller, *Simultaneous Statistical Inference,* McGraw-Hill, New York, 1966.

[29] R.C. Milton, Tables of the equally correlated multivariate normal probability integral, Tech. Report No. 27, Department of Statistics, University of Minnesota, 1963.

[30] R.C. Milton, An extended table of critical values for the Mann – Whitney (Wilcoxon) two-sample statistic, *J. Am. Statist. Assoc.,* 59 (1964), 925-934.

[31] R.C. Milton, *Rank Order Probabilities: Two-Sample Normal Shift Alternatives,* John Wiley, New York, 1970.

[32] J.B. Ofosu, On selection procedures for exponential distributions, *Bull. Math. Statist.,* 16 (1974), 1-9.

[33] S. Panchapakesan, On a subset selection procedure for the most probable event in a multinomial distribution, *Statistical Decision Theory and Related Topics,* Academic Press, New York, (1971), 275-298.

[34]   M.L. Puri — P.S. Puri, Multiple decision procedures based
        on ranks for certain problems in analysis of variance, *Ann. Math.
        Statist.,* 40 (1969), 619-632.

[35]   M.H. Rizvi — G.G. Woodworth,   On selection proce-
        dures based on ranks: counterexamples concerning least favorable
        configurations, *Ann. Math. Statist.,* 41 (1970), 1942-1951.

[36]   R.G.D. Steel, A multiple comparison rank sum test: treat-
        ment versus control, *J. Am. Statist. Assoc.,* 54 (1959), 767-775.

S.S. Gupta

Dept. of Stat. Purdue University, West Lafayette, IN 47907, USA.

G.C. McDonald

Math. Dept. General Motors Research Laboratories, Warren, MI 48090, USA.

COLLOQUIA MATHEMATICA SOCIETATIS JÁNOS BOLYAI

32. NONPARAMETRIC STATISTICAL INFERENCE,

BUDAPEST (HUNGARY), 1980.

LINEAR RANK STATISTICS GENERATED BY UNIFORMLY DISTRIBUTED SEQUENCES

B. GYIRES

1. INTRODUCTION

Let $\{(x_1, \ldots, x_N)\} = R_N$ be the real vector space of dimension $N = m + n$, where m and n are positive integers. Let us denote by $\omega_0^{(N)} \subset R_N$ the set of vectors for which at least two components are equal.

Let the components of the vector $(x_1, \ldots, x_N)$ be pairwise different from each other. If the rearrangement of the components according to order of magnitude is $z_1 < \ldots < z_N$ and numbers $x_k = z_{r_k}$, then we say that x_k has rank r_k with respect to this order.

We shall denote by $\Pi_m^{(N)}$ the set of all $(r_1, \ldots, r_m)$ chosen without repetition from the elements $1, \ldots, N$. Let

$$\omega_{r_1, \ldots, r_m}^{(N)} = \{(x_1, \ldots, x_N) \in R_N \mid x_j \neq x_k, \ j \neq k;$$

$$\text{rank } x_k = r_k \quad (k = 1, \ldots, m)\}, \qquad (r_1, \ldots, r_m) \in \Pi_m^{(N)}.$$

Obviously these sets and $\omega_0^{(N)}$ are mutually disjoint and their union is equal to R_N.

If the joint distribution of the identically distributed random variables $X_1, \ldots, X_N$ is a symmetric and continuous, then ([6], 363, Satz 10)

(1.1) $\qquad P[(X_1, \ldots, X_N) \in \omega_{r_1, \ldots, r_m}^{(N)}] = \dfrac{1}{(n + 1) \ldots (n + m)}.$

The conditions listed are satisfied, if $X_1, \ldots, X_N$ are indepedndent identically distributed random variables with continuous distribution function.

Let the conditions (1.1) be satisfied. In this case the random variable

$$X_{m,n} = r_1 + \ldots r_m \qquad ((r_1, \ldots, r_m) \in \Pi_m^{(N)}),$$

is the well-known Wilcoxon statistics ([7]), which is asymptotically normally distributed if first $n \to \infty$, then $m \to \infty$ ([5]).

Let $f_1, \ldots, f_m$ be continuous functions on the closed interval $[0, 1]$ of bounded variations $V(f_1), \ldots, V(f_m)$, and suppose we are given the points

(1.2) $\qquad x_1^{(j)}, \ldots, x_N^{(j)} \qquad (j = 1, \ldots, m),$

on the open interval $(0, 1)$.

The aim of this paper is to investigate the so called linear rank statistics ([2])

(1.3) $\qquad X_{m,n} = f_1(x_{r_1}^{(1)}) + \ldots + f_m(x_{r_m}^{(m)}),$

if

$$(X_1, \ldots, X_N) \in \omega_{r_1, \ldots, r_m}^{(N)} \qquad ((r_1, \ldots, r_m) \in \Pi_m^{(N)}),$$

and to investigate the asymptotic behaviour of $X_{m,n}$ if first $n \to \infty$, then $m \to \infty$.

Let us denote by $\varphi_{m,n}(t)$ and by $\varphi_N^{(j)}(t)$ the characteristic functions of the random variable $X_{m,n}$ and of the random variable $Y_N^{(j)}$ defined as follows:

(1.4) $P(Y_N^{(j)} = x_k^{(j)}) = \dfrac{1}{N}$ $(k = 1, \ldots, N, \ j = 1, \ldots, m)$,

respectively.

In paper [2] I proved the following theorem (Theorem 1.2):

Theorem 1.1.

$$\left| \varphi_{m,n}(t) - \frac{N^m}{(n+1)\ldots(n+m)} \varphi_N^{(1)}(t) \ldots \varphi_N^{(m)}(t) \right| \leqslant$$

$$\leqslant \frac{N^m}{(n+1)\ldots(n+m)} - 1.$$

2. UNFORMLY DISTRIBUTED SEQUENCES

In this chapter we present the definitions and theorems, which will be used in this paper from the theory of the uniformly distributed sequences.

For a subset E of $(0, 1)$ let the counting function $A(E; N)$ be defined as the number of real numbers $x_k \in (0, 1)$ $(k = 1, \ldots, N)$, for which $x_j \in E$.

The sequence $\{x_k\}_{k=1}^{\infty}$ of numbers $x_k \in (0, 1)$ is said to be uniformly distributed if for every pair a, b of real numbers with $0 \leqslant a < b \leqslant 1$ we have

$$\lim_{N \to \infty} \frac{1}{N} A([a, b); N) = b - a.$$

For the finite sequence $x_k \in (0, 1)$ $(k = 1, \ldots, N)$ the number

$$D_N = \sup_{0 < \alpha \leqslant 1} \left| \frac{1}{N} A([0, \alpha); N) - \alpha \right|$$

is considered, and it is called the discrepancy of the given sequence.

Lemma 2.1 ([4], Corollary 1.1). *The sequence* $x_k \in (0, 1)$ $(k = 1, 2, \ldots)$ *is uniformly distributed if and only if* $\lim_{N \to \infty} D_N = 0$.

Koksma's inequality ([4], Theorem 5.1). *Let f be a function on $[0, 1]$ of bounded variation $V(f)$, and suppose we are given N points $x_1, \ldots, x_N$ in $(0, 1)$ with discrepancy D_N. Then*

$$\left| \frac{1}{N} \sum_{k=1}^{N} f(x_k) - \int_0^1 f(x) \, dx \right| \leqslant V(f) D_N .$$

In the applications for the linear rank statistics generated by uniformly distributed sequences, the following estimations of the discrepancy have an important role:

Lemma 2.2 ([4], Theorem 2.3). *For any infinite sequence in $(0, 1)$ we have*

$$ND_N > c \lg N$$

for infinitely many positive integers N, where $c > 0$ is an absolute constant.

Lemma 2.3 (Le Veque's inequality, [4], Theorem 2.4). *The discrepancy D_N of the finite sequence $x_1, \ldots, x_N$ in $(0, 1)$ satisfies*

$$D_N \leqslant \left(\frac{6}{\pi^2} \sum_{h=1}^{\infty} \frac{1}{h^2} \left| \frac{1}{N} \sum_{n=1}^{N} e^{2\pi i h x_n} \right|^2 \right)^{\frac{1}{3}} .$$

For $0 \leqslant \delta < 1$ and $\epsilon > 0$, a finite sequence $x_1 < x_2 < \ldots < x_N$ in $(0, 1)$ is called an *almost-arithmetic progression* (δ, ϵ), if there exists an η, $0 < \eta \leqslant \epsilon$, such that the following conditions are satisfied:

(a) $0 \leqslant x_1 \leqslant \eta + \delta\eta$;

(b) $\eta - \delta\eta \leqslant x_{n+1} - x_n \leqslant \eta + \delta\eta$ for $1 \leqslant n \leqslant N - 1$;

(c) $1 - \eta - \delta\eta \leqslant x_N < 1$.

Lemma 2.4 ([4], Theroem 3.1). *Let $x_1 < x_2 < \ldots < x_N$ be an almost-arithmetic progression (δ, ϵ) and let η be the positive real number corresponding to the sequence according to the above mentioned definition. Then*

$$D_N \leq \frac{1}{N} + \frac{\delta}{1 + \sqrt{1 - \delta^2}} \qquad \text{for} \quad \delta > 0,$$

$$D_N \leq \min\left(\eta, \frac{1}{N}\right) \qquad \text{for} \quad \delta = 0.$$

A must important class of uniformly distributed mod 1 sequences is given by $\{n\alpha\}_{n=1}^{\infty}$ with α irrational. The discrepancy of $\{n\alpha\}_{n=1}^{\infty}$ will depend on the finer arithmetical properties of α. For further details we refer to [4] (Diophantic approximation, 121-127). Here we mention only one theorem.

Lemma 2.5 ([4], Theorem 3.4). *Suppose the irrational* $\alpha = [a_0, a_1, a_2, \ldots]$ *has bounded partial quotiens, say* $a_i \leq K_N$ *for* $i \geq 1$. *Then the discrepancy* D_N *of* $\{n\alpha\}_{n=1}$ *satisfies*

$$ND_N \leq 3 + \left(\frac{1}{\lg \xi} + \frac{K}{\lg (K + 1)}\right) \lg N,$$

where $\xi = \frac{1}{2}(1 + \sqrt{5})$.

We define the so-called Van der Corput sequence $\{x_n\}_{n=1}^{\infty}$ as follows: For $n \geq 1$, let $n - 1 = \sum_{j=0}^{s} a_j 2^j$ be the dyadic expansion of $n - 1$. Let $x_n = a_j 2^{-(1+j)}$. Then the sequence $\{x_n\}_{n=1}^{\infty}$ is contained in $(0, 1)$.

Lemma 2.6 ([4], Theorem 3.5). *The discrepancy* D_N *of the Van der Corput sequence* $\{x_n\}_{n \equiv 1}^{N}$ *satisfies*

$$ND_N \leq \frac{\lg (N + 1)}{\lg 2}.$$

3. LINEAR RANK STATISTICS BASED ON UNIFORMLY DISTRIBUTED SEQUENCES

Now we turn to the main results of this paper, mentioned in the Introduction.

Lemma 3.1. *If* f *is continuous on* $[0, 1]$ *of bounded variation* $V(f)$, *then*

$$V(e^{itf}) \leq |t| V(f) \qquad (t \in R_1).$$

Proof. Let $0 = x_0 < x_1 < \ldots < x_{n-1} \leqslant x_n = 1$, and let

$$S(x_0, x_1, \ldots, x_n) = \sum_{k=1}^{n} |e^{itf(x_k)} - e^{itf(x_{k-1})}|.$$

After a simple calculation we get

$$|e^{itf(x_k)} - e^{itf(x_{k-1})}| = 2\left|\sin\left(\frac{1}{2}\,[f(x_k) - f(x_{k-1})]t\right)\right|.$$

Since f is continuous on $[0, 1]$, if the numbers $x_0, x_1, \ldots, x_n$ are close enough to each other, then the differences

$$|f(x_k) - f(x_{k-1})| \qquad (k = 1, \ldots, n),$$

are small. Thus

$$2\left|\sin\left(\frac{1}{2}\,[f(x_k) - f(x_{k-1})]t\right)\right| \leqslant |t|\,|f(x_k) - f(x_{k-1})|$$

$$(k = 1, \ldots, n).$$

Using the result, we obtain that the inequality

$$S(x_0, x_1, \ldots, x_n) \leqslant |t|\,V(f)$$

holds, and this is the statement of our Lemma.

By Koksma's inequality from Lemma 3.1 we obtain

Lemma 3.2. *Let f be a continuous on $[0, 1]$ of bounded variation $V(f)$, and suppose N points $x_1, \ldots, x_N$ are given in $(0, 1)$ with discrepancy D_N. If we denote by $\varphi_N(t)$ the characteristic function of the discrete random variable Y_N defined by*

$$P(Y_N = f(x_k)) = \frac{1}{N} \qquad (k = 1, \ldots, N),$$

then

$$\left|\varphi_N(t) - \int_0^1 e^{itf(x)}\,dx\right| \leqslant |t|\,V(f)D_N$$

for $t \in R_1$.

Lemma 3.3. *Let* $f_1, \ldots, f_m$ *be continuous functions on* $[0, 1]$ *of bounded variations* $V(f_1), \ldots, V(f_m)$, *and suppose* N *points are given by* (1.2) *in* $(0, 1)$ *with discrepancies* $D_N^{(1)}, \ldots, D_N^{(m)}$. *If we denote by* $\varphi_N^{(j)}(t)$ *the characteristic function of the random variable* $Y_N^{(j)}$ *defined by* (1.4) $(j = 1, \ldots, m)$, *then*

$$\left| \prod_{j=1}^{m} \varphi_N^{(j)}(t) - \prod_{j=1}^{m} \int_0^1 e^{itf_j(x)} \, dx \right| \leqslant |t| \sum_{j=1}^{m} V(f_j) D_N^{(j)}$$

for $t \in R_1$.

Proof. Using the notations

$$a_j = \varphi_N^{(j)}(t), \quad b_j = \int_0^1 e^{itf_j(x)} \, dx,$$

we get the relation

$$a_1 \ldots a_m - b_1 \ldots b_m =$$

(3.1)
$$= \sum_{j=1}^{m} (a_1 \ldots a_{m-j} a_{m-j+1} b_{m-j+2} \cdots b_m -$$
$$- a_1 \ldots a_{m-j} b_{m-j+1} b_{m-j+2} \cdots b_m),$$

$$(a_0 = b_{m+1} = 1).$$

Since a_j, b_j $(j = 1, \ldots, m)$ are characteristic functions,

$$|a_j| \leqslant 1, \quad |b_j| \leqslant 1 \qquad (j = 1, \ldots, m).$$

Applying these inequalities in (3.1) we get

(3.2) $$|a_1 \ldots a_m - b_1 \ldots b_m| \leqslant \sum_{j=1}^{m} |a_j - b_j|.$$

Applying Lemma 3.2 to the right hand side of (3.2), we get the statement of Lemma 3.3.

From Theorem 1.1 we obtain the following theorem using Lemma 3.3.

Theorem 3.1. *Let* $f_1, \ldots, f_m$ *be continuous functions on* $[0, 1]$ *of bounded variations* $V(f_1), \ldots, V(f_m)$, *and suppose* N *points are given* by (1.2) in $(0, 1)$ with discrepancies $D_N^{(1)}, \ldots, D_N^{(m)}$, resp. Let us denote by $\varphi_{m,n}(t)$ and by $\varphi_N^{(j)}(t)$ the characteristic functions of the random variables $X_{m,n}$ and $Y_N^{(j)}$ defined by (1.3) and (1.4) respectively $(j = 1, \ldots, m)$. Then for $t \in R_1$ we have

(3.3)
$$\left| \varphi_{m,n}(t) - \frac{N^m}{(n+1) \ldots (n+m)} \prod_{j=1}^{m} \int_{0}^{1} e^{it f_j(x)} \, dx \right| \le$$

$$\le \frac{N^m}{(n+1) \ldots (n+m)} \left[1 + |t| \sum_{j=1}^{m} V(f_j) D_N^{(j)} \right] - 1.$$

Theorem 3.2. *Let f_j be continuous function on $[0, 1]$ of bounded variation $V(f_j) > 0$ $(j = 1, \ldots, m)$ and suppose that the sequences $\{x_k^{(j)}\}_{k=1}^{\infty}$ are given in $(0, 1)$ with discrepancy $D_N^{(j)}$ of the points $x_1^{(j)}, \ldots, x_N^{(j)}$ $(j = 1, \ldots, m)$. If the random variable $X_{m,n}$ is defined by (1.3), then*

(3.4) $\qquad X_{m,n} \to f_1(\eta_1) + \ldots + f_m(\eta_m), \quad n \to \infty$

in weak sense if and only if the sequences $\{x_k^{(j)}\}_{k=1}^{\infty}$ $(j = 1, \ldots, m)$ are uniformly distributed on $(0, 1)$, where $\eta_1, \ldots, \eta_m$ are independent random variables uniformly distributed on $(0, 1)$.

Proof of Theorem 3.1. Since

$$\lim_{n \to \infty} \frac{N^m}{(n+1) \ldots (n+m)} = 1,$$

the right hand side of (3.3) has zero limit uniformly in all finite interval $t \in [-T, T]$ if and only if

$$\lim_{N \to \infty} D_N^{(j)} = 0 \quad \text{for} \quad j = 1, \ldots, m.$$

Thus from Lemma 2.1 it follows that

$$\lim_{n \to \infty} \varphi_{m,n}(t) = \prod_{j=1}^{m} \int_{0}^{1} e^{it f_j(x)} \, dx$$

holds uniformly in all finite interval $t \in [-T, T]$ if and only if the sequences $\{x_k^{(j)}\}_{k=1}^{\infty}$ for $j = 1, \ldots, m$ are uniformly distributed on $(0, 1)$. Thus the proof of our theorem is completed.

4. REMARKS

On the basis of the linear rank statistics $X_{m,n}$ defined by the expression (1.3), we can decide to accept or to reject the hypothesis

$$H_0: \mathrm{P}(X < x) = \mathrm{P}(Y < x), \quad x \in R_1,$$

provided that the common distribution function of the random variables X and Y is continuous. Theorem 3.1 gives the bound of the error if we apply the statistics $f_1(\eta_1) + \ldots + f_m(\eta_m)$ with independently and uniformly distributed random variables $\eta_1, \ldots, \eta_m$ on $(0, 1)$ instead of the exact statistics $X_{m,n}$.

Remark 4.2. Theorem 3.2 shows much more. Namely if the sample size of the random variable X (m) is given, and the sample size of Y (n) is large enough and the linear rank statistics $X_{m,n}$ is based on the uniformly distributed sequences $\{x_k^{(j)}\}_{k=1}^{\infty}$ $(j = 1, \ldots, m)$, then the error will be small if we apply the statistics $f_1(\eta_1) + \ldots + f_m(\eta_m)$ instead of $X_{m,n}$. Moreover we can give the rate of convergence in the Theorem 3.2 by the help of Lemmas 2.2 and 2.3 generally, and by the help of Lemmas 2.4, 2.5 and 2.6 in the special cases of uniformly distributed arithmetic progressions, of uniformly distributed mod 1 sequences, and of the Van der Corput sequence respectively on the basis of Theorem 3.1.

Remark 4.3. If it is given the sequence $\{f_j\}_{j=1}^{\infty}$ of continuous functions on $[0, 1]$ of bounded variations $\{V(f_j)\}_{j=1}^{\infty}$, moreover the sequences $\{x_k^{(j)}\}_{k=1}^{\infty}$ $(j = 1, 2, \ldots)$ are defined in $(0, 1)$ with discrepancy $D_N^{(j)}$ of the points $x_1^{(j)}, \ldots, x_N^{(j)}$ $(j = 1, 2, \ldots)$, then under the conditions of Theorem 3.2 (3.4) hold for $m = 1, 2, \ldots$ if and only if the sequences $\{x_k^{(j)}\}_{k=1}^{\infty}$ $(j = 1, 2, \ldots)$ are uniformly distributed on $(0, 1)$.

In this case the question, what is the limit distribution of the statistics (1.3) if first $n \to \infty$, and then $m \to \infty$, is equivalent to the question, what is the limit distribution of the sequence

$$\{f_1(\eta_1) + \ldots + f_m(\eta_m)\}_{m=1}^{\infty},$$

where the random variables $\{\eta_j\}_{j=1}^{\infty}$ are independently and uniformly distributed on $(0, 1)$. This question belongs to the theory of limit distribution of sums of independent random variables ([1]) in case, if all moments of the random variables exist.

Remark 4.4. The author's Theorems 2.1, 2.2, 2.5 of the paper [2], moreover those 1, 2, 3 of paper [3] can trace back to Theorem 3.2.

REFERENCES

[1] B.V. Gnedenko — A.N. Kolmogorov, *Limit distributions for sums of independent random variables,* Addison-Wesley, Reading, Mass., 1954.

[2] B. Gyires, Limit distribution of linear rank statistics, *Publ. Math.,* (Debrecen), 21 (1974), 95-112.

[3] B. Gyires, The construction of linear rank statistics with the help of pseudo random numbers, *Publ. Math.,* (Debrecen), 21 (1974), 225-232.

[4] L. Kuipers — H. Niederreiter, *Uniform distribution of sequences,* John Wiley, New York, 1974, Ch. 2.

[5] A. Rényi, New criteria for the comparison of two samples, *Selected papers of A. Rényi,* Vol. I., Akadémiai Kiadó, Budapest, 1976, 381-401.

[6] L. Schmetterer, *Einführung in die mathematische Statistik,* Springer-Verlag, Wien, 1956.

[7] F. Wilcoxon, Individual comparisons by ranking methods, *Biometrics Bulletin,* 1 (1945), 80-83.

B. Gyires

Kossuth L. University of Debrecen, Hungary.

AN EFFECTIVE ESTIMATE OF NONLINEAR FUNCTIONALS

S.A. HASHIMOV

Many quantitative characteristics of distributions, goodness of statistical estimates, strength of criteria, are expressed by various functionals of the probability density of the following form:

$$(1) \qquad I = I(f) = \int g(f(x), f'(x)) f(x)\, dx$$

where g is some given function, $f(x)$ is the probability density of a random variable X, and $f'(x)$ is the derivative of $f(x)$.

The problem of finding effective estimates for functionals of type (1) was studied in [1].

Let $X_1, X_2, \ldots, X_n$ be independent identically distributed random variables with the common distribution function

$$F(x) = \int_{-\infty}^{x} f(y)\, dy$$

We assume in the sequel that, for some $r \geq 4$, the r-th derivative $f^{(r)}(x)$ of the density $f(x)$ exists.

For estimating $f^{(s)}(x)$ $(s = 0, 1)$, we are going to use the following type of statistics

$$(2) \qquad f_n^{(s)}(x) = \frac{1}{nh^{1+s}} \sum_{i=1}^{n} K^{(s)} \left(\frac{x - x_i}{h} \right) \qquad (s = 0, 1),$$

where $h = h(n) \to 0$ as $n \to \infty$, and

(a) $K(x) = K(-x)$, $\int_0^\infty K(x)\, dx = \frac{1}{2}$,

$$G_2(K) = G_4(K) = \ldots = G_{2j}(K) = 0 \qquad j = \left[\frac{r-1}{2} \right] ,$$

where

$$G_i(K) = \int_0^\infty x^i K(x)\, dx$$

(b) $\int_0^\infty xK'(x)\, dx = \frac{1}{2}$,

$$G_3(K') = G_5(K') = \ldots = G_{2j-1}(K') = 0 \quad \text{if} \quad r \geqslant 5)$$

where

$$G_i(K') = \int_0^\infty x^i K'(x)\, dx.$$

We introduce the operator (cf. [1])

$$Ag(x) = g + \frac{\partial g}{\partial f} f - \frac{d}{dx} \left[\frac{\partial g}{\partial f'} f \right].$$

Now for estimating the functional I we use the statistics

$$I_n = \int_{-k_n}^{k_n} g(f_n(x), f_n'(x)) f_n(x)\, dx$$

where $k_n \to \infty$ as $n \to \infty$, and $f_n^{(s)}(x)$ $(s = 0, 1)$, are defined in (2).

We assume in the sequel the validity of the following five conditions:

Condition A. The function $g(y_1, y_2)$ and its partial derivatives $\dfrac{\partial g}{\partial y_1}, \dfrac{\partial g}{\partial y_2}$ satisfy a global Lipschitz condition in the variable (y_1, y_2),

Condition B

$$\int_{-\infty}^{\infty} \int_{-\infty}^{\infty} |K(t)| \, |Ag(x+th) - Ag(x)|^2 f(x) \, dt dx \to 0, \quad \text{as } h \to 0,$$

Condition C.

$$f^{(r)}(x) \in \text{Lip } 1, \quad (r \geqslant 4), \quad f(\pm k_n) \sim h,$$

Condition D.

$$0 < \sigma^2(f) = D_f^2[Ag(X)] < \infty$$

$$\int_{|x| \geqslant k_n} g(f(x), f'(x)) f(x) \, dx = o(n^{-\frac{1}{2}}).$$

Condition E.

$$\sqrt{n} h^3 \to \infty, \quad \sqrt{n} h^r \to 0, \quad (r \geqslant 4), \quad \sqrt{n} h^{r+1} k_n \to 0.$$

Lemma. *Let* $f^{(r)}(x) \in \text{Lip } 1$. *If* $K(x)$ *and* $K'(x)$ *satisfy Conditions* (a) *and* (b), *then the following estimations hold:*

$$(3) \qquad \sup_{x \in R} |Ef_n^{(s)}(x) - f^{(s)}(x)| = O(h^{r+1-s}) \qquad (s = 0, 1),$$

$$(4) \qquad |f_n(x) - Ef_n(x)| = O_P\left(\frac{f(x)\gamma_n}{\sqrt{nh}}\right),$$

$$(5) \qquad |f_n'(x) - Ef_n'(x)| = O_P\left(\frac{f(x)\gamma_n'}{\sqrt{nh^3}}\right)$$

where $\gamma_n, \gamma_n' \to \infty$ *arbitrarily slowly (but we assume that*

$$h^{\frac{1}{2}}\gamma_n \to 0 \quad \text{and} \quad \frac{(\gamma_n')^2}{\sqrt{nh^3}} \to 0).$$

The proof of the Lemma is simple and we omit it.

Denote by $\mathscr{B}$ the class of all density functions f that satisfy Conditions B, C, D. We provide $\mathscr{B}$ with a topology $\mathscr{T}$ generated by the open base

$$V_\epsilon(f_0) = \left\{ f: \sup_x \sum_{s=0}^1 \left| \frac{f^{(s)}(x)}{f_0^{(s)}(x)} - 1 \right| < \epsilon \right\} \qquad (f_0 \in \mathscr{B}, \ \epsilon > 0).$$

We use the expression (cf. [1]) "$T_n \sim N(0, \sigma^2(f))$" uniformly in (the neighbourhood) V" if for $n \to \infty$ we have

$$\sup_{f \in V} \sup_x \left| P_f\{T_n < x\} - \frac{1}{\sqrt{2\pi}\,\sigma(f)} \int_0^x \exp\left\{ -\frac{y^2}{2\sigma^2(f)} \right\} dy \right| \to 0.$$

We have the following theorem.

Theorem 1. *Assume Conditions* A-E *are satisfied. Then*

$$\sqrt{n}(I_n - I) \sim N(0, \sigma^2(f)),$$

uniformly in some neighbourhood $V \subset \mathscr{T}$ *of the density* f_0.

Proof. We use the following notations:

$$V_1 = \int_{-k_n}^{k_n} g(f, f') f_n(x)\, dx,$$

$$V_2 = \int_{-k_n}^{k_n} A g(f, f') [f_n(x) - E f_n(x)]\, dx,$$

$$V_3 = \int_{-k_n}^{k_n} g(E f_n, E f_n') f_n(x)\, dx,$$

$$V_4 = \int_{-k_n}^{k_n} \left\{ \frac{\partial g(E f_n, E f_n')}{\partial f} [f_n(x) - E f_n(x)] + \right.$$

$$\left. + \frac{\partial g(E f_n, E f_n')}{\partial f'} [f_n'(x) - E f_n'(x)] \right\} E f_n(x)\, dx,$$

$$V_5 = I_n - V_3 - V_4,$$

$$V_6 = \int_{-k_n}^{k_n} \left\{ \frac{\partial g(f,f')}{\partial f} \left[f_n(x) - E f_n(x) \right] + \right.$$

$$\left. + \frac{\partial g(f,f')}{\partial f'} \left[f_n'(x) - E f_n'(x) \right] \right\} E f_n(x) \, dx,$$

$$V_7 = \int_{-k_n}^{k_n} \left[\frac{\partial g(f,f')}{\partial f} f(x) - \frac{d}{dx} \left(\frac{\partial g}{\partial f'} f \right) \right] \left[f_n(x) - E f_n(x) \right] dx.$$

We estimate now the following differences.

By (3), (4) and A we get

(6) $\qquad V_3 - V_1 = O_P(h^r).$

By (3)-(5) and A we find

(7) $\qquad V_4 - V_6 = O_P \left(\dfrac{h^r \gamma_n'}{\sqrt{nh^3}} \right).$

By (3)-(5) and C we have

(8) $\qquad V_6 - V_7 = O_P \left(\dfrac{\gamma_n \sqrt{h}}{\sqrt{n}} \right) + O_P \left(\dfrac{\gamma_n' h^{r - \frac{1}{2}}}{\sqrt{n}} \right).$

By (3)-(5) and A we get

(9) $\qquad V_5 = O_P \left(\dfrac{(\gamma_n')^2}{nh^3} \right).$

Furthermore, by D

(10) $\qquad V_7 - V_2 + V_1 - I = O(h^{r+1} k_n) + o(n^{-\frac{1}{2}}).$

Finally, combining (6)-(10) we find

(11) $\qquad I_n - I = V_2 + o_P \left(\dfrac{1}{\sqrt{n}} \right).$

Now we determine $D^2 V_2$. By Condition B

$$- 405 -$$

$$\mathsf{E} V_2^2 = \frac{1}{n} \int_{-\infty}^{\infty} [Ag(x)]^2 f(x) \, dx +$$

$$+ \frac{n-1}{nh^2} \Big[\int_{-k_n}^{k_n} Ag(x) \mathsf{E}K\Big(\frac{x-x_1}{h}\Big) dx \Big]^2 + o\Big(\frac{1}{n}\Big).$$

Hence we obtain

$$(12) \qquad \mathsf{D}^2 V_2 = \frac{1}{n} \sigma^2(f) + o\Big(\frac{1}{n}\Big) + O\Big(\frac{h^{r+1} k_n}{n}\Big).$$

Thus, V_2 is the sum of independent centraled random variables which, by Condition D, are uniformly square integrable in some neighbourhood V of the distribution F_0. Theorem 1 is proved.

Theorem 2. *Assume Conditions* A-E *are satisfied, and for some* m

$$(13) \qquad \mathsf{E}_f(|g|^m + |Ag|^m) < \infty.$$

Then uniformly in some neighbourhood V *of* f_0

$$\mathsf{E}_f n^{\frac{m}{2}} |I_n - I|^m \to \frac{1}{\sqrt{2\pi}\sigma(f)} \int_{-\infty}^{\infty} |y|^m \exp\Big\{-\frac{y^2}{2\sigma^2(f)}\Big\} dy.$$

Proof. According to (13) and Theorem 1 (cf. [2])

$$\mathsf{E}_f V_2^m \to \frac{1}{\sqrt{2\pi}\sigma(f)} \int_{-\infty}^{\infty} |y|^m \exp\Big\{-\frac{y^2}{2\sigma^2(f)}\Big\} dy.$$

Thus, it is sufficient to show that relations (6)-(10) hold true with $o_p(n^{-\frac{1}{2}})$ changed to $o_m(n^{-\frac{1}{2}})$ (o_m means "o in m-th order mean"). Since

$$\mathsf{E}_f \Big[\int_{-k_n}^{k_n} |f_n(x)| \, dx \Big]^m < \infty,$$

thus

$$\mathsf{E}_f |V_3 - V_1|^m = O(h^{rm}).$$

In view of the relation

$$\mathsf{E}\left[\int_{-k_n}^{k_n} |f_n(x) - \mathsf{E}f_n(x)|\, dx\right]^m = O(1)$$

$$\mathsf{E}\left[\int_{-k_n}^{k_n} |f_n'(x) - \mathsf{E}f_n'(x)|^2\, dx\right]^m = O((nh^2)^{-m})$$

$$\mathsf{E}\,|f_n(x) - \mathsf{E}f_n(x)|^m = O(n^{-\frac{m}{2}} h^{1-m}),$$

we have

$$\mathsf{E}\,|V_4 - V_6|^m = o(h^{rm}),$$

$$\mathsf{E}\,|V_6 - V_7|^m = O(h^{rm}) + O(hn^{-\frac{m}{2}}),$$

$$\mathsf{E}\,|V_5|^m = O((nh^2)^{-m}),$$

$$\mathsf{E}\,|V_7 - V_2 + V_1 - I|^m = O(h^{r+1}k_n)^m) + o(n^{-\frac{m}{2}}).$$

Hence Theorem 2 is proved.

The assertions of Theorems 1 and 2 imply

Corollary. *Assume that the nonnegative loss function l satisfies the condition*

$$l(y) \leqslant C(1 + |y|^m) \qquad (C > 0).$$

Then uniformly in some neighbourhood of f_0

$$\lim_{n \to \infty} \mathsf{E}_f l(\sqrt{n}(I_n - I)) = \frac{1}{\sqrt{2\pi}\,\sigma(f)} \int_{-\infty}^{\infty} l(y) \exp\left\{-\frac{y^2}{2\sigma^2(f)}\right\} dy.$$

Theorem 3. *Assume the conditions of Theorems 1, 2 and the Corollary. Further, let $\Phi_n = \Phi_n(x_1, x_2, \ldots, x_n)$ denote an arbitrary estimate. Then the estimate I_n is locally asymptotically optimal in the class of all estimates Φ_n.*

Theorem 3 can be proved analogously to Theorem 4 in [1].

REFERENCES

[1] B.Ja. Levit, Asymptotical effective estimation of nonlinear functionals, *Problems of information transmission,* 1978, Vol. XIV/3, 65-72 (in Russian).

[2] S.N. Bernstein, *Collected works,* IV., Moscow, Nauka, 1964, 358-363 (in Russian).

S.A. Hashimov
Mathematical Institute, SU-700000 GSP Tashkent, UzbSSR.

COLLOQUIA MATHEMATICA SOCIETATIS JÁNOS BOLYAI

32. NONPARAMETRIC STATISTICAL INFERENCE,

BUDAPEST (HUNGARY), 1980.

NONPARAMETRIC BAYESIAN ESTIMATION OF THE HORIZONTAL DISTANCE BETWEEN TWO POPULATIONS

M. HOLLANDER* — R.M. KORWAR**

1. INTRODUCTION

When F and G are continuous distribution functions, the horizontal distance

$$(1) \qquad \Delta(x) = G^{-1}(F(x)) - x, \qquad x \text{ real,}$$

has been shown by D o k s u m [1], to be a useful measure of the difference, at each x, between F and G. Under suitable regularity, Doksum shows that $\Delta(x)$ is essentially the only function satisfying

$$(2) \qquad X + \Delta(X) \overset{d}{=} Y,$$

where, in (2), X is distributed according to F, Y is distributed according

*Research was supported by the Air Force Office of Scientific Research, AFSC, USAF under Grant AFOSR-81-0038.

**Research was supported by the Air Force Office of Scientific Research, AFSC, USAF under Grant F49620-79-C-0105.

to G, and "$\overset{\text{d}}{=}$" means "has the same distribution as".

When the linear model

$$(3) \qquad F(x) = G(x + \Delta) \quad \text{for all} \quad x,$$

holds, where Δ is a constant, then $\Delta(x) \equiv \Delta$ (and, of course, when $F \equiv G$, $\Delta(x) \equiv 0$).

When one observes a random sample of n X's from F and an independent random sample of m Y's from G, Doksum suggests estimating $\Delta(x)$ by

$$(4) \qquad \hat{\Delta}_N(x) = G_m^{-1}(F_n(x)) - x,$$

where $N = m + n$ and F_n, G_m are the empirical distribution functions based on the X's and Y's, respectively. Doksum also derives a simultaneous confidence band for $\Delta(x)$ and shows that $\sqrt{N}\{\hat{\Delta}_N(x) - \Delta(x)\}$ converges weakly to a Gaussian process.

In this paper we consider the one-sample problem where G *is known* and (just) a random sample of n X's from F is available for estimating $\Delta(x)$. One natural estimator for this problem is the one-sample limit $(m \to \infty)$ of Doksum's estimator [1] $\hat{\Delta}_N$. This one-sample limit is

$$(5) \qquad \hat{\Delta}_n(x) = G^{-1}(F_n(x)) - x.$$

The estimator $\hat{\Delta}_n$ does not utilize prior information about the unknown F. Our approach is Bayesian and leads to an estimator $\tilde{\Delta}_n$ which does use prior information about F.

We assume that F is a *random* distribution function chosen according to Ferguson's [2] Dirichlet process prior (Definition 2.2) with parameter $\alpha(\cdot)$, a completely specified measure on the real line R with the Borel σ-field $\mathscr{B}$. A defect to this approach is that the randomly chosen F will not be continuous (Ferguson's Dirichlet process prior chooses, with probability one, a discrete distribution) and thus the desirability of estimating $\Delta(x)$ is slightly diminished. Nevertheless, in this case $\Delta(x)$ remains a useful measure of the distance between F and G at x, and the resulting estimator $\tilde{\Delta}_n(x)$ combines sample information and prior information in an effective manner.

Our loss function is

$$L(\hat{\Delta}, \Delta) = \int (\hat{\Delta}(x) - \Delta(x))^2 \, dW(x), \tag{6}$$

where $\hat{\Delta}$ is an estimator of Δ and W is a finite measure on $(R, \mathscr{B})$. A general expression for the Bayes estimator $\tilde{\Delta}_n$ is given in Section 3, and explicit expressions for $\tilde{\Delta}_n$ are obtained for the cases when G is

(i) exponential

(ii) uniform.

Furthermore, in the uniform case we derive the Bayes risk of $\tilde{\Delta}_n$.

Section 2 contains preliminaries relating to the Dirichlet process.

2. DIRICHLET PROCESS PRELIMINARIES

This section briefly gives some definitions and theorems associated with the Dirichlet process. For further details the reader is referred to Ferguson [2].

Definition 2.1 (Ferguson). Let $Z_1, \ldots, Z_k$ be independent random variables with Z_j having a gamma distribution with shape parameter $\alpha_j \geqslant 0$ and scale parameter 1 $(j = 1, \ldots, k)$. Let $\alpha_j > 0$ for some j. The Dirichlet distribution with parameter $(\alpha_1, \ldots, \alpha_k)$, denoted by $D(\alpha_1, \ldots, \alpha_k)$, is defined as the distribution of $(Y_1, \ldots, Y_k)$, where $Y_j = \dfrac{Z_j}{\sum\limits_{i=1}^{k} Z_i}$ $(j = 1, \ldots, k)$.

Since $\sum\limits_{i=1}^{k} Y_i = 1$, the Dirichlet distribution is singular with respect to Lebesgue measure in k-dimensional space. If $\alpha_j = 0$, the corresponding Y_j is degenerate at zero. If however $\alpha_j > 0$ for all j, the $(k-1)$-dimensional distribution of $(Y_1, \ldots, Y_{k-1})$ is absolutely continuous with density

$$f(y_1, \ldots, y_{k-1} \mid \alpha_1, \ldots, \alpha_k) = \tag{7}$$

$$= \frac{\Gamma(\alpha_1 + \ldots + \alpha_k)}{\Gamma(\alpha_1) \ldots \Gamma(\alpha_k)} \left(\prod_{j=1}^{k-1} y_j^{\alpha_j - 1} \right) \times$$

$$\times \left(1 - \sum_{j=1}^{k-1} y_j \right)^{\alpha_k - 1} I_S(y_1, \ldots, y_{k-1})$$

where S is the simplex $S = \left\{ (y_1, \ldots, y_{k-1}): y_j \geqslant 0, \sum_{j=1}^{k-1} y_j \leqslant 1 \right\}$.

Definition 2.2 (F e r g u s o n). Let $(X, \mathcal{A})$ be a measurable space. Let α be a non-null finite measure (nonnegative and finitely additive) on $(X, \mathcal{A})$. We say P is a *Dirichlet process* on $(X, \mathcal{A})$ with parameter α if for every $k = 1, 2, \ldots$ and measurable partition $(B_1, \ldots, B_k)$ of X, the distribution of $(P(B_1), \ldots, P(B_k))$ is Dirichlet with parameter $(\alpha(B_1), \ldots, \alpha(B_k))$.

Definition 2.3 (F e r g u s o n). The X-valued random variables $X_1, \ldots, X_n$ constitute a sample of size n from a Dirichlet process P on $(X, \mathcal{A})$ with parameter α if for any $m = 1, 2, \ldots$ and measurable sets $A_1, \ldots, A_m, C_1, \ldots, C_n$,

$$Q\{X_1 \in C_1, \ldots, X_n \in C_n \mid P(A_1), \ldots, P(A_m), P(C_1), \ldots, P(C_n)\} =$$

$$= \prod_{i=1}^{n} P(C_i) \quad \text{a.s.,}$$

where Q denotes probability.

Theorem 2.4 (F e r g u s o n). *Let* P *be a Dirichlet process on* $(X, \mathcal{A})$ *with parameter* α, *and let* $X_1, \ldots, X_n$ *be a sample of size* n *from* P. *Then the conditional distribution of* P *given* $X_1, \ldots, X_n$ *is a Dirichlet process on* $(X, \mathcal{A})$ *with parameter* $\beta = \alpha + \sum_{i=1}^{n} \delta_{X_i}$, *where, for* $x \in X$, $A \in \mathcal{A}$, $\delta_x(A) = 1$ *if* $x \in A$, 0 *otherwise.*

3. A BAYES ESTIMATOR OF THE HORIZONTAL DISTANCE

We suppose that F is chosen according to a Dirichlet process prior on $(R, \mathcal{B})$ with parameter α. With the loss function given by (6), the Bayes estimator for the no-sample problem is found by minimizing the right hand side of (8),

(8) $EL(\tilde{\Delta}, \Delta) = \int E(\tilde{\Delta}(x) - \Delta(x))^2 \, dW(x),$

where the expectation is with respect to F. The estimator is obtained by minimizing $E(\tilde{\Delta}(x) - \Delta(x))^2$ for each x, yielding

(9) $\tilde{\Delta}(x) = E(\Delta(x)) = E\{G^{-1}F(x)\} - x.$

We next evaluate (9) in the cases where

   (i) G is exponential

   (ii) G is uniform.

   3.1. The case where G is exponential. Let $G(x) = 1 - \exp(-\lambda x)$, $x > 0$, and 0 for $x \leqslant 0$, for some $\lambda > 0$. Then

$$G^{-1}(x) = -\frac{\ln(1-x)}{\lambda} \qquad (0 \leqslant x < 1),$$

and (9) reduces to

(10) $\tilde{\Delta}(x) = \dfrac{1}{B(\alpha', \beta')} \displaystyle\int_0^1 \left[-\dfrac{\ln(1-y)}{\lambda} \right] y^{\alpha'-1}(1-y)^{\beta'-1} \, dy - x,$

where $B(\alpha', \beta') = \dfrac{\Gamma(\alpha')\Gamma(\beta')}{\Gamma(\alpha'+\beta')}$. Equation (10) makes use of the fact that for each x, $F(x)$ is distributed according to the beta distribution with parameters $\alpha' = \alpha((-\infty, x])$, $\beta' = \alpha(R) - \alpha'$. (To see this use Definition 2.2 with the measurable partition $B_1 = (-\infty, x], B_2 = R - B_1$.) Thus, for the "no-sample" problem, by expanding $\ln(1-y)$ in a power series, we obtain

$$\tilde{\Delta}(x) = \frac{1}{\lambda B(\alpha', \beta')} \int_0^1 \sum_{j=1}^{\infty} \frac{y^{\alpha'+j-1}(1-y)^{\beta'-1}}{j} \, dy - x =$$

$$= \frac{1}{\lambda} \sum_{j=1}^{\infty} \frac{B(\alpha'+j, \beta')}{jB(\alpha', \beta')} - x.$$

Using Theorem 2.4, the Bayes estimator when a sample $X_1, \ldots, X_n$ is available from F, is

$$(11) \qquad \tilde{\Delta}_n(x) = \frac{1}{\lambda} \sum_{j=1}^{\infty} \frac{B(\alpha'' + j, \beta'')}{jB(\alpha'', \beta'')} - x,$$

where

$$\alpha'' = \alpha((-\infty, x]) + \sum_{i=1}^{n} \delta_{X_i}((-\infty, x]),$$

$$\beta'' = \alpha(R) + n - \alpha''.$$

3.2. The case where G is uniform. Let $G(x) = 0$ for $x < a$, $\dfrac{x - a}{b - a}$ for $a \leqslant x \leqslant b$, and 1 for $x > b$, for some $a < b$. Then (9) reduces to

$$\tilde{\Delta}(x) = \int_0^1 [y(b - a) + a] \frac{y^{\alpha' - 1}(1 - y)^{\beta' - 1}}{B(\alpha', \beta')} \, dy - x =$$

$$= a + (b - a) \frac{B(\alpha' + 1, \beta')}{B(\alpha', \beta')} - x =$$

$$= a + (b - a) \frac{\alpha'}{\alpha' + \beta'} - x = a + (b - a)F_0(x) - x,$$

where

$$F_0(x) = \frac{\alpha((-\infty, x])}{\alpha(R)} \qquad (x \in R),$$

can be interpreted as the "prior guess" at F.

Thus, from Theorem 2.4, when a sample $X_1, \ldots, X_n$ is available from F, the Bayes estimator is

$$(12) \qquad \tilde{\Delta}_n(x) = a + (b - a)\tilde{F}_n(x) - x \qquad (x \in R),$$

where

$$\tilde{F}_n(x) = \frac{\alpha((-\infty, x]) + \sum_{i=1}^{n} \delta_{X_i}((-\infty, x])}{\alpha(R) + n}.$$

The minimum Bayes risk $S(\alpha)$ of $\tilde{\Delta}_n$ (12) can be computed using results of K o r w a r and H o l l a n d e r [3]. K o r w a r and H o l l a n d e r obtained the minimum Bayes risk $R(\alpha)$ of the estimator $\tilde{F}_n$ against weighted squared error loss to be

$$R(\alpha) = \frac{\alpha(R)}{(\alpha(R) + 1)(\alpha(R) + n)} \int F_0(x)(1 - F_0(x)) \, dW(x).$$

(See equation (2.19) of H o l l a n d e r and K o r w a r [3] and replace the m of that equation with n here.) It immediately follows that $S(\alpha) = = (b - a)^2 R(\alpha)$.

We note that we can also directly obtain the risk $T(\alpha)$ (say) of the one-sample limit of D o k s u m 's estimator $\hat{\Delta}_n$ (see equation (5)) with respect to the Dirichlet process prior with parameter α in this case when G is uniform. We find

$$\hat{\Delta}_n(x) = a + (b - a)F_n(x) - x \qquad (x \in R),$$

where $F_n(x)$ is the empirical distribution function of the X's. Using (3.3) of K o r w a r and H o l l a n d e r [3] we obtain $T(\alpha) = = (b - a)^2 \left(1 + \frac{\alpha(R)}{n}\right) R(\alpha)$.

REFERENCES

[1] K. D o k s u m, Empirical probability plots and statistical inference for nonlinear models in the two-sample case, *Ann. Statist.*, 2 (1974), 267-277.

[2] T.S. F e r g u s o n, A Bayesian analysis of some nonparametric problems, *Ann. Statist.*, 1 (1973), 209-230.

[3] R.M. K o r w a r − M. H o l l a n d e r, Empirical Bayes estimation of a distribution function, *Ann. Statist.*, 4 (1976), 581-588.

M. H o l l a n d e r
The Florida State University, Department of Statistics, Tallahassee, Florida, USA.
R. K o r w a r
University of Massachusetts, Department of Mathematics and Statistics, Amherst, Massachusetts, USA.

COLLOQUIA MATHEMATICA SOCIETATIS JÁNOS BOLYAI

32. NONPARAMETRIC STATISTICAL INFERENCE,

BUDAPEST (HUNGARY), 1980.

TWO-SAMPLE PROBLEMS UNDER RANDOM CENSORSHIP

L. HORVÁTH

1. INTRODUCTION

Burke, Csörgő and Horváth [1] introduced a general model for random censorship and proved some strong approximation results for the one sample problem from censored data. In this paper some two sample problems are discussed, and several test statistics are proposed. The strong approximation technique gives rates of convergence for these test statistics.

The general scheme of [1] is as follows. Let X be a real random variable with distribution function $F(t) = P(X < t)$, $t \in R$ and A^1, A^2 be disjoint events, and define the subdistribution functions

$$F^i(t) = P(X < t, A^i) \qquad (i = 1, 2).$$

Imagine that observing X we are interested in not only the behaviour of X, but in the joint behaviour of the pairs (X, A^i) $(i = 1, 2)$. So let $X_1, X_2, \ldots$ be a sequence of independent random variables defined on a probability space $(\Omega, \mathscr{A}, P)$ with distribution function $F(t) = P(X_n < t)$ $(n = 1, 2, \ldots)$. We are observing the sample

$$(X_1, A_1^1), (X_2, A_2^1), \ldots$$

(1.1)

$$(X_1, A_1^2), (X_2, A_2^2), \ldots$$

where $\{X_n, A_n^1, A_n^2\}_{n=1}^{\infty}$ is a sequence of independent replicas of $\{X, A^1, A^2\}$. Let F_n denote the empirical distribution function of $X_1, \ldots, X_n$. Introduce the empirical subdistribution functions

$$F_n^i(t) = \frac{1}{n} \# \{m : 1 \leqslant m \leqslant n, \ X_m < t \text{ and } A_m^i \text{ occurs}\}$$

and the empirical (censored empirical) processes

$$\alpha_n^0(t) = \sqrt{n} (F_n(t) - F(t))$$

$$\alpha_n^i(t) = \sqrt{n} (F_n^i(t) - F^i(t)) \qquad (i = 1, 2).$$

For $x = (x_1, x_2, x_3)$ let $\| x \| = \max(|x_1|, |x_2|, |x_3|)$ be the maximum-norm in R^3. We shall use the main theorem of [1] on the strong approximation of the empirical and the censored empirical processes with a Gaussian vector processes.

Theorem 1.A. (B u r k e, C s ö r g ő and H o r v á t h [1]).

(a) *If the underlying probability space is rich enough, than one can define three sequences of Gaussian processes* $B_n^0(t), B_n^1(t), B_n^2(t)$ *such that for* $\alpha_n(x) = (\alpha_n^0(x_0), \alpha_n^1(x_1), \alpha_n^2(x_2))$ *and* $B_n(x) = (B_n^0(x_0), B_n^1(x_1), B_n^2(x_2))$, $x = (x_0, x_1, x_1)$ *we have*

$$P\Big\{ \sup_{x \in R^3} \| \alpha_n(x) - B_n(x) \| > \frac{M \log n + z}{\sqrt{n}} \Big\} \leqslant K \exp(-Lz)$$

for all real z, *where* $M = 5A_1$, $K = 5A_2$ *and* $L = \dfrac{A_3}{5}$ *with* A_1, A_2 *and* A_3 *of Theorem 3 K o m l ó s, M a j o r and T u s n á d y [8]. Moreover* B_n *is a three-dimensional vector valued Gaussian process having the same covariance structure as* α_n, *namely*

$$E B_n^i(t) = 0 \qquad (i = 0, 1, 2)$$

and for any $i, j = 1, 2$, $i \neq j$,

$$\mathbb{E}B_n^0(t)B_n^0(s) = \min\left(F(t), F(s)\right) - F(t)F(s)$$

$$\mathbb{E}B_n^i(t)B_n^j(s) = -F^i(t)F^j(s)$$

$$\mathbb{E}B_n^i(t)B_n^i(s) = \min\left(F^i(t), F^i(s)\right) - F^i(t)F^i(s)$$

$$\mathbb{E}B_n^i(t)B_n^0(s) = \min\left(F^i(t), F^i(s)\right) - F^i(t)F(s).$$

(b) *If the underlying probability space is rich enough, then one can define a sequence of three-dimensional vector valued Gaussian process* $K(x, n) = (K^0(x_0, n), K^1(x_1, n), K^2(x_2, n))$ *such that we have*

$$\mathsf{P}\left\{ \sup_{x \in R^3} \left\| \alpha_n(x) - \frac{1}{\sqrt{n}} K(x, n) \right\| \frac{\bar{M} \log^2 n + z}{\sqrt{n}} \right\} \leq \bar{K} \exp\left(-\bar{L}z\right)$$

for all real z, *where* $\bar{M} = 5A_4$, $\bar{K} = 5A_5$ *and* $\bar{L} = \dfrac{A_6}{5}$ *with* A_4, A_5, A_6 *of Theorem 4 of* $K\,o\,m\,l\,\acute{o}\,s,\; M\,a\,j\,o\,r$ *and* $T\,u\,s\,n\,\acute{a}\,d\,y$ [8]. *The covariance structure of the process* $K(x, n)$ *is the following:*

$$\mathbb{E}K^i(t, n) = 0 \qquad (i = 0, 1, 2).$$

If $i, j = 1, 2$ *and* $i \neq j$, *then*

$$\mathbb{E}K^0(t, m)K^0(s, n) =$$

$$= \min(m, n)\left[\min\left(F(t), F(s)\right) - F(t)F(s)\right]$$

$$\mathbb{E}K^i(t, m)K^i(s, n) =$$

$$= \min(m, n)\left[\min\left(F^i(t), F^i(s)\right) - F^i(t)F^i(s)\right]$$

$$\mathbb{E}K^i(t, m)K^j(s, n) =$$

$$= \min(m, n)\left[-F^i(t)F^j(s)\right]$$

$$\mathbb{E}K^i(t, m)K^0(s, n) =$$

$$= \min(m, n)\left[\min\left(F^i(t), F^i(s)\right) - F^i(t)F(s)\right].$$

Note that the continuity assumptions on F, F^1, F^2 are not necessary (see [7]). Henceforth, we will assume that we are working on the probability space $(\Omega, \mathscr{A}, \mathsf{P})$ of Theorem 1.A. That is, all the approximation theorem in the sequel are understood to hold on this triale.

We will use the following important theorem.

Theorem 1.B. (Dvoretzky, Kiefer and Wolfowitz [5]).

$$P(\sup_{-\infty < t < \infty} |\alpha_n^0(t)| > z) \leqslant D \exp(-2z^2)$$

for all $z > 0$ and D is an absolute constant.

The limiting process of the empirical process is a Brownian Bridge $B(\cdot)$ for which it is well known that

$$(1.2) \qquad P(\sup_{0 \leqslant t \leqslant 1} |B(t)| > z) \leqslant 2 \exp(-2z^2).$$

In the two sample case (first considered by Efron [6]) we have an other sample

$$(1.3) \qquad \begin{array}{l} (\hat{X}_1, \hat{A}_1^1), (\hat{X}_2, \hat{A}_2^1), \ldots \\[6pt] (\hat{X}_1, \hat{A}_1^2), (\hat{X}_2, \hat{A}_2^2), \ldots \end{array}$$

with the corresponding distribution functions $\hat{F}, \hat{F}^1, \hat{F}^2$. We assume that the sample of (1.1) independent from the sample of (1.3). We make a comparison between the two samples by general Chernoff – Savage tests. These tests were considered in the censored data problem at first by Efron [6]. His limit distribution result will be reproduced in Theorem 3.1 (with a rate of convergence by an appropriate choice of the A's and the score functions. The general model (with another appropriate choice of the A's) makes us able also to handle the case when the data are censored on the left (cf. Theorem 3.2). Moreover, we can also consider the case when $X_n = \hat{X}_n$ ($n = 1, 2, \ldots$), i.e., when there are two possible censorship on the same data (Theorem 3.3). All these results (together with the accompanying rates of convergence) are derived from the main Theorem 2.1 proved in Section 2.

2. MAIN THEOREM

Consider, the bivariate functions R and S defined on the unit square. Suppose that the partial derivatives of R and S are exist. Set $N = m + n,$

$$\Psi_N = \int_{-\infty}^{\infty} R(F_n^1(t), F_n^2(t))\, dS(\hat{F}_m^1(t), \hat{F}_m^2(t))$$

and

$$\Psi = \int_{-\infty}^{\infty} R(F^1(t), F^2(t))\, dS(\hat{F}^1(t), \hat{F}^2(t))$$

Θ_N denotes the following random variable

$$\Theta_N = \sqrt{\frac{m}{N}} \int_{-\infty}^{\infty} R_x(F^1(t), F^2(t)) B_n^1(t)\, dS(\hat{F}^1(t), \hat{F}^2(t)) +$$

$$+ \sqrt{\frac{m}{N}} \int_{-\infty}^{\infty} R_y(F^1(t), F^2(t)) B_n^2(t)\, dS(\hat{F}^1(t), \hat{F}^2(t)) -$$

$$- \sqrt{\frac{n}{N}} \int_{-\infty}^{\infty} S_x(\hat{F}^1(t), \hat{F}^2(t)) \hat{B}_m^1(t)\, dR(F^1(t), F^2(t)) -$$

$$- \sqrt{\frac{n}{N}} \int_{-\infty}^{\infty} S_y(\hat{F}^1(t), \hat{F}^2(t)) \hat{B}_m^2(t)\, dR(F^1(t), F^2(t)).$$

The random variable Θ_N has normal distribution. The expectation of Θ_N is 0, and we can compute $E\Theta_N^2$ by the help of the given covariance structure of $B_n^1, B_n^2, \hat{B}_m^1, \hat{B}_m^2$. From the independence of (1.1) and (1.3) it follows that the approximating processes (B_n^1, B_n^2) and $(\hat{B}_m^1, \hat{B}_m^2)$ are also independent.

We also introduce the rate sequence

$$r(n, m, T) = 2v_1(n, m) + 10v_1(m, n) + v_2(n, m) + 5v_2(m, n) +$$

$$+ 4v_3(n, m) + 4v_4(N) + 4v_5(n, m),$$

where

$$v_1(n, m) = 2T^2 C \sqrt{\frac{m}{N}} \frac{1}{\sqrt{n}} \log n,$$

$$v_2(n, m) = 8T^2 \epsilon \sqrt{\frac{m}{N}} \frac{1}{\sqrt{n}} \log N,$$

$$v_3(n, m) = 4T^2 \sqrt{\frac{n}{N}} [\sqrt{2\epsilon} + 2 + \epsilon] \frac{1}{N^{\frac{1}{3}}} \sqrt{\log N},$$

$$v_4(N) = 2T^2 \epsilon \frac{\log N}{\sqrt{N}},$$

$$v_5(n, m) = 4\sqrt{\frac{N}{n}} \sqrt{\frac{\epsilon}{2}} T^2 C \frac{(\log N)^{\frac{3}{2}}}{N^{\frac{1}{3}}} +$$

$$+ T^2 \left(\epsilon + \sqrt{\frac{2}{3} \epsilon + \epsilon^2} \right) \frac{\log N}{\sqrt{N}},$$

$\epsilon > 0$ is an arbitrary constant and $C = M + \frac{\epsilon}{L}$. The constant T depends only on the functions R, S and we define it exactly in the following theorem. If $\delta < \frac{m}{N} < 1 - \delta$, $0 < \delta < 1$, then we have

$$r(n, m, T) = O\left(\frac{(\log N)^{\frac{3}{2}}}{N^{\frac{1}{3}}} \right).$$

The above (at the first sight) complicated notation will considerably shorten the space of proofs later.

Set

$$Q_N = 6Kn^{-\epsilon} + 6Km^{-\epsilon} + 20DN^{-\epsilon} + 188N^{-\epsilon}.$$

Theorem 2.1. *Suppose that R and S have continuous partial derivatives up to the second order and let T denote the common bound of all of these functions. Then*

$$P\left\{ \left| \sqrt{\frac{mn}{N}} (\Psi_N - \Psi) - \Theta_N \right| > r(n, m, T) \right\} \leq Q_N.$$

Proof. By Taylor expansion

$$\sqrt{\frac{mn}{N}} (\Psi_N - \Psi) - \Theta_N =$$

$$= \sqrt{\frac{m}{N}} \int_{-\infty}^{\infty} R_x(F^1(t), F^2(t))(\alpha_n^1(t) - B_n^1(t)) \, dS(\hat{F}^1(t), \hat{F}^2(t)) +$$

$$+ \sqrt{\frac{m}{N}} \int_{-\infty}^{\infty} R_y(F^1(t), F^2(t))(\alpha_n^2(t) - B_n^2(t)) \, dS(\hat{F}^1(t), \hat{F}^2(t)) -$$

$$- \sqrt{\frac{n}{N}} \int_{-\infty}^{\infty} S_x(\hat{F}^1(t), \hat{F}^2(t))(\hat{\alpha}_m^1(t) - \hat{B}_m^1(t))\, dR(F^1(t), F^2(t)) -$$

$$- \sqrt{\frac{n}{N}} \int_{-\infty}^{\infty} S_y(\hat{F}^1(t), \hat{F}^2(t))(\hat{\alpha}_m^2(t) - \hat{B}_m^2(t))\, dR(F^1(t), F^2(t)) +$$

$$+ \frac{1}{2}\sqrt{\frac{mn}{N}} \int_{-\infty}^{\infty} R''(\tau_1(t)) \times$$

$$\times \begin{pmatrix} F_n^1(t) - F^1(t) \\ F_n^2(t) - F^2(t) \end{pmatrix} \begin{pmatrix} F_n^1(t) - F^1(t) \\ F_n^2(t) - F^2(t) \end{pmatrix} dS(\hat{F}_m^1(t), \hat{F}_m^2(t)) -$$

$$- \frac{1}{2}\sqrt{\frac{mn}{N}} \int_{-\infty}^{\infty} S''(\tau_2(t)) \times$$

$$\times \begin{pmatrix} \hat{F}_m^1(t) - \hat{F}^1(t) \\ \hat{F}_m^2(t) - \hat{F}^2(t) \end{pmatrix} \begin{pmatrix} F_m^1(t) - F^1(t) \\ F_m^2(t) - F^2(t) \end{pmatrix} dR(F_n^1(t), F_n^2(t)) +$$

$$+ \sqrt{\frac{m}{N}} \int_{-\infty}^{\infty} R_x(F^1(t), F^2(t)) \times$$

$$\times \alpha_n^1(t)\, d(S(\hat{F}_m^1(t), \hat{F}_m^2(t)) - S(\hat{F}^1(t), \hat{F}^2(t))) +$$

$$+ \sqrt{\frac{m}{N}} \int_{-\infty}^{\infty} R_y(F^1(t), F^2(t)) \times$$

$$\times \alpha_n^2(t)\, d(S(\hat{F}_m^1(t), \hat{F}_m^2(t)) - S(\hat{F}^1(t), \hat{F}^2(t))) =$$

$$= q_{N1} + \ldots + q_{N8},$$

where R_x and R_y denote the first order partial derivatives of R with respect to its first and second variable, while R'' is the second order total derivative of R, and the notations for S are analogous. Handling the above eight terms there are only three cases which differ from each other essentially. These different cases are studied in the following lemmas. The estimation of $q_{N1} + \ldots + q_{N4}$ will follow from Lemma 2.1

Lemma 2.1.

$$P\left\{\sqrt{\frac{m}{N}} \left| \int_{-\infty}^{\infty} R_x(F^1(t), F^2(t)) \times \right. \right.$$

$$\left. \left. \times (\alpha_n^1(t) - B_n^1(t)) \, dS(\hat{F}^1(t), \hat{F}^2(t)) \right| > v_1(n, m) \right\} \leqslant K n^{-\epsilon}.$$

Proof. For a function $f(t)$, let $V(f)$ denote the total variation of $f(t)$ on the whole line. The probability in the lemma is not greater than

$$P\left\{\sqrt{\frac{m}{N}} \, T \sup_{-\infty < t < \infty} |\alpha_n^1(t) - B_n^1(t)| V(S(\hat{F}^1(t), \hat{F}^2(t)) > \right.$$

$$\left. > v_1(n, m) \right\} \leqslant K n^{-\epsilon}$$

by the inequality $V(S(\hat{F}^1(t), \hat{F}^2(t)) \leqslant 2T$ and Theorem 1.A.

We estimate the terms $q_{N5} + q_{N6}$ by the following

Lemma 2.2.

$$P\left\{\sqrt{\frac{mn}{N}} \left| \int_{-\infty}^{\infty} R''(\tau_1(t)) \times \right. \right.$$

$$\left. \left. \times \begin{pmatrix} F_n^1(t) - F^1(t) \\ F_n^2(t) - F^2(t) \end{pmatrix} \begin{pmatrix} F_n^1(t) - F^1(t) \\ F_n^2(t) - F^2(t) \end{pmatrix} dS(\hat{F}_m^1(t), \hat{F}_m^2(t)) \right| > \right.$$

$$\left. > 2v_2(n, m) \right\} \leqslant 4DN^{-\epsilon}.$$

Proof. The total variation of $S(\hat{F}_m^1(t), \hat{F}_m^2(t))$ is less than or equal to $2T$. Therefore the probability in the lemma is not greater than

$$P(\sup_{-\infty < t < \infty} |\alpha_n^1(t)| > \sqrt{2\epsilon \log N}) +$$

$$+ P(\sup_{-\infty < t < \infty} |\alpha_n^1(t)||\alpha_n^2(t)| > 2\epsilon \log N) +$$

$$+ P(\sup_{-\infty < t < \infty} |\alpha_n^2(t)| > \sqrt{2\epsilon \log N}) \leqslant 4DN^{-\epsilon},$$

using equality (3.2) in [1] and Theorem 1.B.

Lemma 2.3.

$$P\left\{\sqrt{\frac{m}{N}} \left| \int_{-\infty}^{\infty} R_x(F^1(t), F^2(t)) \times \right. \right.$$

$$\left. \times \alpha_n^1(t) \, d(S(\hat{F}_m^1(t), \hat{F}_m^2(t)) - S(\hat{F}^1(t), \hat{F}^2(t))) \right| >$$

$$> 4v_1(m, n) + 2v_2(m, n) + 2v_3(n, m) +$$

$$\left. + 2v_4(N) + 2v_5(n, m) \right\} \leqslant$$

$$\leqslant 6DN^{-\epsilon} + 2Km^{-\epsilon} + 94N^{-\epsilon} + 2Kn^{-\epsilon}.$$

Proof. We use the proof of Theorem 4.2 in [1]. Integrating by parts and using the Taylor expansion, the probability in the lemma is not greater than the sum

$$P\left\{\frac{1}{\sqrt{N}} \left| \int_{-\infty}^{\infty} S_x(\hat{F}^1(t), \hat{F}^2(t))\hat{\alpha}_m^1(t) \, dR_x(F^1(t), F^2(t))\alpha_n^1(t) \right| > \right.$$

$$\left. > 2v_1(m, n) + v_3(n, m) + v_4(N) + v_5(n, m) \right\} +$$

$$+ P\left\{\frac{1}{\sqrt{N}} \left| \int_{-\infty}^{\infty} S_y(\hat{F}^1(t), \hat{F}^2(t)) \times \right. \right.$$

$$\left. \times \hat{\alpha}_m^2(t) \, dR_x(F^1(t), F^2(t))\alpha_n^1(t) \right| >$$

$$\left. > 2v_1(m, n) + v_3(n, m) + v_4(N) + v_5(n, m) \right\} +$$

$$+ P\left\{\frac{1}{2}\sqrt{\frac{mn}{N}} \left| \int_{-\infty}^{\infty} S''(\tau_3(t)) \times \right. \right.$$

$$\times \begin{pmatrix} \hat{F}_m^1(t) - \hat{F}^1(t) \\ \hat{F}_m^2(t) - \hat{F}^2(t) \end{pmatrix} \begin{pmatrix} \hat{F}_m^1(t) - \hat{F}^1(t) \\ \hat{F}_m^2(t) - \hat{F}^2(t) \end{pmatrix} dR_x(F^1(t), F^2(t)) \times$$

$$\left. \times (F_n^1(t) - F^1(t)) \right| > 2v_2(m, n) \right\} = p_{1N} + p_{2N} + p_{3N}.$$

Using Lemma 2.2, we obtain $p_{3N} \leqslant 4DN^{-\epsilon}$, because

$$V(R_x(F^1(t), F^2(t))(F_n^1(t) - F^1(t))) \leqslant 4T.$$

The estimations of the probabilities p_{1N} and p_{2N} are similar, hence we study only p_{1N}.

$$p_{1N} \leqslant P\left\{\frac{1}{\sqrt{N}} \left| \int_{-\infty}^{\infty} S_x(\hat{F}^1(t), \hat{F}^2(t)) \times \right.\right.$$

$$\left. \times \hat{B}_m^1(t) \, dR_x(F^1(t), F^2(t)) \alpha_n^1(t) \right| >$$

$$> v_3(n, m) + v_4(N) + v_5(n, m)\right\} +$$

$$+ P\left\{\sqrt{\frac{n}{N}} \left| \int_{-\infty}^{\infty} S_x(\hat{F}^1(t), \hat{F}^2(t)) \times \right.\right.$$

$$\left. \times (\hat{\alpha}_m^1(t) - \hat{B}_m^1(t)) \, dR_x(F^1(t), F^2(t))(F_n^1(t) - F^1(t)) \right| >$$

$$> 2v_1(m, n)\right\}$$

and the second probability is majorized by $Km^{-\epsilon}$ using Theorem 1.A.

Divide the real line into $2[N^{\frac{2}{3}}]$ parts ($[\cdot]$ denotes the integer part function) by the points

$$t'_j = \inf\left\{s: F(s) \geqslant \frac{j}{[N^{\frac{2}{3}}]}\right\}$$

$$\qquad (j = 0, 1, \ldots, [N^{\frac{2}{3}}]),$$

$$t''_j = \inf\left\{s: \hat{F}(s) \geqslant \frac{j}{[N^{\frac{2}{3}}]}\right\}$$

and let $\{t_j\}$ denote the union of the pointsets $\{t'_j\}, \{t''_j\}$. We will now approximate $S_x(\hat{F}^1(t), \hat{F}^2(t))\hat{B}_m^1(t)$ by a jump process $\tilde{B}_m^1(t)$ defined as

$$\tilde{B}_m^1(t) = S_x(\hat{F}^1(t_j), \hat{F}^2(t_j))\hat{B}_m^1(t_j) \quad \text{if} \quad t_j \leqslant t < t_{j+1}.$$

The difference of the two processes is majorized by

$$|\tilde{B}_m^1(t) - S_x(\hat{F}^1(t), \hat{F}^2(t))\hat{B}_m^1(t)| \leqslant$$

$$\leqslant |S_x(\hat{F}^1(t), \hat{F}^2(t)) - S_x(\hat{F}^1(t_j), \hat{F}^2(t_j))| |\hat{B}_m^1(t)| +$$

$$+ |S_x(\hat{F}^1(t_j), \hat{F}^2(t_j))| |\hat{B}_m^1(t) - \hat{B}_m^1(t_j)|.$$

By the inequality of C s ö r g ő and R é v é s z ([3], [1]) we have

$$P\left\{\max_{t_j} \sup_{t_j \leqslant t < t_{j+1}} |S_x(\hat{F}^1(t), \hat{F}^2(t)) - S_x(\hat{F}^1(t_j), \hat{F}^2(t_j))| \times\right.$$

$$\left. \times |\hat{B}_m^1(t_j)| > \frac{2T\sqrt{\epsilon \log N}}{N^{\frac{2}{3}}\sqrt{2}}\right\} +$$

$$+ P\left\{\max_{t_j} \sup_{t_j \leqslant t < t_{j+1}} |\hat{B}_m^1(t_j) - \hat{B}_m^1(t)| > \frac{\sqrt{(2+\epsilon)\log N}}{N^{\frac{1}{3}}}\right\} \leqslant$$

$$\leqslant 2N^{-\epsilon} + 41N^{-\epsilon}.$$

Therefore, using (1.2),

$$P\left\{\frac{1}{\sqrt{N}} \left| \int_{-\infty}^{\infty} S_x(\hat{F}^1(t), \hat{F}^2(t))\hat{B}_m^1(t)\, dR_x(F^1(t), F^2(t))\alpha_n^1(t) \right| > \right.$$

$$\left. > v_3(n,m) + v_4(N) + v_5(n,m))\right\} \leqslant 43N^{-\epsilon} +$$

$$+ P\left\{\frac{1}{\sqrt{N}} \sup_{-\infty < t < \infty} |\tilde{B}_m^1(t)| \times \right.$$

$$\times \sum_{j=0}^{2[N^{\frac{2}{3}}]-1} |R_x(F^1(t_{j+1}), F^2(t_{j+1}))\alpha_n^1(t_{j+1}) -$$

$$\left. - R_x(F^1(t_j), F^2(t_j))\alpha_n^1(t_j) \right| > v_4(N) + v_5(n,m)\right\} \leqslant 45N^{-\epsilon} +$$

$$+ P\left\{\frac{T\sqrt{\epsilon \log N}}{\sqrt{2N}} \sup_{-\infty < t < \infty} |\alpha_n^1(t)| 2T > v_4(N)\right\} +$$

$$+ P\left\{\frac{T^2\sqrt{\epsilon \log N}}{\sqrt{2N}} \sum_{j=0}^{2[N^{\frac{2}{3}}]-1} |\alpha_n^1(t_{j+1}) - \alpha_n^1(t_j)| > v_5(n,m)\right\} \leqslant$$

$$\leqslant 45N^{-\epsilon} + DN^{-\epsilon} + Kn^{-\epsilon} + 2N^{-\epsilon}.$$

The last inequality we obtain from the proof of Theorem 4.1 in [1]. So Lemma 2.3 is proved.

This, together with Lemma 2.1 and Lemma 2.2 gives Theorem 2.1.

3. APPLICATIONS

The first application is the Kaplan – Meier model, the most important example for random censorship from the right. Let $X_1^0, \ldots, X_n^0$ be a sequence of independent identically distributed random variables with a continuous distribution function $F^0(t) = P(X_1^0 < t)$ $(t \in R)$. In the most frequent applications, t is a time parameter and $X_1^0, \ldots, X_n^0$ are the survival times of n individuals in a life table. Then $F^0(0) = 0$, but this is not assumed here. The X_i^0 are censored on the right by Y_i $(1 \leqslant i \leqslant n)$, a sequence of independent random variables, independent from the $X_1^0, \ldots, X_n^0$ sequence, with common continuous distribution function $H(t) = P(Y_1 < t)$ $(t \in R)$. One can observe

(3.1) $X_j = \min (X_j^0, Y_j)$

and

(3.2) $A_j^1 = \{\omega: X_j^0(\omega) = X_j(\omega)\}$.

Consequently, $X_1, \ldots, X_n$ is a random sample with distribution function F satisfying

(3.3) $1 - F(t) = (1 - F^0(t))(1 - H(t)$ $(t \in R)$.

The uncensored observations $(A_j^1$ occurs) have subdistribution function

(3.4) $F^1(t) = \int_{-\infty}^{t} (1 - H(s))\, dF^0(s)$,

and the censored subdistribution function is

(3.5) $F^2(t) = \int_{-\infty}^{t} (1 - F^0(s))\, dH(s)$.

Our null hypothesis, which we wish to test, is $F^0 = \hat{F}^0$, that is, the random variables X_i^0, X_j^0 have the same distribution. It is desirable to choose such functions R, S that the value of Ψ depends only on the null hypothesis. This is realized if

$$\Psi^1 = \int_{-\infty}^{\infty} (2\hat{F}^1(t) + \hat{F}^2(t))\, dF(t) + \int_{-\infty}^{\infty} F^2(t)\, d\hat{F}(t).$$

If $F^0 = \hat{F}^0$, then, by (3.3), (3.4) and (3.5),

$$\Psi^1 = \int_{-\infty}^{\infty} (1 - F^0(t))(1 - H(t))(1 - \hat{H}(t)) \, d\hat{F}^0(t) +$$

$$+ \int_{-\infty}^{\infty} (1 - \hat{F}^0(t))(1 - H(t))(1 - \hat{H}(t)) \, dF^0(t) +$$

$$+ \int_{-\infty}^{\infty} (1 - F^0(t))(1 - H(t))(1 - \hat{F}^0(t)) \, d\hat{H}(t) +$$

$$+ \int_{-\infty}^{\infty} (1 - \hat{F}^0(t))(1 - \hat{H}(t))(1 - F^0(t)) \, dH(t) =$$

$$= P(\min(X^0, Y, \hat{Y}) > \hat{X}^0) + P(\min(\hat{X}^0, Y, \hat{Y}) > X^0) +$$

$$+ P(\min(X^0, Y, \hat{X}^0) > \hat{Y}) + P(\min(X^0, \hat{Y}, \hat{X}^0) > Y) = 1.$$

The statistic

$$\Psi_N^1 = \int_{-\infty}^{\infty} (2\hat{F}_m^1(t) + \hat{F}^2(t)) \, d\hat{F}_n(t) + \int_{-\infty}^{\infty} F_n^2(t) \, d\hat{F}_m(t)$$

has been proposed independently by G e h a n and G i l b e r t (according to E f r o n [6]) as a reasonable extension of Wilcoxon's statistic to the case of censored data. Let Θ_N^1 denote the corresponding normal random variable

$$\Theta_N^1 = 2\sqrt{\frac{n}{N}} \int_{-\infty}^{\infty} \hat{B}_m^1(t) \, dF(t) + \sqrt{\frac{n}{N}} \int_{-\infty}^{\infty} \hat{B}_m^2(t) \, dF(t) -$$

(3.6)
$$- \sqrt{\frac{m}{N}} \int_{-\infty}^{\infty} (B_n^1(t) + B_n^2(t)) \, d(2\hat{F}^1(t) + \hat{F}^2(t)) +$$

$$+ \sqrt{\frac{m}{N}} \int_{-\infty}^{\infty} B_n^2(t) \, d\hat{F}(t) - \sqrt{\frac{n}{N}} \int_{-\infty}^{\infty} (\hat{B}_m^1(t) + \hat{B}_m^2(t)) \, dF^2(t).$$

From Theorem 2.1 we obtain

Theorem 3.1. *In the Kaplan -- Meier model*

$$P\left\{ \left| \sqrt{\frac{mn}{N}} (\Psi_N^1 - 1) - \Theta_N^1 \right| > r(n, m, 1) + r(m, n, 2) \right\} \leqslant 2Q_N,$$

and

$$\sup_{-\infty < x < \infty} \left| P\left(\sqrt{\frac{mn}{N}} \, (\Psi_N^1 - 1) < x \right) - P(\Theta_N^1 < x) \right| \le$$

$$\le (2\pi E(\Theta_N^1)^2)^{\frac{1}{2}} (r(n, m, 1) + r(m, n, 2)) + 2Q_N.$$

In the second application we consider the model of random censorship on the left. A general theory was formulated in [4] for left censorship problems based on one basic sequence $\{X_j^0\}$. In this model

$$X_j = \max (X_j^0, Y_j), \quad A_j^1 = \{\omega \colon X_j^0(\omega) = X_j(\omega)\},$$

the random variables X_j^0, Y_j are as in the Kaplan – Meier model. Here we have

$$F(t) = F^0(t)H(t) \qquad (t \in R),$$

(3.7)
$$F^1(t) = \int_{-\infty}^{t} H(s) \, dF^0(s),$$

$$F^2(t) = \int_{-\infty}^{t} F^0(s) \, dH(s).$$

In the corresponding two-sample problem we are given another two sequences $\{\hat{X}_j^0\}, \{\hat{Y}_j\}$ with $\hat{X}_j = \max (\hat{X}_j^0, \hat{Y}_j)$. The variant of the Gehan – Gilbert statistic for the case of censorship from the left (in obvious notation) is

$$\Psi_N^2 = \int_{-\infty}^{\infty} F_n(t) \, d(2\hat{F}_m^1(t) + \hat{F}_m^2(t)) + \int_{-\infty}^{\infty} \hat{F}_m(t) \, dF_n^2(t).$$

If our null hypothesis, $F^0 = \hat{F}^0$, is satisfied, then

$$\Psi^2 = \int_{-\infty}^{\infty} F(t) \, d(2\hat{F}^1(t) + \hat{F}^2(t)) + \int_{-\infty}^{\infty} \hat{F}(t) \, dF^2(t) =$$

$$= \int_{-\infty}^{\infty} H(t)F^0(t)\hat{H}(t) \, d\hat{F}^0(t) + \int_{-\infty}^{\infty} H(t)\hat{F}^0(t)\hat{H}(t) \, dF^0(t) +$$

$$+ \int_{-\infty}^{\infty} F^0(t)H(t)\hat{F}^0(t) \, d\hat{H}(t) + \int_{-\infty}^{\infty} F^0(t)\hat{H}(t)\hat{F}^0(t) \, dH(t) =$$

$$= P(\max(Y_i, X_i^0, \hat{Y}_i) < \hat{X}_i^0) + P(\max(Y_i, \hat{X}_i^0, \hat{Y}_i) < X_i^0) +$$

$$+ P(\max(X_i^0, Y_i, \hat{X}_i^0) < \hat{Y}_i) + P(\max(\hat{X}_i^0, \hat{Y}_i, X_i^0) < Y_i) =$$

$$= 1.$$

For the corresponding normal variable Θ_N^2, playing the role of Θ_N^1, we have

$$\Theta_N^2 = \sqrt{\frac{m}{N}} \int_{-\infty}^{\infty} (B_n^1(t) + B_n^2(t)) \, d(2\hat{F}^1(t) + \hat{F}^2(t)) -$$

$$- 2\sqrt{\frac{n}{N}} \int_{-\infty}^{\infty} \hat{B}_m^1(t) \, dF(t) - \sqrt{\frac{n}{N}} \int_{-\infty}^{\infty} \hat{B}_m^2(t) \, dF(t) +$$

$$+ \sqrt{\frac{n}{N}} \int_{-\infty}^{\infty} (\hat{B}_m^1(t) + \hat{B}_m^2(t)) \, dF^2(t) - \sqrt{\frac{m}{N}} \int_{-\infty}^{\infty} B_n^2(t) \, d\hat{F}(t).$$

Theorem 3.2. *In the case of censored data from the left*

$$P\left(\left|\sqrt{\frac{mn}{N}} (\Psi_N^2 - 1) - \Theta_N^2\right| > v(n, m, 2) + v(m, n, 1)\right) \leqslant 2Q_N,$$

and

$$\sup_{-\infty < x < \infty} \left| P\left(\sqrt{\frac{mn}{N}} (\Psi_N^2 - 1) < x\right) - P(\Theta_N^2 < x) \right| \leqslant$$

$$\leqslant (2\pi E(\Theta_N^2)^2)^{\frac{1}{2}} (v(n, m, 2) + v(m, n, 1)) + 2Q_N.$$

These theorems and Theorem 2.1 are extensions of Chernoff – Savage statistics to the case of the censored data. If $R(x, y) = f(x + y)$ and $S(x, y) = g(x + y)$, then we get back the original Chernoff – Savage statistics.

There are many kinds of statistics to test the null hypothesis of the equality of subdistribution functions. In this case $X_n = \hat{X}_n$ and A_n^1, $\hat{A}_n^1 = A_n^2$ are arbitrary disjoint events. Introduce the following Chernoff – Savage statistics

$$\Psi_N^3 = \int_{-\infty}^{\infty} R(F_n^1(t), F_m^2(t)) \, dS(F_n^1(t), F_m^2(t)),$$

$$\Psi^3 = \int\limits_{-\infty}^{\infty} R(F^1(t), F^2(t))\, dS(F^1(t), F^2(t)),$$

and the approximating variable

$$\Theta_N^3 = \sqrt{\frac{m}{N}} \int\limits_{-\infty}^{\infty} R_x(F^1(t), F^2(t)) K^1(t, n)\, dS(F^1(t), F^2(t)) +$$

$$+ \sqrt{\frac{n}{N}} \int\limits_{-\infty}^{\infty} R_y(F^1(t), F^2(t)) K^2(t, m)\, dS(F^1(t), F^2(t)) -$$

$$- \sqrt{\frac{m}{N}} \int\limits_{-\infty}^{\infty} S_x(F^1(t), F^2(t)) K^1(t, n)\, dR(F^1(t), F^2(t)) -$$

$$- \sqrt{\frac{n}{N}} \int\limits_{-\infty}^{\infty} S_y(F^1(t), F^2(t)) K^2(t, m)\, dR(F^1(t), F^2(t)).$$

The strong approximation technique, which we used in the previous cases, can be applied to this problem as well. Using part (b) of Theorem 1.A instead of part (a), the following result is obtained by a repetition of the proof of Theorem 2.1.

Theorem 3.3. *If there exists a constant* δ, $0 < \delta < 1$, *such that* $\delta < \dfrac{m}{N} < 1 - \delta$, *then*

$$\sup_{-\infty < x < \infty} \left| P\left(\sqrt{\frac{mn}{N}}\, (\Psi_N^3 - \Psi^3) < x \right) - P(\Theta_N^3 < x) \right| =$$

$$= O\left(\frac{(\log N)^{\frac{5}{2}}}{N^{\frac{1}{3}}} \right).$$

Acknowledgement. I would like to express my sincere thanks to Professor S. Csörgő for introducing me to this field of research and for helpful conversations.

REFERENCES

[1] M.D. Burke – S. Csörgő – L. Horváth, Strong approximation of some biometric estimates under random censorship, *Z. Wahrscheinlichkeitstheorie verw. Geb.*, 56 (1981), 87-112.

[2] H. Chernoff – I.R. Savage, Asymptotic normality and efficiency of certain nonparametric test statistics, *Ann. Math. Statist.*, 29 (1958), 972-997.

[3] M. Csörgő – P. Révész, *Strong approximations in probability and statistics*, Academic Press, New York, 1980.

[4] S. Csörgő – L. Horváth, Random censorship from the left, *Studia Sci. Math. Hungar.*, (to appear).

[5] A. Dvoretzky – J. Kiefer – J. Wolfowitz, Asymptotic minimax character of the sample distribution function and of the multinomial estimator, *Ann. Math. Statist.*, 27 (1956), 642-669.

[6] B. Efron, The two-sample problem with censored data, *Proc. Fifth Berkeley Symp. Math. Statist. Prob.*, 4 (1967), 831-853.

[7] L. Horváth, Dropping continuity assumptions in random censorships models, *Periodica Math. Hungar.*, (submitted).

[8] J. Komlós – P. Major – G. Tusnády, An approximation of partial sums of independent r.v.'s and the sample d.f. I., *Z. Wahrscheinlichkeitstheorie verw. Geb.*, 32 (1975), 113-132.

L. Horváth

Bolyai Inst. of József A. University, Aradi vértanuk tere 1, Szeged, Hungary.

COLLOQUIA MATHEMATICA SOCIETATIS JÁNOS BOLYAI

32. NONPARAMETRIC STATISTICAL INFERENCE,

BUDAPEST (HUNGARY), 1980.

ON BOUNDED LENGTH SEQUENTIAL CONFIDENCE INTERVAL FOR PARAMETER IN REGRESSION MODEL BASED ON RANKS

M. HUŠKOVÁ

1. INTRODUCTION

Consider the simple linear regression model:

$$(1.1) \qquad Y_i = \Delta_0 + \Delta_1 c_i + X_i \qquad (i = 1, 2, \ldots)$$

where c_i are known regression constants Δ_0, Δ_1 are parameters and X_i are independent identically distributed random variables with continuous distribution function.

There is a couple of papers (e.g. [2], [4], [6], [7], [8], [14], [16]), where a sequential procedure for confidence interval for parameter Δ_1, such that the length of this interval is bounded from above by $2d$ $(2d > 0$ given) and the confidence coefficient is $1 - \alpha$, is proposed and some asymptotic properties are studied.

The sequential procedure based on Wilcoxon rank statistics (for location and shift parameter) was proposed and studied by G e e r t s e m a [6]. G h o s h and S e n [7] developed the sequential procedure based on a general class of regression rank statistics, showed that the confidence coefficient is asymptotically equal $1 - \alpha$ (for $d > 0$) and derived the asymptotic properties of the sample size $N(d)$. Namely, they proved that

$$\lim_{d \to 0} N(d) = \infty \quad \text{a.s.,}$$

$$\lim_{d \to 0} d^2 N(d) \quad \text{and} \quad \lim_{d \to 0} E N(d) d^2$$

exist and finite.

Jurečková [14] investigated the asymptotic distribution of properly standardized $\sqrt{N(d)}$ corresponding to the mentioned procedure based on Wilcoxon rank statistics in the two sample location model.

Here we will consider the assumptions slightly different from those of Ghosh and Sen — both type of assumptions include "smooth unbounded" scores, namely, widely used normal scores. We prove that the assertion of Ghosh and Sen [7] holds true here too, moreover, we show that the asymptotic distribution of properly standardized $\sqrt{N(d)}$ is asymptotically normal.

To prove the results we need to show the almost sure convergence of a rank statistics residual process to some constant and the weak convergence of a properly standardized rank statistics residual process to some Gaussian process (Section 3 and Section 4). The first convergence was already proved by Ghosh and Sen [7] under different assumptions. The second one was showed for Wilcoxon rank statistics by Jurečková [13] and Antille [2].

2. ASYMPTOTIC PROPERTIES OF STOPPING RULE

Consider the model described in the previous section. Denote

$$(2.1) \qquad S_N(Y_i - \Delta c_i) = \sum_{i=1}^{N} c_{iN} \varphi \left(\frac{R_{iN}^{\Delta}}{N + 1} \right),$$

where c_i are known constants, φ is a nondecreasing scoregenerating function defined on $(0, 1)$ and R_{iN}^{Δ} is the rank of $Y_i - \Delta c_i$ among

$$Y_1 - \Delta c_1, Y_2 - \Delta c_2, \ldots, Y_N - \Delta c_N,$$

$$c_{iN} = \frac{c_i - \bar{c}_N}{\sqrt{\sum_{j=1}^{N} (c_j - \bar{c}_N)^2}}, \qquad \bar{c}_N = \frac{1}{N} \sum_{i=1}^{N} c_i.$$

The confidence interval is determined by

(2.2)

$$\hat{\Delta}_N^+ = \inf \left\{ \Delta; \, S_N(Y_i - \Delta c_i) < -\Phi^{-1} \left(1 - \tfrac{\alpha}{2}\right) \times \right.$$

$$\left. \times \left(\int_0^1 \left(\varphi(u) - \int_0^1 \varphi(z) \, dz \right)^2 du \right)^{\frac{1}{2}} \right\},$$

(2.3)

$$\hat{\Delta}_N^- = \sup \left\{ \Delta; \, S_N(Y_i - \Delta c_i) > \Phi^{-1} \left(1 - \tfrac{\alpha}{2}\right) \times \right.$$

$$\left. \times \left(\int_0^1 \left(\varphi(u) - \int_0^1 \varphi(z) \, dz \right)^2 du \right)^{\frac{1}{2}} \right\},$$

where Φ^{-1} is the quantile function corresponding to the standard normal distribution, $1 - \alpha$ is the confidence coefficient.

It is known (see e.g. [12]) that $S_N(Y_i - \Delta c_i)$ is nonincreasing in Δ with probability 1 and thus $\hat{\Delta}_N^- \geqslant \hat{\Delta}_N^+$.

Now we define the sequential procedure: for every $d > 0$, let $N(d)$, stopping variable, be the smallest positive integer $N \geqslant N_0$ (initial sample size) for which $\hat{\Delta}_N^+ - \hat{\Delta}_N^- \leqslant 2d$. The confidence interval is then

$$I_\alpha(N(d)) = (\hat{\Delta}_{N(d)}^-, \hat{\Delta}_{N(d)}^+).$$

We formulate the assumptions.

Assumptions A. $X_1, X_2, \ldots$ are independent identically distributed random variables with distribution function F, density f and the first derivative of f satisfying:

$$\left| \frac{f'(F^{-1}(u))}{f(F^{-1}(u))} \right| \leqslant K_1 (r(u))^{\frac{1}{2} - \epsilon}$$

for all $u \in (0, 1)$, some $K_1 > 0$, $\epsilon > 0$, for every $\eta > 0$ there exists $\delta > 0$ such that $\int (r(F(x)))^{1-\eta} dF(x + \delta) < + \infty$, where $r(u) = \dfrac{1}{u(1-u)}$.

Assumption B. The function φ is defined on $(0, 1)$, nondecreasing and has the second derivative φ'' satisfying:

$$|\varphi''(u)| \leqslant K_2 (r(u))^2, \qquad u \in (0, 1) \quad \text{for some} \quad K_2 > 0.$$

Assumption C. The constants $c_1, c_2, \ldots$ satisfy for $N \to \infty$:

$$\frac{1}{\sqrt{N}} \max_{1 \leqslant i \leqslant N} (c_i - \bar{c}_N)^2 \to 0, \qquad \frac{1}{N} \sum_{i=1}^{N} (c_i - \bar{c}_N)^2 \to \lambda_1 > 0,$$

$$\frac{1}{N} \sum_{i=1}^{N} (c_i - \bar{c}_N)^4 \to \lambda_2, \qquad \frac{1}{N} \sum_{i=1}^{N} (c_i - \bar{c}_N)^8 \to \lambda_3.$$

Remark. The Assumption A is slightly stronger than the usual assumption that Fisher's information is finite. But still the most of the standard distributions satisfy it. Assumption B is Chernoff – Savage type assumption. As for the Assumption C it is stronger than Noether's condition but the most frequent case ($\max_{1 \leqslant i \leqslant N} c_i^2 = O(1)$) is included.

Now, we formulate the main theorem of this paper.

Theorem 2.1. *Under Assumptions A, B, C it holds*

(i) $N(d)$ *is nonincreasing in* d $(d > 0)$;

(ii) $P(N(d) < + \infty) = 1$ *for any* $d > 0$;

(iii) $\lim\limits_{d \to 0} N(d) = + \infty$ *a.s.*;

(iv) $\lim\limits_{d \to 0} N(d) d^2 = \dfrac{\gamma^2 (F, \varphi)}{\lambda_1}$ *a.s., where*

$$\gamma(F, \varphi) = \Phi^{-1} \left(1 - \frac{\alpha}{2}\right) \frac{\left(\int_0^1 \left(\varphi(u) - \int_0^1 \varphi(z)\, dz\right)^2 du\right)^{\frac{1}{2}}}{\int \varphi'(F(x)) f^2(x)\, dx};$$

(v) $\lim\limits_{d \to 0} d^2 E N(d) = \dfrac{\gamma^2 (F, \varphi)}{\lambda_1}$;

(vi) $\quad \lim_{d \to 0} P(\Delta_1 \in I_\alpha(N(d))) = 1 - \alpha;$

(vii) *the asymptotic distribution of* $\sqrt{\dfrac{\lambda_1}{\lambda_2}} \dfrac{\gamma(F, \varphi)}{d} \left\{ \dfrac{d\sqrt{N(d)}\lambda_1}{\gamma(F, \varphi)} - 1 \right\}$

is normal for $d \to 0$ *with parameters* 0 *and* $\dfrac{\sigma^2}{\left(\int \varphi'(F(x)) f^2(x)\, dx \right)^2}$,

where σ^2 *is given by* (4.2) *below.*

3. ALMOST SURE ASYMPTOTIC LINEARITY OF RANK STATISTICS

Consider the random process $\{T_N(\Delta): \Delta \in (-M, M)\}$, where M is a positive constant and

(3.1) $\qquad T_N(\Delta) = S_N(X_i + \Delta d_{iN}) - S_N(X_i),$

(3.2) $\qquad S_N(X_i + \Delta d_{iN}) = \sum_{i=1}^{N} c_{iN}\, \varphi \left(\dfrac{R_{iN}^\Delta}{N+1} \right)$

with $c_{1N}, \dots, c_{NN}$ and $d_{1N}, \dots, d_{NN}$ being known constants, the function φ defined on $(0, 1)$ and R_{iN}^Δ being the rank of $X_i + \Delta d_{iN}$ among $X_1 + \Delta d_{1N}, \dots, X_N + \Delta d_{NN}$.

The process $\{T_N(\Delta): \Delta \in (-M, M)\}$ is sometimes called the rank statistics residual process. Under mild conditions Jurečková [12] proved the weak convergence of this process to some constant. This property is called the first order asymptotic linearity of the corresponding rank statistics. Ghosh and Sen [7] strenghtened this statement to the almost sure convergence (under stronger conditions). Here we establish this result under conditions different from those considered by Ghosh and Sen. In the next section we will investigate the weak convergence of $\{T_N(\Delta): \Delta \in (-M, M)\}$ properly standardized to some Gaussian process which is often called the second order asymptotic linearity of the corresponding rank statistics.

Theorem 3.1. *Let assumptions A and B be satisfied. Let the sequences of vectors* $\{(c_{1N}, \dots, c_{NN})\}_{N=1}^{\infty}$ *and* $\{(d_{1N}, \dots, d_{NN})\}_{N=1}^{\infty}$ *fulfil:*

$$\sum_{i=1}^{N} c_{iN} = 0, \quad \sum_{i=1}^{N} c_{iN}^2 = 1, \quad \sum_{i=1}^{N} d_{iN} = 0, \quad \sum_{i=1}^{N} d_{iN}^2 = 1,$$

$$\lim_{N \to \infty} N \max_{1 \leqslant i \leqslant N} c_{iN}^2 d_{iN}^2 = 0, \quad \lim_{N \to \infty} \max_{1 \leqslant i \leqslant N} c_{iN}^2 = 0,$$

$$\lim_{N \to \infty} \max_{1 \leqslant i \leqslant N} d_{iN}^2 = 0,$$

$$\lim_{N \to \infty} N \sum_{i=1}^{N} c_{iN}^2 d_{iN}^2 = \lambda_2^*, \quad \lim_{N \to \infty} N^3 \sum_{i=1}^{N} c_{iN}^4 d_{iN}^4 = \lambda_4^*,$$

$$(c_{iN} - c_{jN})(d_{iN} - d_{jN}) \geqslant 0 \quad \text{for all} \ i, j.$$

Then there exists $\kappa_1 > 0$ and $\kappa_2 > 0$ such that

$$P\Big(\sup_{|\Delta| \leqslant M} \Big| T_N(\Delta) - \Delta \sum_{i=1}^{N} c_{iN} d_{iN} \int \varphi'(F(x)) f^2(x) \, dx \Big| \geqslant$$

$$\geqslant \frac{\log N}{N^{\kappa_1}} \Big) = O\Big(\frac{1}{N^{1+\kappa_2}} \Big).$$

Hence,

$$\sup_{|\Delta| \leqslant M} \Big| T_N(\Delta) - \Delta \sum_{i=1}^{N} c_{iN} d_{iN} \int \varphi'(F(x)) f^2(x) \, dx \Big| \to 0 \quad a.s.,$$

as $N \to \infty$.

Proof. According to Theorem 2.1 in [12] $T_N(\Delta)$ is monotone function in Δ. Thus (see [12]) it suffices to prove that there exist constants $B > 0$ and $N_0 > 0$ such that for $N > N_0$ and for $|\Delta| \leqslant M \log N$

$$P\Big(\Big| T_N(\Delta) - \Delta \sum_{i=1}^{N} c_{iN} d_{iN} \int \varphi'(F(x)) f^2(x) \, dx \Big| \geqslant \frac{1}{N^{\kappa_1}} \Big) \leqslant$$

$$\leqslant \frac{B}{N^{1+\kappa_2}}.$$

Decompose $T_N(\Delta)$ as follows:

$$(3.3) \qquad \begin{aligned} T_N(\Delta) &= T_{N1}(\Delta) + Q_N(\Delta) + V_{N1}(\Delta) - V_{N1}(0) + \\ &\quad + V_{N2}(\Delta) - V_{N2}(\Delta), \end{aligned}$$

where

$$T_{N1}(\Delta) = \sum_{i=1}^{N} c_{iN} [\varphi(F(X_i + \Delta d_{iN}))I\{F(X_i + \Delta d_{iN}) \in A_N\} -$$

$$- \varphi(F(X_i))I\{F(X_i) \in A_N\}] +$$

(3.4)
$$+ \sum_{i=1}^{N} \sum_{j=1}^{N} \frac{c_{iN}}{N} \int (u(x - X_j) - F(x)) \times$$

$$\times \varphi'(F(x))I\{F(x) \in A_N\} d(F(x - \Delta d_{iN}) - F(x)),$$

$$Q_N(\Delta) = \sum_{i=1}^{N} c_{iN} \left[\left(\frac{R_{iN}^{\Delta}}{N+1} - F(X_i + \Delta d_{iN}) \right) \times \right.$$

$$\times \varphi'(F(X_i + \Delta d_{iN}))I\{F(X_i + \Delta d_{iN}) \in A_N\} -$$

(3.5)
$$\left. - \left(\frac{R_{iN}}{N+1} - F(X_i) \right) \varphi'(F(X_i))I\{F(X_i) \in A_N\} \right] -$$

$$- \sum_{i=1}^{N} \sum_{j=1}^{N} \frac{c_{iN}}{N} \int (u(x - X_j) - F(x)) \times$$

$$\times \varphi'(F(x))I\{F(x) \in A_N\} d(F(x - \Delta d_{iN}) - F(x)),$$

(3.6)
$$V_{N1}(\Delta) = \sum_{i=1}^{N} c_{iN} \varphi \left(\frac{R_{iN}^{\Delta}}{N+1} \right) I\{F(X_i + \Delta d_{iN}) \in A_N\},$$

$$V_{N2}(\Delta) = \frac{1}{2} \sum_{i=1}^{N} c_{iN} \left(\frac{R_{iN}^{\Delta}}{N+1} - F(X_i + \Delta d_{iN}) \right) \times$$

(3.7)
$$\times I\{F(X_i + \Delta d_{iN}) \in A_N\} \times$$

$$\times \varphi'' \left(F(X_i + \Delta d_{iN})\lambda_N + (1 - \lambda_N) \frac{R_{iN}^{\Delta}}{N+1} \right),$$

where

$$u(x) = \begin{cases} 1 & \text{if } x \geqslant 0, \\ 0 & \text{if } x < 0 \end{cases}$$

and $I\{A\}$ denotes the indicator of a set A, $0 < \alpha < \dfrac{\epsilon}{\epsilon + \dfrac{3}{2}}$ (see Assumption A), $A_N = (N^{\alpha-1}, 1 - N^{\alpha-1})$, $|\lambda_N| \leq 1$. This decomposition will be also useful in the next section.

Before we show that all terms in (3.3) are sufficiently small we give a short review of elementary results that we use repeatedly. Assumptions A and B imply that there exists a constant K^* such that

(3.8)
$$f(F^{-1}(u)) \leq K^*(r(u))^{-\frac{1}{2}-\epsilon} \qquad (u \in (0, 1)),$$
$$|f'(F^{-1}(u))| \leq K^*(r(u))^{-2\epsilon}$$

(3.9) $\quad |\varphi'(u)| \leq K^* r(u), \quad |\varphi(u)| \leq K^* (\log r(u)) \qquad (u \in (0, 1)).$

Further, the integrals

(3.10) $\quad \int (r(F(x)))^{\frac{1}{2}-\frac{\eta}{2}} f'(x + z) \, dx,$

(3.11) $\quad \int (r(F(x)))^{1-\eta} \, dF(x + z)$

are uniformly bounded for $|z| \leq \delta$, where η, δ are from Assumption A. Also, for $\beta > 0$

(3.12) $\quad \int (r(F(x)))^{\beta} I\{F(x) \in A_N\} dF(x + z) = O(N^{(1-\alpha)(\beta-1)})$

holds uniformly for $|z| \leq \delta$.

Lemma 1 in [4] implies that there exists some constant $D > 0$, such that

(3.13)
$$P\left(\sup_{i; F(X_i + \Delta d_{iN}) \notin A_N} \frac{1}{N+1} \left| R_{iN}^{\Delta} - E(R_{iN}^{\Delta} \mid X_i) \right| \geq \right.$$
$$\left. \geq DN^{-1+\frac{3\alpha}{4}} \right) = O\left(\frac{1}{N^2}\right),$$

$$P\left(\sup_{i;\,F(X_i + \Delta d_{iN}) \in A_N} \frac{1}{\sqrt{N+1}} \times\right.$$

(3.14)
$$\left.\times \frac{|R_{iN}^\Delta - E(R_{iN}^\Delta \mid X_i)|}{\sqrt{F(X_i + \Delta d_{iN})(1 - F(X_i + \Delta d_{iN}))}} \geq D \log N\right) =$$

$$= O\left(\frac{1}{N^2}\right).$$

Now we will treat the members in (3.3). We start with $V_{N1}(\Delta)$. There exists a constant D_1 such that

$$V_{N1}(\Delta) = D_1 \sum_{i=1}^{N} Z_{iN}(\Delta),$$

where $Z_{iN}(\Delta) = |c_{iN}| (\log N) I\{F(X_i + \Delta d_{iN}) \notin A_N\}$ are independent random variables with

$$E Z_{iN}(\Delta) = 2 (\log N) \max (N^{\alpha-1}, F(F^{-1}(N^{\alpha-1}) + \Delta d_{iN})) |c_{iN}|$$

$$\text{(i.e. } E \sum_{i=1}^{N} Z_{iN}(\Delta) = O(N^{\alpha - \frac{1}{2}} \log N)),$$

$$E \exp\{t(Z_{iN}(\Delta) - E Z_{iN}(\Delta))\} \leq$$

$$\leq \exp\{t^2 e c_{iN}^2 \max (N^{\alpha-1}, F(F^{-1}(N^{\alpha-1}) - \Delta d_{iN}))\}$$

for $|t| \leq \dfrac{1}{\max\limits_{1 \leq i \leq N} |c_{iN}|}$. Applying the exponential inequality to

$\sum\limits_{i=1}^{N} Z_{iN}(\Delta)$ (see (3.4.15 in [16]) we obtain

$$P\left(|V_{N1}(\Delta)| \geq \frac{1}{N^{\kappa_1}}\right) \leq$$

$$\leq P\left(D_1 \left|\sum_{i=1}^{N} (Z_{iN}(\Delta) - E Z_{iN}(\Delta))\right| \geq \frac{3}{N^{\kappa_1}}\right) =$$

$$= o(e^{-N^{1-\alpha+2\kappa_1}})$$

for $\dfrac{1}{2} - \alpha < \kappa_1 < \dfrac{1}{2} - \dfrac{\alpha}{2}$.

As for $V_{N2}(\Delta)$, according to (3.14) it remains to study $V_{N1}(\Delta)I\{B_N\}$, where

$$B_N = \left\{ \max \frac{|R_{iN}^\Delta - E(R_{iN}^\Delta \mid X_i)|}{\sqrt{F(X_i + \Delta d_{iN})(1 - F(X_i + \Delta d_{iN}))}} \leqslant \right.$$

$$\left. \leqslant D\,(\log N)\sqrt{N}\right\}$$

with maximum extended over all $1 \leqslant i \leqslant N$ such that $F(X_i + \Delta d_{iN}) \in$ $\in A_N$. Obviously, there exists a constant D_2 such that

$$|V_{N2}(\Delta)|I\{B_N\} \leqslant D_2 \sum_{i=1}^N |c_{iN}| \frac{(\log N)^2}{N} r(F(X_i + \Delta d_{iN})) \times$$

$$\times I\{F(X_i + \Delta d_{iN}) \in A_N\}.$$

On the right hand side there is a sum of independent random variables and we can proceed similarly as with $V_{N1}(\Delta)$. Thus for some $\kappa_1 > 0$ and $\kappa_2 > 0$

$$P\left(|V_{N2}(\Delta)| \geqslant \frac{1}{N^{\kappa_1}}\right) = O\left(\frac{1}{N^{1+\kappa_2}}\right).$$

Now, we turn to $T_{N1}(\Delta)$ which is the sum of independent random variables. By direct computations we get

$$E(T_{N1}(\Delta) - E\,T_{N1}(\Delta))^4 = O\left(\frac{1}{N^{1+\alpha}}\right)$$

which gives the needed result for $T_{N1}(\Delta) - E\,T_{N1}(\Delta)$.

The result concerning $Q_N(\Delta)$ follows from the proofs of Lemma 4.3 and Lemma 4.4. Q.E.D.

4. SECOND ORDER ASYMPTOTIC LINEARITY OF RANK STATISTICS

We will study again the rank statistics residual process $\{T_N(\Delta): \Delta \in (-M, M)\}$ given by (3.1). We prove that this properly standardized process converges weakly to a Gaussian process (the realization of $T_N(\Delta)$ can be considered as a right-continuous function in Δ). This convergence is of its own importance. Besides the application presented in this paper there is an application in the asymptotic representation of R-estimators.

Theorem 4.1. *Let Assumptions A and B be satisfied. Let the sequences of vectors* $\{(c_{1N}, \ldots, c_{NN})\}_{N=1}^{\infty}$ *and* $\{(d_{1N}, \ldots, d_{NN})\}_{N=1}^{\infty}$ *fulfil:*

$$\sum_{i=1}^{N} c_{iN} = 0, \quad \sum_{i=1}^{N} c_{iN}^2 = 1, \quad \lim_{N \to \infty} \max_{1 \leqslant i \leqslant N} c_{iN}^2 = 0,$$

$$\sum_{i=1}^{N} d_{iN} = 0, \quad \sum_{i=1}^{N} d_{iN}^2 = 1, \quad \lim_{N \to \infty} \max_{1 \leqslant i \leqslant N} d_{iN}^2 = 0,$$

$$\lim_{N \to \infty} \max_{1 \leqslant i \leqslant N} \frac{c_{iN}^2 d_{iN}^2}{\sum\limits_{j=1}^{N} c_{jN}^2 d_{jN}^2} = 0, \quad \lim_{N \to \infty} \sum_{i=1}^{N} c_{iN} d_{iN} = b > 0.$$

Then the process $\left\{ \dfrac{T_N(\Delta) - \mathsf{E}\, T_N(\Delta)}{\sqrt{\sum\limits_{i=1}^{N} c_{iN}^2 d_{iN}^2}}, \; \Delta \in (-M, M) \right\}$ *converges weakly*

to the process $\{\Delta X; \; \Delta \in (-M, M)\}$, *where* X *is the random variable normally distributed with*

(4.1) $\qquad \mathsf{E} X = 0, \quad \operatorname{var} X = \sigma^2,$

(4.2)
$$\sigma^2 = \operatorname{var} \left\{ \varphi'(U) f(F^{-1}(U)) - \right.$$
$$\left. - b \int (u(F(x) - U) - F(x)) \varphi'(F(x)) f'(x) \, dx \right\}$$

where U *is distributed uniformly on* $(0, 1)$.

The assertion remains true if we replace $\mathsf{E} T_N(\Delta)$ *by*

$$\mu_N(\Delta) = \sum_{i=1}^{N} c_{iN} \int (\varphi F(x + \Delta d_{iN}) - \varphi(F(x))) \, dF(x).$$

Remark. If the assumptions of Theorem 4.1 are satisfied and, moreover, $\varphi''(u)$ and $f'(x)$ are continuous, then $\mathsf{E} T_N(\Delta)$ can be replaced by

$$\mu_N^*(\Delta) = \Delta \sum_{i=1}^{N} c_{iN} d_{iN} \int \varphi'(F(x)) f^2(x)\, dx +$$

$$+ \frac{\Delta^2}{2} \sum_{i=1}^{N} c_{iN} d_{iN}^2 \int (\varphi(F(x)))''\, dF(x).$$

Remark. The assertion of Theorem 4.1 remains true if we omit the assumption of monotonicity of φ.

Now we proceed to state the modification of Theorem 4.1 that gives a simple form of the asymptotic center of the process

$$T_N^*(\Delta) = S_N(X_i + \Delta d_{iN}^*) - S_N(X_i),$$

where $S_N(X_i + \Delta d_{iN}^*)$ is defined by (3.2) and

$$d_{iN}^* = d_{iN} \Big(\sum_{j=1}^{N} c_{iN}^2 d_{iN}^2 \Big)^{\frac{\gamma}{2}} \qquad 0 < \gamma < \frac{\epsilon}{\epsilon + \frac{3}{2}}$$

with ϵ from Assumption A.

Theorem 4.2. *Under assumptions of Theorem 4.1 the process*

$$\left\{ (T_N^*(\Delta) - \Delta \sum_{i=1}^{N} c_{iN} d_{iN}^* \frac{\int \varphi'(F(x)) f^2(x)\, dx}{\sqrt{\sum_{i=1}^{N} c_{iN}^2 d_{iN}^{*2}}} ,\ \Delta \in (-M, M) \right\}$$

converges weakly to the process $\{\Delta X;\ \Delta \in (-M, M)\}$ *satisfying* (4.1).

Proof of Theorem 4.1. In accordance with results of Ch. III of [3] it suffices to prove the weak convergence of finite dimensional distributions of (3.1) to the corresponding distributions of $\{G(\Delta),\ \Delta \in (-M, M)\}$ and the tightness of (3.1).

We use the decomposition (3.3), Notice that $T_{N1}(\Delta)$ is the sum of independent random variables and is the "leading term" and the other terms are "remainder terms".

In Lemmas 4.3-4.4 we show that the remainder terms are sufficiently small, the statements can be summarized as follows:

(4.3) $\qquad \lim_{N \to \infty} P\Big(|T_N(\Delta) - T_{N1}(\Delta) - (E T_N(\Delta) - E T_{N1}(\Delta))| \geq$

$$\geqslant \sqrt{\sum_{i=1}^{N} c_{iN}^2 d_{iN}^2}\ \Bigg) = 0$$

uniformly in $\Delta \in (-M, M)$.

We will study $T_{N1}(\Delta)$. Let $\Delta_1 < \ldots < \Delta_q$ be a fixed set of parameter values and let $(\lambda_1, \ldots, \lambda_q)$ be a fixed vector from R_q. We show that

(4.4) $$\lim_{N \to \infty} \mathrm{var} \Big(\sum_{r=1}^{q} \lambda_r T_{N1}(\Delta_r) \Big) = \sigma^2 \Big(\sum_{r=1}^{q} \lambda_r \Delta_r \Big)^2.$$

This will follow from these convergences:

(4.5)
$$\frac{1}{(\Delta_r d_{iN})^2} \int (\varphi(F(x + \Delta_r d_{iN})) - \varphi(F(x)))^2 I\{F(x) \in A_N\} \to$$

$$\to \int \varphi'^2(F(x)) f^2(x)\, dF(x), \quad \text{as } N \to \infty \qquad (i = 1, \ldots, N)$$

(4.6)
$$\frac{1}{\Delta_r^2 d_{jN} d_{iN}} \int\int (F(\min(x, y)) - F(x) F(y)) \varphi'(F(x)) \varphi'(F(y)) \times$$

$$\times I\{F(x) \in A_N, F(y) \in A_N\} \times$$

$$\times (f(x - \Delta_r d_{iN}) - f(x))(f(x - \Delta_r d_{jN}) - f(x))\, dx\, dy \to$$

$$\to \int\int (F(\min(x, y)) - F(x) F(y)) \times$$

$$\times \varphi'(F(x)) \varphi'(F(y)) f'(x) f'(y)\, dx\, dy \quad \text{as } N \to \infty.$$

The assumptions imply

(4.7)
$$\lim_{N \to \infty} \frac{1}{(\Delta_r d_{iN})^2} (\varphi(F(x + \Delta_r d_{iN})) - \varphi(F(x)))^2 I\{F(x) \in A_N\} =$$

$$= (\varphi'(F(x)) f(x))^2, \quad \text{a.s.}$$

Further, for $\Delta_r d_{iN} > 0$

$$\frac{1}{(\Delta_r d_{iN})^2} \int (\varphi(F(x + \Delta_r d_{iN})) - \varphi(F(x)))^2 I\{F(x) \in A_N\}\, dF(x) \leqslant$$

$$\leqslant \frac{1}{\Delta_r d_{iN}} \int_0^{\Delta_r d_{iN}} \Big(\int (\varphi'(F(x + y)) f(x + y)))^2\, dF(x) \Big)\, dy.$$

We get a similar inequality for $\Delta_r d_{iN} < 0$ and, hence

$$\varlimsup_{N \to \infty} \frac{1}{(\Delta_r d_{iN})^2} \int (\varphi(F(x + \Delta_r d_{iN})) - \varphi(F(x)))^2 \times$$

$$\times I\{F(x) \in A_N\} \, dF(x) \leqslant \int (\varphi'(F(x))f(x))^2 \, dF(x)$$

which together with (4.7) and Fatou's lemma implies (4.5). The convergence (4.6) can be showed similarly.

Since $\sum_{r=1}^{q} \lambda_r T_{N1}(\Delta_r)$ is the sum of independent random variables

the asymptotic normality of $\sum_{r=1}^{q} \lambda_r T_N(\Delta_r)$ follows from (4.3), (4.4) and Theorem V. 1.2 of [9].

As for the tightness of the process $T_N(\Delta)$, (4.3)-(4.4) imply that for fixed numbers $\Delta \in (-M, M)$ and $\delta \in (0, 1)$ and $m = 1, 2, \ldots,$

$$\varlimsup_{N \to \infty} P \left(\min \left(\left| T_N \left(\Delta + \frac{j\delta}{m} \right) - T_N \left(\Delta + \frac{i\delta}{m} \right) \right|, \right. \right.$$

$$\left. \left. \left| T_N \left(\Delta + \frac{k\delta}{m} \right) - T_N \left(\Delta + \frac{j\delta}{m} \right) \right| \right) \geqslant \lambda \right) \leqslant$$

$$\leqslant \left[\frac{k-i}{m} \right]^2 \frac{\sigma^3 \delta^2}{\lambda^2},$$

where $0 \leqslant i \leqslant j \leqslant k \leqslant m$ $(m = 1, 2, \ldots)$ and σ^2 is given by (4.2). Thus by Lemma 3.1 in [13] the process $\{T_N(\Delta), \Delta \in (-M, M)\}$ is tight.

From the proof of Lemmas 4.3 and 4.4 easily follows that $E T_N(\Delta)$ can be replaced by $\mu_N(\Delta)$. Q.E.D.

Proof of Remark. It suffices to prove

$$(4.8) \qquad |\mu_N^*(\Delta) - \mu_N(\Delta)| = o \left(\sqrt{\sum_{i=1}^{N} c_{iN}^2 d_{iN}^2} \right).$$

The convergence

$$\lim_{N \to \infty} \{\varphi''(F(x + \Delta d_{iN}))f^2(x + \Delta d_{iN})) +$$

$$+ \varphi'(F(x + \Delta d_{iN}))f(x + \Delta d_{iN}))\} =$$

$$= \varphi''(F(x))f^2(x) + \varphi'(F(x))f(x), \quad \text{a.s.},$$

and assumptions together with Lebesgue's theorem ensure the validity of (4.8). Q.E.D.

Proof of Theorem 4.2. If we go trough the proof of Theorem 4.1 we find that all steps are valid and it remains prove that

$$\left| E\left\{ \sum_{i=1}^{N} c_{iN} (\varphi(F(X_i + \Delta d_{iN}^*)) - \varphi(F(X_i))) \right\} - \Delta d_{iN}^* \int (\varphi'(F(x))f^2(x) \, dx) \right| = o\left(\sum_{i=1}^{N} c_{iN} d_{iN}^{*2} \right).$$

This easily follows applying the Taylor expansion to $\varphi(F(X_i + \Delta d_{iN}))$ and noticing that the integral

$$\int |\varphi''(F(x + y))f^2(x + y) + \varphi'(F(x + y))f(x + y)| \, dF(x)$$

is uniformly bounded for $|y| \leqslant \delta$. Q.E.D.

We finish this section proving several lemmas concerning the remainder terms in the decomposition (3.3).

First, we notice that for arbitrary random variables $\{(S_{N1}, \ldots \ldots, S_{NN})\}_{N=1}^{\infty}$ and $0 < \alpha < \dfrac{\epsilon}{\epsilon + \dfrac{3}{2}}$ it holds

(4.9) $\quad \lim_{N \to \infty} P\left(\sum_{i=1}^{N} S_{Ni} I\{F(X_i) \in A_N, F(X_i + \Delta d_{iN}) \notin A_N\} \neq 0 \right) = 0.$

(4.10) $\quad \lim_{N \to \infty} P\left(\sum_{i=1}^{N} S_{Ni} I\{F(X_i) \notin A_N, F(X_i + \Delta d_{iN}) \in A_N\} \neq 0 \right) = 0.$

Actually, for $\Delta d_{iN} > 0$

(4.11) $\quad P\left(\sum_{i=1}^{N} S_{Ni} I\{F(X_i) \notin A_N, F(X_i + \Delta d_{iN}) \in A_N\} \neq 0 \right) \leqslant$

$$\leqslant \sum_{\substack{i=1 \\ d_{iN} > 0}}^{N} P(F(X_i) \notin A_N, F(X_i + \Delta d_{iN}) \in A_N) =$$

$$= O(\Delta \max (N^{\frac{\alpha}{2} - \epsilon - \alpha\epsilon}, D_N)),$$

where we used the fact that $F(X_i) < N^{\alpha - 1}$ implies

(4.12) $\quad F(X_i + \Delta d_{iN}) < F(X_i) + \Delta d_{iN} K^* N^{\frac{(\alpha - 1)(1 + \epsilon)}{2}} + (\Delta d_{iN})^2 D_N$

$D_N \to 0$, as $N \to \infty$.

Lemma 4.3. *Under assumptions of Theorem* 4.1 *it holds*

$$\lim_{N \to \infty} \mathsf{P}\left(|V_{N2}(\Delta) - \mathsf{E} V_{N2}(\Delta)| \geqslant \sqrt{\sum_{i=1}^{N} c_{iN}^2 d_{iN}^2} \right) = 0,$$

$$\lim_{N \to \infty} \mathsf{P}\left(|V_{N1}(\Delta) - V_{N1}(0) - \mathsf{E}(V_{N1}(\Delta) - V_{N1}(0))| \geqslant \right.$$

$$\left. \geqslant \sqrt{\sum_{i=1}^{N} c_{iN}^2 d_{iN}^2} \right) = 0,$$

where $V_{N1}(\Delta)$, $V_{N2}(\Delta)$ *are given by* (3.6) *and* (3.7), *respectively.*

Proof. Let us start with $V_{N1}(\Delta) - V_{N1}(0)$. Applying the Taylor expansion and noticing (3.8) we have

$$|V_{N1}(\Delta) - V_{N1}(0)| \leqslant$$

(4.13)
$$\leqslant \sum_{i=1}^{N} K^* |c_{iN}| |R_{iN}^{\Delta} - R_{iN}| I\{F(X_i) \notin A_N\} +$$

$$+ \sum_{i=1}^{N} |c_{iN}| K^* (\log N) I\{F(X_i) \notin A_N, F(X_i + \Delta d_{iN}) \in A_N\} =$$

$$= J_1 + J_2.$$

By (4.10)

$$\lim_{N \to \infty} \mathsf{P}(J_2 \neq 0) = 0.$$

Again using the Taylor expansion and (3.8) we obtain

$$\mathsf{E}\,|J_1| \leqslant \sum_{i=1}^{N} |c_{iN}| \sum_{\substack{i=1 \\ j \neq 1}}^{N} K^* \mathsf{E}\,|F(X_i + \Delta(d_{iN} - d_{jN})) - F(X_i)| \times$$

(4.14)
$$\times I\{F(X_i) \in A_N\} =$$

$$= O\left(\Delta \max N^{-\frac{1}{2} - \epsilon + \frac{3\alpha}{2} + \epsilon\alpha}, \sqrt{\sum_{i=1}^{N} c_{iN}^2 d_{iN}^2} \, N^{\alpha - 2\epsilon + 2\alpha\epsilon}\right).$$

As for $V_{N2}(0)$, similarly as in the proof of Lemma 4 in [11] we derive the inequality

$$\mathsf{E}\{(R_{1N} - \mathsf{E}(R_{1N} \mid X_1))^4 \mid X_i\} \leqslant \frac{N}{r(F(X_1))} + \frac{3N^2}{(r(F(X_1)))^2}$$

which together with convexity of the function $(r(u))^\gamma$, $\gamma > 0$, implies

$$\mathsf{E}(V_{N2}(0))^2 \leqslant$$

$$\leqslant \frac{1}{2} \mathsf{E}\left(\frac{R_{1N}}{N+1} - F(X_1)\right)^4 \varphi''^2 \left(\frac{\lambda_{1N} R_{1N}}{N+1} + (1 - \lambda_{1N}) F(X_1)\right) \times$$

$$\times I\{F(X_1) \in A_N\} \leqslant$$

$$\leqslant \frac{K^{*2}}{2} \mathsf{E}\left(\frac{R_{1N}}{N+1} - F(X_1)\right)^4 \left[\lambda_N \left(r\left(\frac{R_{1N}}{N+1}\right)\right)^4 + \right.$$

$$\left. + (1 - \lambda_N)(r(F(X_1)))^4\right] I\{F(X_1) \in A_N\} = O\left(\frac{1}{N^{1+\frac{\alpha}{2}}}\right).$$

The needed statement for $V_{N2}(\Delta)$ follows from contiguity

$$\left\{\prod_{i=1}^{N} f(x_i + \Delta d_{iN})\right\}_{N=1}^{\infty}$$

to

$$\left\{\prod_{i=1}^{N} f(x_i)\right\}_{N=1}^{\infty}.$$

Q.E.D.

Lemma 4.4. *Under assumptions of Theorem 4.1 it holds*

$$\lim_{N \to \infty} \mathsf{P}\left(|Q_N(\Delta)| \leqslant \sqrt{\sum_{i=1}^{N} c_{iN}^2 d_{iN}^2}\right) = 0,$$

where $Q_N(\Delta)$ is given by (3.5).

Proof. After simple considerations we find that it remains to treat

$$(4.15) \qquad Q_{N1}(\Delta) = \sum_{i=1}^{N} \sum_{\substack{i=1 \\ j \neq 1}}^{N} h_N(X_i, X_j),$$

where

$$h_N(X_i, X_j) =$$

$$(4.16) \qquad \begin{aligned} &= \frac{c_{iN}}{N+1} \Big[(u(X_i - X_j + \Delta(d_{iN} - d_{jN})) - F(X_i + \Delta d_{iN})) \times \\ &\qquad\qquad \times \varphi'(F(X_i + \Delta d_{iN}))I\{F(X_i + \Delta d_{iN}) \in A_N\} - \\ &\qquad - (u(X_i - X_j) - F(X_i))\varphi'(F(X_i))I\{F(X_i) \in A_N\} - \\ &\qquad - \int (u(x - X_j) - F(x))\varphi'(F(x)) \times \\ &\qquad\qquad \times I\{F(x) \in A_N\}d(F(x - \Delta d_{iN}) - F(x)) \Big]. \end{aligned}$$

Clearly,

$$E(h_N(X_i, X_j) \mid X_i) =$$

$$(4.17) \qquad \begin{aligned} &= (F(X_i + \Delta(d_{iN} - d_{jN})) - F(X_i + \Delta d_{iN})) \times \\ &\qquad \times \varphi'(F(X_i + \Delta d_{iN}))I\{F(X_i + \Delta d_{iN}) \in A_N\}\frac{c_{iN}}{N+1}, \end{aligned}$$

$$E(h_N(X_i, X_j) \mid X_j) =$$

$$(4.18) \qquad \begin{aligned} &= \int (u(x - X_j - \Delta d_{jN}) - u(x - X_j)) \times \\ &\qquad \times \varphi'(F(x))I\{F(x) \in A_N\}dF(x - \Delta d_{iN})\frac{c_{iN}}{N+1}. \end{aligned}$$

By elementary considerations we obtain

$$(4.19) \qquad \mathrm{var}\{Q_{N1}(\Delta)\} \leqslant$$

$$\leqslant 6 \sum_{i=1}^{N} \sum_{j=1}^{N} E(h_N(X_i, X_j) - E(h_N(X_i, X_j) \mid X_i))^2 +$$

$$+ 3 \sum_{i=1}^{N} \text{var} \left\{ \sum_{j=1}^{N} \mathsf{E}(h_N(X_i, X_j) \mid X_i) \right\} +$$

$$+ 3 \sum_{j=1}^{N} \text{var} \left\{ \sum_{i=1}^{N} \mathsf{E}(h_N(X_i, X_j) \mid X_j) \right\}.$$

We estimate the terms separately. We start with the first one:

$$\sum_{i=1}^{N} \sum_{j=1}^{N} \mathsf{E}(h_N(X_i, X_j) - \mathsf{E}(h_N(X_i, X_j) \mid X_i))^2 \leqslant$$

$$\leqslant 3 \sum_{i=1}^{N} \sum_{j=1}^{N} \frac{c_{iN}^2}{(N+1)^2} \times$$

$$\times \mathsf{E}[(u(X_i - X_j + \Delta d_{iN}) - F(X_i + \Delta d_{iN})) \times$$

$$\times \varphi'(F(X_i + \Delta d_{iN})) I\{F(X_i + \Delta d_{iN}) \in A_N\} -$$

$$- (u(X_i - X_j) - F(X_i))\varphi'(F(X_i)) I\{F(X_i) \in A_N\}]^2 +$$

$$+ 3 \sum_{i=1}^{N} \sum_{j=1}^{N} \frac{c_{iN}^2}{(N+1)^2} \times$$

$$\times \mathsf{E}[(u(X_i - X_j + \Delta(d_{iN} - d_{jN})) - u(X_i - X_j + \Delta d_{iN}) -$$

$$- F(X_i + \Delta(d_{iN} - d_{jN})) + F(X_i + \Delta d_{iN})) \times$$

$$\times \varphi'(F(X_i + \Delta d_{iN})) I\{F(X_i + \Delta d_{iN}) \in A_N\}]^2 +$$

$$+ 3 \sum_{i=1}^{N} \sum_{j=1}^{N} \frac{c_{iN}^2}{(N+1)^2} \mathsf{E}\left[\int (u(x - X_j) - F(x))\varphi'(F(x)) \times \right.$$

$$\left. \times I\{F(x) \in A_N\} d(F(x - \Delta d_{iN}) - F(x)) \right]^2 =$$

$$= 3J_3 + 3J_4 + 3J_5.$$

For J_3 we can write (computing the expectation with respect to X_j) for $\Delta d_{iN} > 0$

(4.20) $$J_3 = \frac{1}{(N+1)^2} \sum_{i=1}^{N} \sum_{j=1}^{N} c_{iN}^2 \{ \int F(x)\varphi'(F(x)) I\{F(x) \in A_N\} \times$$

$$\times [-(1 - F(x + \Delta d_{iN}))\varphi'(F(x + \Delta d_{in})) \times$$

$$\times \, I\{F(x + \Delta d_{iN}) \in A_N\} + (1 - F(x))\varphi'(F(x))] \, dF(x)\big\} \, +$$

$$+ \int (1 - F(x + \Delta d_{iN}))\varphi'(F(x - \Delta d_{iN})) \times$$

$$\times \, I\{F(x + \Delta d_{iN}) \in A_N\} \times$$

$$\times \, [F(x + \Delta d_{iN})\varphi'(F(x + \Delta d_{iN})) -$$

$$- F(x)\varphi'(F(x))I\{F(x) \in A_N\}] \, dF(x).$$

By assumptions and (4.9)

$$\Big| \int F(x)\varphi'(F(x))I\{F(x) \in A_N\} \times$$

$$\times \, [(1 - F(x + \Delta d_{iN}))\varphi'(F(x + \Delta d_{iN})) \times$$

$$\times \, I\{F(x + \Delta d_{iN}) \in A_N\} - (1 - F(x))\varphi'(F(x))] \, dF(x) \Big| \leqslant$$

$$\leqslant K^{*2} \int r(F(x)) \times$$

$$\times \, I\{F(x) < 1 - N^{\alpha-1}, F(x + \Delta d_{iN}) > 1 - N^{\alpha-1}\} \, dF(x) \, +$$

$$+ \int |F(x)\varphi'(F(x)) \times$$

$$\times \, I\{F(x) > N^{\alpha-1}, F(x + \Delta d_{iN}) < 1 - N^{\alpha-1}\} \times$$

$$\times \int_0^{\Delta d_{iN}} ((1 - F(x + y))\varphi''(F(x + y)) \times$$

$$\times \, f(x + y) - f(x + y)\varphi'(F(x + y))) \, dy \Big] \, dF(x) =$$

$$= o((\Delta d_{iN})^2 N^{1-\alpha}) + O(\Delta d_{iN} N^{\frac{1}{2(1-\alpha)}}).$$

The second integral in (4.20) can be treated similarly and so does the case $\Delta d_{iN} < 0$. Finally, we get

$$J_3 = O\Big(\frac{\Delta}{N^{\frac{\alpha}{2}}} \sum_{i=1}^N c_{iN}^2 d_{iN}^2\Big).$$

Computing the expectation with respect to X_j we have

$$J_4 = \frac{1}{(N + 1)^2} \sum_{i=1}^N \sum_{j=1}^N c_{iN}^2 \int |F(x) - F(x - \Delta d_{jN})| \times$$

$$\times \ (1 - |F(x) - F(x - \Delta d_{jN})|) \times$$

$$\times \ \varphi'^2(F(x))I\{F(x) \in A_N\} \, dF(x - \Delta d_{iN}) \leqslant$$

$$\leqslant \frac{1}{(N+1)^2} \sum_{i=1}^{N} \sum_{j=1}^{N} c_{iN}^2 \int [|\Delta d_{jN}|(r(F(x)))^{\frac{3}{2} - \epsilon} K^{*2} +$$

$$+ \ (\Delta d_{jN})^2 K^* \frac{3}{2} (r(F(x)))^2 \, I\{F(x) \in A_N\}] \, dF(x) =$$

$$= O\left(\frac{\Delta}{N^{1 + \frac{\alpha}{2}}}\right),$$

where we used the simple inequality

$$|F(x) - F(x - \Delta d_{jN})| \leqslant$$

$$\leqslant |\Delta d_{jN}| K^*(r(F(x)))^{-\frac{1}{2} - \epsilon} + \frac{1}{2} K^*(\Delta d_{jN})^2.$$

As for J_5 we can write

$$\mathsf{E}\left[\int_0^{-\Delta d_{iN}} \left[\int (u(x - X_j) - F(x))\varphi'(F(x)) \times \right.\right.$$

$$\left.\left. \times \ I\{F(x) \in A_N\}f'(x + z) \, dxdz\right]\right]^2 =$$

$$= \int_0^{-\Delta d_{iN}} \int_0^{-\Delta d_{iN}} \int\int (F(\min(x, y)) - F(x)F(y)) \times$$

$$\times \ \varphi'(F(x))\varphi'(F(y))I\{F(x) \in A_N, F(y) \in A_N\} \times$$

$$\times \ f'(x + z)f'(y + v) \, dvdzdxdy \leqslant$$

$$\leqslant \left[\int_0^{-\Delta d_{iN}} \int \sqrt{r(F(x))} \, K^* f'(x + z) \, dzdx\right]^2 = o((\Delta d_{iN})^2 N^{\frac{\alpha}{2}}).$$

Hence,

$$J_5 = o\left(\Delta^2 \sum_{i=1}^{N} c_{iN}^2 d_{iN}^2 N^{-1 + \frac{\alpha}{2}}\right).$$

Now we turn to the second member in (4.19).

After some computations we receive

$$\sum_{i=1}^{N} \text{var}\left\{\sum_{j=1}^{N} E(h_N(X_i, X_j) \mid X_j)\right\} \leqslant \frac{1}{(N+1)^2} \sum_{i=1}^{N} c_{iN}^2 \times$$

$$\times E\left[\sum_{j=1}^{N} (F(X_i + \Delta(d_{iN} - d_{jN})) - F(X_i + \Delta d_{iN})) \times\right.$$

$$\left. \times \varphi'(F(X_i + \Delta d_{iN}))I\{F(X_i + \Delta d_{iN}) \in A_N\}\right]^2 \leqslant$$

$$\leqslant \frac{1}{(N+1)^2} \sum_{i=1}^{N} c_{iN}^2 \int (2(\Delta d_{iN})^2 f^2(x) + \Delta^2 K^{*2}) \times$$

$$\times \varphi'^2(F(x))I\{F(x) \in A_N\} \, dF(x - \Delta d_{iN})) =$$

$$= O\left(\frac{\Delta}{N} \sum_{i=1}^{N} c_{iN}^2 d_{iN}^2\right) + O\left(\frac{\Delta^2}{N^{1+\alpha}}\right).$$

Finally, we show that

$$\sum_{j=1}^{N} \text{var}\left\{\sum_{i=1}^{N} E(h_N(X_i, X_j) \mid X_j)\right\} = O\left(\frac{1}{N} \max\left(\frac{1}{N}, \sum_{i=1}^{N} d_{iN}^4\right)\right).$$

Obviously,

$$\sum_{j=1}^{N} \text{var}\left\{\sum_{i=1}^{N} E(h_N(X_i, X_j) \mid X_j)\right\} \leqslant$$

$$\leqslant \sum_{j=1}^{N} E\left(\sum_{i=1}^{N} E(h_N(X_i, X_j) \mid X_j)\right)^2 =$$

$$= \frac{1}{(N+1)^2} \sum_{\substack{j=1 \\ }}^{N} \sum_{\substack{i=1 \\ i \neq j}}^{N} \sum_{\substack{m=1 \\ m \neq j}}^{N} c_{iN} c_{mN} \times$$

$$\times \int \int \varphi'(F(x))\varphi'(F(y))I\{F(x) \in A_N, F(y) \in A_N\} \times$$

$$\times E(u(x - X_j - \Delta d_{jN}) - u(x - X_j)) \times$$

$$\times (u(y - X_j - \Delta d_{jN}) - u(y - X_j)) \times$$

$$\times [dF(x)dF(y) + dF(y)d(F(x - \Delta d_{iN}) - F(x)) +$$

$$+ dF(x)d(F(y - \Delta d_{mN}) - F(y)) +$$

$$+ d(F(x - \Delta d_{iN}) - F(x))d(F(y - \Delta d_{mN}) - F(y))].$$

Noticing that for $\Delta d_{jN} > 0$

$$\mathsf{E}\,(u(x - X_j) - u(x - X_j - \Delta d_{jN}))\,\times$$

$$\times\,(u(y - X_j) - u(y - X_j - \Delta d_{jN})) \leqslant$$

$$\leqslant \begin{cases} \min\,((F(x) - F(x - \Delta d_{jN})), (F(y) - F(y - \Delta d_{jN}))) \\ \hspace{4cm} \text{if } |x - y| < \Delta d_{jN} \\ 0 \quad \text{otherwise} \end{cases}$$

and

$$\frac{1}{(N + 1)^2} \sum_{\substack{i=1 \\ i \neq j}}^{N} \sum_{\substack{i=1 \\ m \neq j}}^{N} \sum_{m=1}^{N} c_{iN} c_{mN}\,\times$$

$$\times\,\int\int \varphi'(F(x))\varphi'(F(y))I\{F(x) \in A_N, F(y) \in A_N\}\,\times$$

$$\times\,\mathsf{E}\,(u(x - X_j) - u(x - X_j - \Delta d_{jN}))\,\times$$

$$\times\,(u(y - X_j) - u(y - X_j - \Delta d_{jN}))\,dF(x)\,dF(y) = o\left(\frac{\Delta^2}{N^{1 + \alpha}}\right)$$

it suffices to show that the integral

$$\int\int_{|x - y| < \Delta d_{jN}} \min\,((F(y) - F(y - \Delta d_{jN})),$$

(4.21)
$$(F(x) - F(x - \Delta d_{jN})))\,\times$$

$$\times\,\varphi'(F(x))\varphi'(F(y))I\{F(x) \in A_N, F(y) \in A_N\}\,\times$$

$$\times\,f'(x + z)f'(y + v)\,dy\,dx = O((\Delta d_{jN})^2 \max\,(1, d_{jN}^2 N))$$

uniformly for $|z| \leqslant |d_{iN} \Delta|$, $|v| \leqslant |\Delta d_{mN}|$.

Decompose the set $\{(x, y), |x - y| < |d_{jN} \Delta|\}$ as follows.

$$\{(x, y), |x - y| < d_{jN} \Delta\} =$$

$$= \left\{(x, y), x \leqslant y < x + \Delta d_{jN}, F(x) < \frac{1}{3}\right\} \cup$$

$$\cup \left\{(x, y), y < x < y + \Delta d_{jN}, F(x) < \frac{1}{2}\right\} \cup$$

$$\cup \left\{(x, y), |x - y| < \Delta d_{jN}, \frac{1}{3} \leqslant F(x) \leqslant \frac{2}{3}\right\} \cup$$

$$\cup \left\{(x, y), x \leqslant y < x + \Delta d_{jN}, F(x) > \frac{2}{3}\right\} \cup$$

$$\cup \left\{(x, y), y < x < y + \Delta d_{jN}, F(x) > \frac{2}{3}\right\}$$

Obviously, regarding the properties of the function $r(u)$,

$$\int\int\limits_{\{y < x < y + \Delta d_{jN}, F(x) < \frac{1}{3}\}} \min\left((F(x + \Delta d_{jN}) - F(x)),\right.$$

$$(F(y + \Delta d_{jN}) - F(y))) \times$$

$$\times |\varphi'(F(x))\varphi'(F(y))| I\{F(x) \in A_N, F(y) \in A_N\} \times$$

$$\times f'(x + z)f'(x + v) \, dx \, dy =$$

$$= o\left(\Delta d_{jN} \int_0^{\Delta d_{jN}} \left[\int (r(F(y)))^2 f(y + z) I\{F(y) \in A_N\} \, dy\right] dz\right) =$$

$$= o((\Delta d_{jN})^2 N^{(1-\alpha)(1+\frac{\alpha}{2})}).$$

The integrals over the other sets can be treated similarly. Q.E.D.

5. PROOF OF THEOREM 2.1

We give here only the sketch of the proof for there exists a standard technique that enable us to derive the assertion of Theorem 2.1 from the Theorem 3.1 and Theorem 4.2. This technique is described in [7] and [14] in details. It remains to prove the following three lemmas.

Lemma 5.1. *Under assumptions of Theorem 2.1 it holds*

$$(\hat{\Delta}_N^+ - \hat{\Delta}_N^-) \sqrt{\sum_{i=1}^N (c_i - \bar{c}_N)^2}$$

converges in probability to $2\gamma(F, \varphi)$ *as* $N \to \infty$ *and the asymptotic* distribution of

$$(\hat{\Delta}_N^+ - \Delta_1) \sqrt{\sum_{i=1}^{N} (c_i - \bar{c}_N)^2}$$

is normal with parameters

$$\left(\gamma(F, \varphi), \frac{\left(\int_0^1 \left(\varphi(u) - \int_0^1 \varphi(y)\, dy \right)^2 du \right)^{\frac{1}{2}}}{\left(\int \varphi'(F(x)) f^2(x)\, dx \right)^2} \right).$$

The proof is the same as that of Lemma 5.2 in [13]. Q.E.D.

Lemma 5.2. *Under assumptions of Theorem 2.1 there exist positive constants* κ_1, κ_2, K_1 *and* K_2 *and* N_0 *such that for* $N \geqslant N_0$

$$P\left(\left| \sqrt{\sum_{i=1}^{N} (c_i - \bar{c}_N)^2} \, (\hat{\Delta}_N^+ - \hat{\Delta}_N^-) - 2\gamma(F, \varphi) \right| \geqslant \frac{K_1 \log N}{N^{\kappa_1}} \right) \leqslant$$

$$\leqslant \frac{K_2}{N^{1+\kappa_2}}.$$

Proof. The assertion follows from Theorem 3.1 in the same way as Lemma 4.2 in [7]. Q.E.D.

Lemma 5.3. *Under assumptions of Theorem 2.1, the asymptotic distribution of*

$$\frac{\sum_i (c_i - \bar{c}_N)^2}{\sqrt{\sum_i (c_i - \bar{c}_N)^4}} \left(\frac{\hat{\Delta}_N^+ - \hat{\Delta}_N^-}{2\gamma(F, \varphi)} \sqrt{\sum_i (c_i - \bar{c}_N)^2} - 1 \right)$$

is normal with parameters

$$\left(0, \frac{\sigma^2}{\left(\int \varphi'(F(x)) f^2(x)\, dx \right)^2} \right),$$

where σ^2 *is given by* (4.2).

Proof. Denote

$$\rho_N = \frac{\displaystyle\sum_{i=1}^{N}(c_i - \bar{c}_N)^4}{\left(\displaystyle\sum_{j=1}^{N}(c_i - \bar{c}_N)^2\right)^2}.$$

Theorem 4.2 implies that for any $\Delta_3 \neq \Delta_2$

$$\lim_{N\to\infty} \mathsf{P}\left(\frac{1}{\rho_N^{\frac{1}{2}+\frac{\gamma}{2}}\sigma}\left[S_N(X_i + \Delta_2 c_{iN}(\rho_N)^{\frac{\gamma}{2}}) - \right.\right.$$

$$- S_N(X_i - \Delta_3 c_{iN}(\rho_N)^{\frac{\gamma}{2}}) - (\Delta_2 - \Delta_3)(\rho_N)^{\frac{\gamma}{2}} \times$$

$$\left.\left.\times \int \varphi'(F(x))f^2(x)\,dx\right] \leqslant y\right) = \Phi\left(\frac{y}{\Delta_2 - \Delta_3}\right),$$

where

$$c_{iN} = \frac{c_i - \bar{c}_N}{\sqrt{\displaystyle\sum_{j=1}^{N}(c_i - \bar{c}_N)^2}}.$$

Applying the procedure in Ch. III [2] and Lemma 5.1 we get

$$\lim_{N\to\infty} \mathsf{P}\left(\frac{1}{\rho_N^{\frac{1}{2}+\frac{\gamma}{2}}\sigma}\left[S_N(X_i - (\hat{\Delta}_N^+ - \Delta_1)c_i(\rho_N)^{\frac{\gamma}{2}}) - \right.\right.$$

(5.1)
$$- S_N(X_i - (\hat{\Delta}_N^- - \Delta_1)c_i(\rho_N)^{\frac{\gamma}{2}}) -$$

$$- (\hat{\Delta}_N^- - \hat{\Delta}_N^+)\sqrt{\sum_i(c_i - \bar{c}_N)^2}\,(\rho_N)^{\frac{\gamma}{2}} \times$$

$$\left.\left.\times \int \varphi'(F(x))f^2(x)\,dx\right] \leqslant y\right) = \Phi\left(\frac{y}{2\gamma(F,\varphi)}\right).$$

Further, Theorem 4.2 implies

$$\lim_{N \to \infty} P\left(\max_{|\Delta - \Delta_1| \sqrt{\sum_i (c_i - \bar{c}_N)^2} \geq C} \frac{1}{(\rho_N)^{\frac{\gamma}{2}}} \times \right.$$

$$\times \left| \left[S_N(X_i - (\Delta - \Delta_1)c_i(\rho_N)^{\frac{\gamma}{2}}) - \right. \right.$$

$$\left. - S_N(X_i) + (\Delta - \Delta_1) \sqrt{\sum_i (c_i - \bar{c}_N)^2} \, (\rho_N)^{\frac{\gamma}{2}} \times \right.$$

$$\left. \left. \times \int \varphi'(F(x))f^2(x)\, dx \right] \right| \geq \epsilon \right) = 0$$

and Theorem 3.1 in [12] ensures

$$\lim_{N \to \infty} P\left(\max_{|\Delta - \Delta_1| \sqrt{\sum_i (c_i - \bar{c}_N)^2} \geq C} \left| S_N(X_i - (\Delta - \Delta_1)c_i) - \right. \right.$$

$$\left. - S_N(X_i) + (\Delta - \Delta_1) \sqrt{\sum_i (c_i - \bar{c}_N)^2} \times \right.$$

$$\left. \left. \times \int \varphi'(F(x))f^2(x)\, dx \right| \geq \epsilon \right) = 0$$

for any $\epsilon > 0$ and $C > 0$.

From the last two relations and Lemma 5.1 can be concluded

$$\lim_{N \to \infty} P\left(\frac{1}{\rho_N^{\frac{\gamma}{2}}} \left| [S_N(X_i - (\hat{\Delta}_N^+ - \Delta_1)c_i(\rho_N)^{\frac{\gamma}{2}}) - \right. \right.$$

$$- S_N(X_i - (\hat{\Delta}_N^- - \Delta_1)c_i(\rho_N)^{\frac{\gamma}{2}}) - S_N(X_i - (\hat{\Delta}_N^+ - \Delta_1)c_i) -$$

$$\left. \left. - S_N(X_i - (\hat{\Delta}_N^- - \Delta_1)c_i)] \right| \geq \epsilon \right) = 0$$

for any $\epsilon > 0$. This together with the definition of $\hat{\Delta}_N^+$ and $\hat{\Delta}_N^-$ ensures that

$$\frac{1}{\rho_N^{\frac{\gamma}{2}}} [S_N(X_i - (\hat{\Delta}_N^+ - \Delta_1)c_i(\rho_N)^{\frac{\gamma}{2}}) -$$

$$- S_N(X_i - (\hat{\Delta}_N^- - \Delta_1)c_i(\rho_N)^{\frac{\gamma}{2}})]$$

converges in probability to $-2\Phi^{-1}\left(1 - \frac{\alpha}{2}\right)$. Utilizing this fact in (5.1) we obtain the assertion. Q.E.D.

Lemma 5.4. *For every* $\eta_1 > 0$ *and* $\eta_2 > 0$ *there exists* η_3 *and* N_* *such that for* $N > N_*$

$$P\left(\sup_{|N - N'| \leqslant N\eta_3} \sqrt{\sum_i (c_i - \bar{c}_N)^2} \, |\hat{\Delta}_N^{\pm} - \hat{\Delta}_{N'}^{\pm}| > \eta_1\right) < \eta_2\right).$$

Proof. The assertion follows from the martingale property similarly as Lemma 4.4 in [7]. Q.E.D.

REFERENCES

[1] J. Anscombe, Large sample theory of sequential estimation, *Proc. Camb. Phil. Soc.*, 48 (1952), 51-58.

[2] A. Antille, Linearité asymptotique d'une statistique de rang, *Z. Wahrscheinlichkeitstheorie verw. Geb.*, 24 (1972), 309-324.

[3] P. Billingsley, *Convergence of probability measures*, Wiley, New York, 1968.

[4] V. Dupač – J. Hájek, Asymptotic normality of simple linear rank statistics under alternatives, II, *Ann. Math. Statist.*, 40 (1969), 1992-2017.

[5] Y.S. Chow – H. Robbins, On the asymptotic theory of fixed – width sequential confidence intervals for the mean, *Ann. Math. Statist.*, 36 (1965), 457-462.

[6] J.C. Geertsema, Sequential confidence intervals based on ranks, *Ann. Math. Statist.*, 41 (1970), 1016-1026.

[7] M. Ghosh – P.k. Sen, On bounded length confidence interval for regression coefficient based on a class of rank statistics, *Sankhya A*, 34 (1972), 33-52.

[8] L.J. Gleser, On the asymptotic theory on fixed-size sequential confidence bounds for linear regression parameters, *Ann. Math. Statist.,* 36 (1965), 463-467.

[9] J. Hájek − Z. Šidák, *Theory of rank tests,* Academia, Prague, 1967.

[10] M. Hušková, The rate of convergence of simple linear rank statistics under hypothesis and alternatives, *Ann. Statist.,* 5 (1977), 658-670.

[11] M. Hušková, The Berry − Esseen theorem for rank statistics, *Comment. Math. Univ. Carol.,* 20 (1979), 399-415.

[12] J. Jurečková, Asymptotic linearity of a rank statistic in regression parameter, *Ann. Math. Statist.,* 40 (1969), 1889-1900.

[13] J. Jurečková, Central limit theorem for Wilcoxon rank statistics process, *Ann. Statist.,* 1 (1973), 1046-1060.

[14] J. Jurečková, Bounded-length sequential confidence intervals for regression and location parameter, *Proceedings of the second Prague symposium on asymptotic statistics,* Prague, (1979), 239-250.

[15] H.L. Koul, Asymptotic behavior of a class of confidence regions based on ranks in regression, *Ann. Math. Statist.,* 42 (1971), 466-476.

[16] L. Weiss − J. Wolfowitz, Optimal, fixed length, non-parametric sequential confidence limits for a translation parameter, *Z. Wahrscheinlichkeitstheorie verw. Geb.,* 24 (1972), 203-209.

[17] V. Petrov, *Sums of independent random variables,* Springer-Verlag, Berlin, 1972.

M. Hušková

Dept. of Stat. Charles Univ., Sokolovská 83, CS-18600 Prague 8, Czechoslovakia.